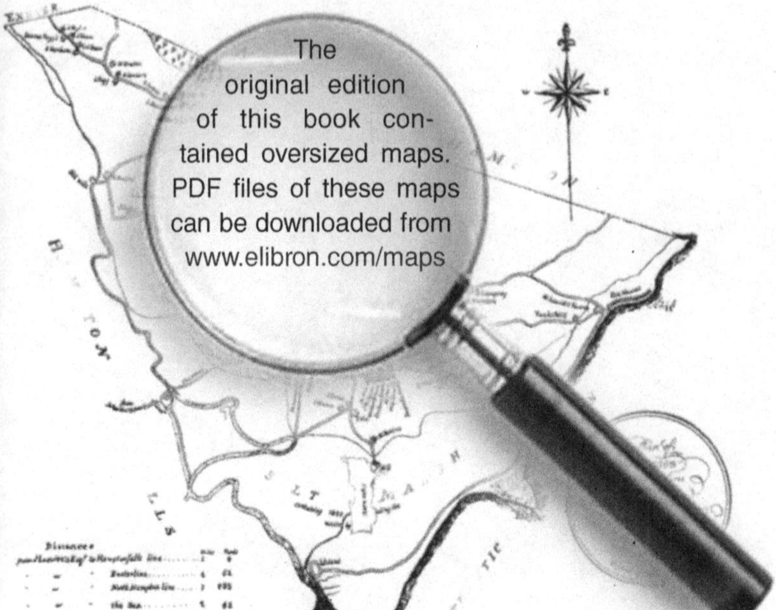
The
original edition
of this book con-
tained oversized maps.
PDF files of these maps
can be downloaded from
www.elibron.com/maps

ROUTES

IN

ABYSSINIA

Elibron Classics
www.elibron.com

Elibron Classics series.

ISBN 1-4021-5883-1 (paperback)
ISBN 1-4021-0632-7 (hardcover)

This Elibron Classics Replica Edition is an unabridged facsimile of the edition published in 1867 by Harrison and Sons, London.

ROUTES

IN

ABYSSINIA.

Presented to the House of Commons, in pursuance of their Address dated November 26, 1867.

LONDON:
PRINTED FOR HER MAJESTY'S STATIONERY OFFICE,
BY HARRISON AND SONS, ST. MARTIN'S LANE,
Printers in Ordinary to Her Majesty.

1867.

PREFACE.

THE object of this compilation is to collect together the information on the routes in Abyssinia which is scattered through the works of different travellers. This information has been arranged in the following order:—

General description of the country and of the different routes by which it can be entered.

Short outline of the nature of the Government, the religion and character of the inhabitants, the currency, the military system of the country, and the career and character of the present Emperor Theodore.

Reference to the Portuguese expedition of 1541, and to the places of entrance into the country which have been adopted by travellers since the 16th century.

Detailed account of the routes leading from Massowah and Ansley Bay to Gondar and Magdala.

Extracts from works of various travellers bearing upon the different lines of roads.

A map, compiled from the records of travellers, is given at the end. The orthography of names of places, &c., in Abyssinia is so indefinite, hardly any two travellers agreeing, that it has been found impossible to avoid, in all cases, discrepancies of spelling between the maps and the text. Abyssinian villages and towns appear to be often of a very temporary nature, and those recorded by one traveller are often not mentioned by the next one who follows the same route; some of the places laid down may therefore be no longer in existence, and others may have sprung up. The data also for laying down many of them are of very doubtful accuracy.

Compiled at the Topographical and Statistical Department of the War Office, by Lieut.-Colonel A. C. Cooke, C.B., R.E.

HENRY JAMES, Colonel R.E.,
Director.

INDEX.

RETURN to an Address of the Honourable the House of Commons, dated November 26, 1867; *for*—

"Copy of a Pamphlet and Appendices relating to the Routes in Abyssinia."

GENERAL DESCRIPTION OF THE COUNTRY OF ABYSSINIA AND OF THE DIFFERENT ROUTES LEADING INTO IT.

ABYSSINIA is often represented on maps as bounded on the east by the Red Sea. This, however, is an error. Abyssinia proper consists of a high mountainous table-land, the eastern boundary of which may be considered roughly as following the 40th degree of east longitude. Between this mountainous region and the sea there is a tract of arid, low-lying, waterless country, inhabited by the savage Danakil tribes; this region at Massowah is only a few miles broad, but it widens out to 200 or 300 miles at Tajurrah. In climate, inhabitants, soil, cultivation, &c., these two regions are totally opposite, the highlands being salubrious, temperate, generally well watered, and traversed by paths in every direction, whilst the low country is arid, waterless, with few exceptions trackless and uncultivated.

This inhospitable region effectually cuts off the highlands from all communication with the sea, except at three points, Massowah and its neighbourhood on the north, Amphilla Bay about 100 miles to the south-east, and Tajurrah on the south.* From these three points the roads into the interior are of a very different character, for at Massowah they very soon climb the eastern boundary of the highlands,† and continue along the elevated land, whilst from Tajurrah they have to traverse 200 or 300 miles, and from Amphilla Bay from 50 to 100 miles, of the low lying country before they reach the highlands.

The western and southern boundaries of Abyssinia are very undefined, but they may be taken roughly as coterminous with the edge of the highlands, as shewn on the map in the margin of the route map at the end.

The different character of the high and low country has an important bearing on the nature of the transport, for, whilst in the latter camels are chiefly used, these animals are useless in the highlands, where they are replaced by asses, mules, oxen, and men.‡

* Travellers speak of other tracks known to the natives, but none have been explored.

† A circle drawn with the centre at Massowah and Ailet on the circumference (a radius of 20 miles), would apparently sweep the spurs of the mountains where they descend into the plains. Camels can, however, go much further, as to the foot of the Taranta pass, about 40 or 50 miles.

‡ That camels are useless as beasts of burden on the highlands seems beyond a doubt. They are invariably changed at the foot of the Taranta Passes for bullocks. Major Harris on approaching Ankobar, the capital of Shoa, from Tajurrah, says:—"As well from the steepness of the rugged mountains of Abyssinia, which towered overhead, as from the pinching climate of their wintry summits, the camel becomes useless as a beast of burden; and none being ever taken beyond the frontier, many of the Widasma's retinue now gazed at these ungainly quadrupeds

There is also a way of approach by Suakim to Kassala, an Egyptian town on the north-west frontier of Abyssinia. From Kartoum and other places on the Nile there are also roads into the interior of the country.

1. *Roads from Massowah.*—After traversing 30 or 40 miles of hot, low-lying desert country, the main road from Massowah to Gondar ascends the mountains by the Taranta pass. This pass is described by all travellers as one of great difficulty, up which in some places loaded animals can with difficulty make their way. It can be avoided by going round to the west, where there are at least two paths, one of which was followed by Ferret and Galinier, on their return to Massowah, in 1842, and the other by Mansfield Parkyns, on his way to Adowa, in 1843. Neither of these travellers describe the same difficulties on reaching the high land by those routes as are experienced in the Taranta pass. M. Münzinger has also lately given a description of two paths by which the table-land can be reached in the direction of Kaya Khor. Of this line Dr. Beke, writing in 1842, says "after quitting the table-land of Serawé at Gura, I descended from Kaiyakhor to Massowah by a gradual and easy road, well watered, and occupying two days and a-half, very gentle travelling. This is so much superior to the steep way up the Taranta, that it is now generally chosen by Europeans." The Taranta pass may be also avoided by following up the Haddas stream to Tohonda; this road was traversed by Krapf, who states that it is an easy one and that it might be made available for camels. From Ansley Bay Tohonda and Senafe may also be reached.

When the high land is reached, the country for the whole distance to Gondar and Magdala appears to present alternations of fertile land, rocky barren tracts, mountain passes and defiles. The table-land appears to be from 5,000 to 8,000 feet above the sea, and the mountains rise from it to the right of 10,000 and 15,000 feet, the whole country being broken and tossed about in a remarkable degree. With the exception of the first 50 or 60 miles after leaving Massowah, there seems to be generally no want of water. There are numerous paths between the villages, but none that deserve the name of roads. Mansfield Parkyns says on this head, "For some distance after passing the church we continued in the great Gondar road. This appellation may give an idea of macadamising, with footpaths alongside, fences, &c., but here the high road is only a track worn by use, and a little larger than the sheep paths, from the fact of more feet passing over it. The utmost labour bestowed on any road in this country is when some traveller, vexed with a thorn that may happen to strike his face, draws his sword and cuts off the spray. Even

for the first time." In all views of the table land of Abyssinia, of which many exist in the works of Lefebvre, Salt, Lord Valentia, &c., a camel is never depicted, mules and horses, on the contrary, entering frequently into the landscape, whilst in sketches taken in the low countries the camel is an invariable feature. The only exception to this rule I have ever found is in the case of the journey of Don Alonzo Mendez from Amphilla Bay to Fremona, in which, after having reached the highlands by the Senafe Pass, he still speaks of his "camel drivers."

this is rarely done; and I have been astonished at seeing many highways, and even some of the most used, rendered almost impassable by the number of thorns which are allowed to remain spread across them." The track he was following when he penned these observations, is the great caravan road from Gondar, the capital of Abyssinia, to the Red Sea. Ferret and Galinier thus alludes to the roads;—"Le chemin que nous suivons est un des plus fréquentés de l'Abyssinie; c'est par là que les caravanes se rendent à la mer et qu'elles en reviennent. Ne vous y trompez pas cependant, ceci ne ressemble guère à un de nos grands chemins, à ces routes commerciales qui traversent notre France; un sentier tout simplement rien qu'un sentier. Le pied des voyageurs l'a tracé lentement, et il serpente au hasard sur les flancs des collines, au milieu des prairies; çà et là il côtoie quelques arbres épineux; prenez garde alors, vous courez risque d'y déchirer vos vêtements et vos jambes."

Besides the main road from Massowah to Gondar by Adowa, there is also, according to Dr. Beke, a second great caravan road from north to south of Abyssinia, from Massowah, through Antalo and Socota. Antalo (about latitude 13¼°, longitude 39½°) may be reached either through Addigraht by the route followed by Rüppell and Ferret and Galinier, or through Adowa, as travelled by Beke, or by an intermediate route followed by Salt. Magdala, according to Dr. Beke* and Dr. Blanc, can be more easily reached from the direction of Socota than from that of Devra-Tabor.

There are numerous streams and rivers and high mountain passes to be traversed on these highlands. Their water system is peculiar. By a glance at the map at the end of the volume it will be seen that the watershed runs very near their eastern edge, from which it results that all the great rivers have their rise near the eastern side of the table land, and then take a westerly course. The principal of these are the Mareb, Taccazze, Atbara, and Abai or Blue Nile. The Takazze crosses the principal caravan road from Massowah to Gondar and is a most formidable obstacle. It is described by Parkyns, who crossed it in the rainy season, as nearly equal in volume to the Thames, at Greenwich, but resembling in rapidity the Rhone when it leaves Lake Geneva; in the dry season it is fordable; Bruce describes it as flowing in a "prodigious valley." Ferret and Galinier say: "the ravine in which it runs is one of the steepest possible. At Tchellatchekenne† it is not less than 2,000 feet deep, the breadth of the river is about 60 feet, and its depth in March not more than 3 or 4 feet, it is then fordable almost everywhere. In the rainy season it overflows its bank, and its depth is often from 15 to 18 feet." Parkyns says that mountains, almost impregnable by nature, are common in this country.

One important feature of Abyssinia is the existence of

* Letter to Secretary of State for War, 9th August, 1867.

† Chilachikan on Map.

ranges of snow-topped mountains in the provinces of Samen and Lasta, which stretch at right angles to the roads leading from the north towards Gondar and Magdala, and the passes through which are described as of the most formidable character. The two principal passes leading to Gondar are the Lamalmon and Selki. Of the first, Ferret and Galinier say: "We found ourselves in the Lamelmon, whose summit, 8,000 feet above the sea, sinks into the plains of Waggara. All at once an immense gulf opened under our feet. We got off our mules, and, leading them by the bridle, descended the mountains by a precipitous path, blocked with fragments of rock, which threatened at every step to precipitate us into the abyss on our left."

Bruce says of the same pass: "We were now ascending the Lamelmon through a very narrow road, or rather path, for it scarcely was two feet wide anywhere. It was a spiral, winding up the side of a mountain, always on the very brink of a precipice. Torrents of water, which in winter carry prodigious stones down the side of the mountains, had divided the path in several places, and opened to us a view of that dreadful abyss below which few heads can bear to look upon. We were here obliged to unload our baggage, and, by slow degrees, crawl up the hill, carrying them little by little upon our shoulders round those chasms where the road was intersected."

Combes and Tamisier thus speak of the Selki Pass: "We then abandoned the banks of the river to mount by an infernal path. . . . At length we arrived at the top of the prodigious mass. On every side still more colossal mountains raised their heads, between which was the path which we had to follow. . . . . After marching two hours we passed the gate called Sancaber, closing the finest and perhaps the only road in Abyssinia, carried along the side of a high mountain, inaccessible from the top to the bottom."

Mr. Dufton, in the journey from Massowah to Devra Tabor, which he describes in a letter published in the *Times* of the 14th September, appears to have taken a more easterly route than that of the Selki pass, but necessarily traversing a portion of the same range of mountains in the Province of Samen, which, as has been before stated, is one of the most rugged in Abyssinia. His remarks on it will probably apply to the whole of the mountainous regions of Samen and Lasta, and are as follows:— "The main features of the route are its ruggedness, making the use of the camel totally impossible; the narrowness of the path over a greater portion of it necessitating the army marching in single file; the salubrity of the climate, this being in general cool and agreeable; the abundance of clear cold water in the mountain torrents, and the magnificence of the scenery passed through. The whole distance,* allowing for sinuosities, is about 400 miles, which the army could not do in less than 40 days." He adds in conclusion that he does not think the route "is a practicable one for the march of an army."

* From Massowah to Debra Tabor.

Mr. Pearce appears to have struck a mountain range after passing Lake Ashangi, as he talks of intense cold, with hoar frost on the ground, and when near Socota, the capital of Lasta, he says, "This province is extremely mountainous throughout, and forms an almost impenetrable barrier between the two great divisions of Abyssinia, generally comprehended under the name of Amhara and Tigré, two passes only existing through the mountains, which are easily commanded by a small number of troops."

Krapf when near Lake Ashangi in about latitude 12½°, says, "our passage was sometimes extremely difficult and narrow. The banks of the mountains had sometimes the appearance of high walls of rocks, a step from which would cause certain death."

Combes and Tamisier, speaking of Samen, say: "It is the highest land in Abyssinia; its mountains are almost continually covered with snow on their highest tops;" and of Lasta, "It is very mountainous, and entire armies have been known to perish there of snow. Its inhabitants are warlike."

These highlands are very healthy. They are stated by Parkyns to "enjoy probably as salubrious a climate as any country on the face of the globe. The heat is by no means oppressive, a fine light air counteracting the power of the sun. At certain seasons of the year the low valleys, as of the Mareb and Taccazy, especially the former, are much to be feared from the malaria which prevails and which brings on, in persons exposed to its influence, most terrible inflammatory fevers, of which four cases out of five are fatal."* Harris thus speaks of the highlands of Shoa: "As if by the touch of the magician's wand, the scene now passes in an instant from parched and arid waste to the green and lovely highlands of Abyssinia, presenting one sheet of rich and thriving cultivation. Each fertile knoll is crowned with its peaceful hamlet—each rural vale traversed by its crystal brook, and teeming with herds and flocks. The cool mountain zephyr is redolent of eglantine and jasmine, and the soft green turf, spangled with clover, daisies, and buttercups, yields at every step the aromatic fragrance of the mint and thyme." Krapf when travelling near Lake Haik says: "The climate in these plains is beautiful, neither too hot nor too cold; the air being always refreshed by the winds blowing from the mountains. There is plenty of water poured out from the veins of the neighbouring mountains."

There is a dry and a rainy season in these highlands. The latter extends from May to September, and, with the exception of a few showers in the beginning of November, little more rain falls during the year. Bruce gives the rainfall from a register kept at Gondar and Koscam, as follows:—

* Bruce says of the Taccazy: "From the falling of the first rains in March till November, it is death to sleep in the country adjoining to it, both within and without its banks; the whole inhabitants retire and live in villages on the top of the neighbouring mountains."

	Gondar, 1770. inches.	Koscam, 1771. inches.
March and April . .	·039	·749
May	2·717	2·501
June	4·307	6·388
July	10·089	14·360
August	15·569	10·019
September	2·834	7·338

After which it rained but little more, except at the beginning of November. During the rainy season the rain almost invariably falls in the afternoon and night. From Bruce's diary of the weather during the rainy months, it appears that there was hardly a single instance of rain falling between 6 a.m. and noon.*

The temperature of the highlands is mild, but at great elevations the cold is severe.

Bruce gives a series of daily observations at Gondar, 7,420 feet above the sea, from which the following is extracted :—

Month.	6 A.M.	Noon.
January .	61 to 66 degrees.	69 to 75 degrees.
February .	63 to 70 ,,	69 to 76 ,,
March . .	56 to 70 ,,	70 to 83 ,,
April . .	59 to 73 ,,	67 to 85 ,,
May . .	60 to 74 ,,	73 to 80 ,,
June . .	60 to 65 ,,	63 to 69 ,,
July . .	55 to 61 ,,	55 to 65 ,,
August . .	55 to 58 ,,	58 to 63 ,,
September .	50 to 67† ,,	61 to 68 ,,
October .	56 to 61 ,,	63 to 69 ,,
November .	59 to 61 ,,	60 to 72 ,,
December .	59 to 63 ,,	67 to 72 ,,

Harris gives the following as the temperature of Ankobar, the capital of Shoa, which is in the hill country, 8,200 feet above the sea.

Month.	Mean of Month.	Extreme maximum.	Extreme minimum.
January .	52 degrees.	65	41
February .	54·6 ,,	66	46
March . .	57·2 ,,	69	46
April . .	55·2 ,,	62	46
May . .	59·7 ,,	67	51
June . .	62·1 ,,	69	52
July . .	58·1 ,,	69	51
August .	55·8 ,,	63	47
September .	55·3 ,,	63	46
October .	52·1 ,,	62	44
November .	51·9 ,,	60	43
December .	51·8 ,,	61	41

* There seems no reason to suppose that the highlands, as a rule, are unhealthy in the rainy season. Doctor Blanc writes, March 31st, 1867—"I believe the rainy season would be the most favourable time for a campaign in this country. The rain by itself is not very severe ; the sky being cloudy, the weather is always agreeable and cool. There is no sickness during that period ; it is before and after that certain localities are unhealthy."

† It is doubtful whether this is not a wrong entry in Bruce's diary and intended for 57.

Ferret and Galinier give the following as the temperatures of different places along the road, from Massowah to Adowa:—

Place.	Day.	Hour.	Temp.
Massowah	2nd to 8th Nov.	Noon	91 to 93
Arkiko	10th Nov. .	"	91½
Oueah	11th " . .	"	91
Valley of Hammamo .	12th " .	"	91
Do. Haddas . .	13th " .	"	82
Foot of Taranta Pass .	14th " .	"	80
Top of do. . .	15th " .	"	61
Dixah	17th " .	"	78
Adowa	2nd Dec. .	"	71
"	" .	midnight	56
Axum	10th Jan. .	Noon	75
"	15th " .	midnight	64
"	5th Feb. .	Noon	71
Intetchaou, near Adowa .	6th June .	"	77
" " .	16th to 24th June	"	75 (mean)
" " .	25th July .	"	65
" " .	31st " .	"	68
" " .	5th August .	"	64
" " .	9th " .	"	69

Some of the higher mountains are covered with snow. Bruce asserted that snow was unknown in Abyssinia, but there is abundant evidence that he was wrong, from the writings of Gobat, Ferret and Galinier, Mansfield Parkyns, and Salt.

It may be mentioned, as bearing on the question of the necessary equipment for an army operating in the highlands, that Steudner, who, in March, 1862, accompanied the army of the Emperor of Abyssinia, which he estimated at 20,000 men, states that the officers had tents and the men made themselves grass huts, called "nogos," into which they crowded for warmth. Several travellers have also recorded the necessity of sleeping under cover in many parts of the highlands on account of the cold.

Massowah is very hot. Ferret and Galinier found the temperature in the shade, in November, to be 88° at 9 a.m., 93° at noon, 91½° at 3 p.m., and 86° at 9 p.m., and they state that in July they experienced a temperature of 127° in the shade, and this is confirmed by Mansfield Parkyns.

The distance from Massowah to Magdala or Debra Tabor is 350 or 400 miles, and to Gondar about 300 miles. From Ansley Bay the distance is about 15 miles less.

The works, from which extracts bearing on the different routes from Massowah and its neighbourhood have been made, will be found by referring to the index and map given at page 39.

2. *Roads from Tajurrah.*—The main road goes from Tajurrah to Ankobar the capital of Shoa. Near Alloolo a branch road turns off in the direction of Magdala.

These roads traverse, for a distance of 200 or 300 miles, the low country which exists between the highlands and the sea. Between Tajurrah and the point where they diverge, there lies the "Tekama," a series of volcanic rocks enclosing a salt lake, the passes through which are of the most frightful nature. Harris traversed it at the worst period of the year, June 1841, and thus describes it. "It is an iron bound waste, which, at this unauspicious season of the year (June) opposes difficulties almost overwhelming in the path of the traveller. Setting aside the total absence of water and forage throughout a burning tract of fifty miles, its manifold intricate mountain passes, barely wide enough to admit the transit of a loaded camel, the bitter animosity of the wild bloodthirsty tribes by which they are infested, and the uniform badness of the road, if road it may be termed, everywhere beset with the huge jagged blocks of lava, and intersected by perilous acclivities and descents—it is no exaggeration to state that the stifling sirocco which sweeps across the unwholesome salt flat during the hotter months of the year could not fail, within 48 hours, to destroy the hardiest European adventurer."

D'Héricourt traversed this road at a more favourable time of year, October 1842, but does not give a much more favourable account. He says: "Je crois pouvoir le dire, sans qu'on m'accuse de vouloir exagérer les difficultés de mes travaux, pour en rehausser le mérite, il y a peu de voyages plus fatigants pour l'esprit et le corps, plus périlleux à la fois et plus monotones que de parcourir les déserts des Adels. Le Major Harris, un des hommes les plus expérimentés dans les voyages Africains, en a gardé une impression semblable, et l'a rendue dans sa relation avec les couleurs les mieux senties et les plus justes. Lui même, lorsque je le rencontrai plus tard dans le Choa, ne pouvait revenir de son étonnement lorsque je lui ai raconté que j'avais tenté tout seul, et pour la seconde fois, une expédition si peu attrayante. Au moins, dans d'autres pays, les magnificences de la nature sont une compensation aux périls que vous bravez, un délassement aux fatigues que vous vous imposez: c'est la nature qui fait du désert des Adels le plus affreux des séjours. Le pays des Adels, que l'on met un mois à parcourir, le pays des Adels est une région montueuse, tourmentée par le travail volcanique à un point qu'on ne saurait rendre. Aucune eau fécondante ne parcourt les brûlants replis de cette terre ravagée en tout sens par les feux souterrancs et embrasée par le soleil des tropiques."

The rest of the road from Aloolo to Ankobar is for the most part very destitute of water and forage. Of the road which branches off towards Magdala, I have not been able to find any description, but there is no reason to believe that the country which it traverses, differs materially from that through which the main road passes. Krapf, writing October 1st, 1866, says that from Aussa this road follows the bank of the noble river Hawash, abounding in wood, grass, and wild animals, as far as the junction of the Berkona, thence to Lake Haik. In one of the itineraries, however, given by Lefebvre, the banks of the Hawash are said to be "frequented by wandering tribes, but they make no stay there, because the air is unhealthy."

The temperature along this road in summer frequently ranges over 100° in the shade at noon. D'Héricourt during his journey in September gives the temperature of Ambabo at 92½°, and of the salt lake to the south-west of Tajurrah at 92½°; the time of day is not stated. The same traveller gives the temperature at the Hawash River as 91°. The rainy season on this low country is in the winter, at the contrary time of year, that is to say, to that in the highlands; D'Héricourt states that the rain only falls between seven and nine o'clock in the evening, and that the rest of the day and night are fine.

The distance from Tajurrah to Magdala, by Aussa, is about 300 miles, and by Ankobar about 400.

The works from which extracts bearing on the routes from Tajurrah have been made will be found by referring to the index and map given at page 39.

3. *Route from Amphilla Bay by Lake Ashangi (about lat. 12½°, long. 39½°).*—This route is advocated by Colonel Merewether, who writes on 28th February, 1867:—

"On my way up from Aden I called in at Amphilla Bay, and found it admirably suited for the debarkation of troops; good anchorage; water good and plentiful; and one of the easiest and best roads comes down from Abyssinia to it—a road constantly used by caravans (camel), who take backwards and forwards the salt obtained near Amphilla. The people in the country are simple and friendly disposed, and there are no Egyptian troops there."

And on April 29, 1867:—

"But I am disposed to think, that should a force be sent into Abyssinia, the best way of approaching Begemeder and the Emperor Theodorus will be by landing the troops at Amphilla, and marching through the Azubo Galla country to the neighbourhood of Lake Ashangi, and then entering Abyssinia Proper, through Lasta and Wadela, both of which are in most active opposition to the Emperor, and thus avoiding Tigré and its high table land, where camel forage is reported difficult to be obtained always.

"The Azubo Gallas are independent, their country is said to be well supplied with grain, grass, water, and camel forage, and the people well disposed. I am having full inquiries made about this route. Reference to Keith Johnston's map of Upper Nubia and Abyssinia will show that Amphilla and Tajooro are exactly equi-distant from Magdala. The former, however, though further by sea from Aden than the latter, has the advantage of leading at once into a suitable country; whereas the latter, in the first 200 miles, opens into a country belonging to the most inhospitable, worst dispositioned people possible, where water and supplies are both very scarce."

And again on 1st May, 1867:—

"But I am strongly of opinion that the best line of operation will be through the Azubo Galla country, where there are no rivers of any consequence, where the people are independent, and where supplies and forage for the baggage animals would be found in abundance, than through Hamozeyn, Adowa, and Sokota, where there are rivers and large ravines, and where, though supplies are plentiful, forage for the baggage cattle is uncertain. The further information I am collecting regarding the route from Amphilla will prove the value of this opinion."

This road was travelled by the Jesuit priest Don Alonzo Mendez in May and June, 1625, and by Mr. Coffin, Mr. Salt's companion, in January, 1810, both of whom have given descriptions of it (see index and map page 39). The distance from Amphilla to the pass by which the Abyssinian table land is reached appears to be about 100 miles. The greater part of this tract is of the usual character of the low country between the mountains and the sea, hot, arid, and deficient in water. About 50 miles from the coast there is a very remarkable feature, consisting of a great plain of solid salt, about 40 or 50 miles long, and 10 or 20 miles broad, and two or three feet thick, which is cut into blocks by the Abyssinians, and used as money. The pass to the highlands is described to be as high as the Taranta pass, but not nearly so difficult. Don Alonzo Mendez traversed this road at the worst period of the year. He says of the first portion of it: "Eating very little besides rice we had with us, meeting no town to furnish us with provisions, and the heat so violent that it melted the wax in our boxes; without any shade but that of briers which did us more harm than good, lying on the hard ground, and drinking brackish water of a very ill scent, and sometimes but very little of that, &c."

M. Lefebvre attempted to make a journey from Atebidera towards the salt plain, apparently in the direction of Amphilla Bay, extracts from which are given at page 198. His account of the great heat tallies with that of Don Alonzo Mendez. He attempted the journey at the worst time of year (June) and he was told that "an Abyssinian, much more a white man could not support the heat; the Taltals* themselves often fell, struck by congestion of the brain." He persisted in his journey, but, before he came in sight of the salt plain, he was struck down by the sun, and was obliged to retreat precipitately, narrowly escaping with his life.

The distance from Amphilla Bay to Magdala is about 40 or 50 miles less than from Massowah.

4. *Route from Suakim to Kassala, Metemmah, and Gondar.*—Suakim is a port of the Red Sea, belonging to the Egyptians, from whence they keep up their communications with their frontier post at Kassala, where they have a considerable force. It is

* The inhabitants of the plains.

stated by Sir S. Baker to be from 16 to 20 days' journey from the latter for a laden camel. That traveller also states that plenty of camels, and the necessary water skins for the journey across the desert, can be procured there, but the water is brackish, and a large supply of Nile water would therefore have to be brought from Suez. Throughout the desert route fodder for the camels is afforded by numerous Mimosas, and water is found every second or third day.

This route was travelled by Mr. Hamilton in 1854, who gives a minute account of it, extracts from which are given at page 130 (marked AA on map page 39). He suffered considerably from heat and want of water in some places, but he made the journey at the worst time of year. The town of Kassala is fortified, and from 6,000 to 8,000 Egyptian troops are usually, according to Sir S. Baker, quartered in the district. It is situated on the Gash or Mareb, which, although dry at some periods of the year, affords an unlimited supply of good water from wells dug in its sandy bed. During the dry season, from 15th November to 1st June, the climate is healthy, but at all other seasons the country is extremely dangerous.* A peculiar fly appears with the first rains, that destroys all domestic animals.

From Kassala there is a route to Adowa by the Basé country, which was traversed by Münzinger in 1861-2. The usual route to Gondar is by Metemmah. This was the one followed by Mr. Rassam. There appears to be also a road from 'Sofie, half way between Kassala and Metemmah, which leads by the Takazze into the Massowah-Gondar road; this was followed by Mansfield Parkyns in the summer of 1845; also, for some portion, by Baker in 1861-2. From Metemmah there seem to be two roads to Gondar, one to the north, the other to the south. The former was followed by Bruce on his return home from Abyssinia in 1771-2 (marked Z 1 on map page 39). He gives a very graphic account of it, extracts from which are given at page 134. Between Gondar and Shelkin, or Tcherkin, he found the country well wooded and watered, generally passing two or three streams a day. After passing Tcherkin, the route lay through enormous forests, roamed over by elephants and other wild beasts. As they approached Metemmah, the forests opened out into a park-like country. They complained a good deal of the heat.

The lower road was traversed by Krapf in May, 1855 (marked Z 2 on map page 39). The journey from Gondar to Metemmah occupied ten days, including two days' halt on the road; for the first five days the route lay through the mountainous country of Abyssinia; they then descended into a vast plain, interspersed with forests. The population of Metemmah is stated to be 1,500. A large market is held there.

* Mr. Rassam, alluding to Kassala, writes—"Cholera and deadly fever were the scourge of the place from the month of July to October; and even while I was there (he left on 9th November, 1865), nearly one-tenth of the garrison was laid up with one kind of disease or other."

It may be observed that the whole route from Suakim, by Kassala and Metemmah, to Gondar, is through the low country, until within four or five days' journey of Gondar, when the Abyssinian highlands are reached. The distance from Suakim to Gondar is about 600 or 700 miles.

Dr. Beke also mentions a road "running westward from the sea coast at Raheita, just within the straits of Babelmandeb;" and he states that the road by Senafé may be reached from Harena, in Hawakil Bay.

The following short account of the different harbours which communicate with the routes that have been mentioned, has been furnished by the Hydrographer to the Admiralty.

Massowah.—Eight or ten of the largest ships with double the number of smaller ones could be securely moored in the harbour. There is also a good harbour called Daha-leah, larger than Massowah, about a mile to the north. In both of these the water is quite smooth. The fresh water supply is from tanks in the island of Massowah. There is fresh water also at Daha-leah. Fresh water is not abundant either here or in any other part of the Red Sea, but the supply at Massowah could probably be increased by digging wells on the main land.

There is a pier with facilities for landing on Massowah island which is connected with the main land by low wet ground about a mile in length. The rainy season is from November to March.

It is 380 miles from Aden, and 290 from Perim. The navigation for the greater part of the way is clear and safe, and for the whole way in the daytime for carefully navigated ships.

Ansley Bay.—The water is inconveniently deep. It is described as presenting a remarkable contrast in point of fertility to other spots, bordered by low land producing rich pastures. Fresh water may be procured. Although not so convenient a spot for naval operations as Massowah, yet, if other advantages preponderated in favour of it, it is more than probable that no great difficulties would be encountered on account of the ship's anchorage, landing, &c.

Suakin.—The harbour is very small, the approach is studded with dangers, and there is no outside anchorage. It is extremely hot. Thermometer in May ranges from 89° to 91°, in June from 93° to 97°, very much less vegetation than at Massowah. Water procured from wells.

Amphilla is described as the most miserable spot on the coast of Abyssinia. In regard to anchorage, facilities for landing, &c., it is not to be compared to Massowah; according to the Admiralty charts it is very circumscribed and intricate.

Tajurrah is quite unsafe and exposed to the north-east monsoon, as well as the southerly winds, and it is probable that ships could not lie there, nor a landing be effected very often.

There is no other spot but Massowah and its immediate neighbourhood where ships could lie safely for any time, and where troops and munitions of war could be disembarked with celerity and safety.

The following are some of the principal towns of Abyssinia:—

Gondar, in Amhara.—The capital of the kingdom. This town is stated by Heuglin, 1862, to have contained from 6,000 to 7,000 inhabitants, but it is said to have been within the last two or three years totally destroyed by the Emperor Theodore.

Debra Tabor, in Amhara.—Formerly a small village. It is now a place of considerable size, and the residence of the Emperor Theodore. Near Debra Tabor is Gaffat, where the European workmen of the Emperor reside, and which may be considered as his arsenal.

Adowa,* the capital of Tigré. This is the second city in the kingdom. It is stated by MM. Ferret and Galinier to have contained in 1840 not more than 4,000 inhabitants. Heuglin, in 1862, put the population at 6,000. The miserable nature of Abyssinian towns may be judged of by the description given by Mansfield Parkyns, in 1843, of this, the second city of the Empire. He says,—

"I own I rather expected to see columns or obelisks, if not an acropolis, on some of the neighbouring hills. Judge, then, of my astonishment when, on arriving at this great city, the capital of one of the most powerful kingdoms of Æthiopia, I found nothing but a large straggling village of huts, some flat-roofed, but mostly thatched with straw, and the walls of all of them built of rough stones, laid together with mud, in the rudest possible manner. Being wet, moreover, with the rain, the place presented the most miserably dirty appearance."

Mr. Dufton, who visited Adowa, puts its population at 10,000.

Antalo.—The capital of Enderta, and one of the principal towns of Tigré. It is said by Ferret and Galinier to contain from 200 to 300 houses.

Chelicut, near Antalo.—This town is said by Lefebvre and Ferret and Galinier to contain about 3,000 inhabitants.

Sokota, the capital of Laag and Wasta, is a place of considerable size.

Dixan, in Tigré.—This is the first town that is met with after surmounting the Taranta passes. Ferret and Galinier say of it,—

"A group of wretched huts, scattered irregularly on the top of a barren mountain, a miserable village, containing about 1,500 souls, Christians and Mussulmans, there is Dixah."

Tzazega.—The capital of Hamazen, said by Heuglin to contain from 1,500 to 2,000 inhabitants.

* Near Adowa is Axum, the ancient capital, among whose ruins, obelisks, churches, and Greek and Abyssinian inscriptions still bear witness to its former importance.

Abbiaddg.—The capital of Tembien. Mr. Dufton says that the place is Mahomedan, and has an occasional market, and that, when he entered it, about 2,000 people were assembled in the market place.

Addigraht.—The capital of Agame.

Hawzen, or Aouissienne.—The capital of Haramat.

Mota.—A large town in Godjam, said by Dr. Beke, to contain 3,000 inhabitants.

Ankobar.—The capital of Shoa.

Angolala.—This is a place of considerable size in Shoa. It is said to contain from 3,000 to 4,000 inhabitants.

Aliya Amba.—This is a large market town in Shoa, and is said by D'Héricourt to contain from 2,000 to 3,000 inhabitants.

GOVERNMENT OF ABYSSINIA.

ABYSSINIA is one of the most ancient monarchies in the world, and has been governed from time immemorial by an Emperor. The form of Government and its military spirit are feudal. Each chief holding the rank of Dejajmatch is entire master of all sources of revenue within his territory, and has practically full power of life and death. His feudal subjection consists in the obligation to send from time to time presents to his superior, and to follow him to war with as large a force as he can muster. It has resulted from this that the great feudatories have become practically independent rulers, and that the Emperor has been for many years, until the accession of Theodore, a mere puppet in the hands of one or other of them.

Doctor Beke thus describes the relations which subsist between the Emperor and the chiefs:—

"Abyssinia," he says, "is an hereditary monarchy, under the sway of an Emperor claiming descent from Solomon, king of Israel, and the Queen of Sheba. Though this parentage is a mere fiction, there are few Christian sovereigns who can show a more illustrious lineage than the Emperors of Ethiopia, whose progenitors received the Christian faith, and possessed a native version of the Holy Scriptures as early as the fourth century, when the now civilized nations of Europe were in a state of barbarism.

"The occupiers of the throne of their once absolute ancestors have for a considerable time past been mere puppets in the hands of one or the other of their powerful vassals; the form having been kept up of nominating an Emperor of the ancient line of Solomon, who, however, has remained a prisoner in his palace at Gondar, his sole revenue consisting of a small stipend and the tolls of the weekly markets of that city.

"Since the commencement of the present century, the seat of government and the person of the Emperor have remained, though with occasional interruptions, in the hands of the chiefs of a powerful tribe of Yedgu (Edgow) Gallas, who for three generations have been able to secure to themselves the dignity of Ras or Vizier of the empire—that is to say, to become its sovereigns in everything but in name. This sovereignty within the central portion of the empire has, however, been far from giving them the command over the surrounding provinces. On the contrary each ruler of a province has mostly acted as an independent sovereign, and if at any time he has found himself strong enough to march upon the capital, he has done so, placed upon the throne another puppet Emperor, and been by him appointed Ras or Vizier, till a rival stronger than himself could turn him out and take his place. Under such circumstances it is not to be wondered at that there should be at one time half a dozen titular Emperors, and that the Governor of each of the principal provinces should have assumed the title of Ras, and continued to bear it, even when no longer in power."

The three principal provinces of the empire are Tigré, Amhara, in which Gonda the capital is situated, and Shoa. The governors of these have all at different times assumed the title of Ras. Three other provinces of some importance are Lasta and Waag, whose capital is Sokota, and which are ruled over by the rebel Gobazie, Godjam, to the south of Lake Tsana, and Kwara, to the west of the same lake, the birth-place of the Emperor Theodore.

The two provinces of Tigré and Shoa have generally been in a state of rebellion from, or acknowledged independence of, the central power at Gondar. The geographical position of Tigré enhances its political importance, for, as the communications between Gondar and the sea at Massowah necessarily lead through Tigré, it follows that that province holds, so to speak, the gate of the capital. The language of the province of Tigré also differs from that of Amhara, as in the former the Gheez and in the latter the Amharic dialect is spoken. Between the two provinces there have been almost constant wars.

The province of Shoa is almost separated from that of Amhara by the Wollo Gallas, a Mahomedan tribe. Shoa has been for a long time virtually independent, and has been governed by a hereditary line of princes, to one of whom the Indian Government sent a special embassy under Major Harris in 1841.

RELIGION AND CHARACTER OF THE ABYSSINIANS.

THE greater part of the inhabitants of Abyssinia are Christians of the Coptic sect; they hold many of the tenets of the Roman Catholic faith, such as priestly absolution, fasts, worship of saints, conventualism, &c. They also follow many Mahomedan practices, such as circumcision, ceremonial uncleanness, abstaining from meat prohibited by the law of Moses, &c., and in their laws and customs Jewish institutions are often traceable. Many of their names betray also a Jewish, or at all events biblical origin, such as Debra Tabor (Mount Tabor), Debra Libanos (Mount Lebanon), Antiokia, &c. Their Christianity has degenerated into little more than form, and they are described as priest ridden and superstitious to the last degree. The head of the church is a Bishop or "Abuna," who is consecrated by the Patriarch of Alexandria, and by whom the Emperors of Abyssinia are crowned. Much weight is attached to this ceremony, as may be gathered from the fact that the Ras of Tigré in 1840 thought it worth while to go to great expense to send a mission to Alexandria to have an Abuna appointed, in the hopes that he would consecrate an Emperor who would be a puppet in his hands, and give him the supreme command. The present Abuna is out of favour with Theodore for the alleged cause, that he imparted to the missionaries his doubts of the authenticity of the descent of Theodore's mother from the Queen of Sheba. That the authority, however, of the Abuna is still recognized, seems probable from the fact that it is stated in a recent letter from the captives that if the rebel Gobazie were to take Magdala and get himself crowned by the Abuna he would be recognized as Emperor by the Abyssinians.

Of the religious character of the people Mr. Dufton says:—"Christian liberty is entirely unknown, as the people are bound down to unmeaning forms and ceremonies, and the observance of fasts which extend over two-thirds of the year. Their calendar is crammed full of saints, and the days of the year by no means suffice for them all, so that they have morning celebrations and evening celebrations. One cannot wonder at this when their latitudinarism leads them to commemorate Balaam and his ass, Pontius Pilate and his wife, and such like doubtful saints. In addition to the heroes of the Bible and Apocryphal books, they have many local saints, who have at various times astonished Abyssinia by their miracles and prodigies .... Apart from such traditional excrescences, the Abyssinians are orthodox in their belief, the grand truths of our religion being received alike by them as by us; but, being void of that charity which edifieth, their knowledge has only tended to puff them up, and the intolerance with which they look upon their Mahomedan and Jewish neighbours is even greater than that of those people themselves towards Christians .... What renders their pride the more offensive, is that the Mahomedans and Jews are in every way their superiors, possessing with an equal amount of intelligence, far greater mechanical genius, and superior habits of industry. All the manufacturers of cotton cloths are Moslems; all the builders and artizans are Jews. But pride is not their only fault, they are deceitful, lying, insincere; their breasts are seldom stirred by generosity towards others, or in gratitude for benefits received, and, added to all, they are inhospitable."

Mansfield Parkyns gives the same account. He says of the Abyssinians: "Bigoted and prejudiced in the extreme, they will not eat of the meat slaughtered by any one but a Christian. They are extremely superstitious in their belief of miracles, and the interposition of the saints, the names of some of whom are continually in their mouths. Their fasts are more numerous perhaps than those of any other Christian people, more than two thirds of the year being assigned to abstinence, &c."

Dr. Gobat, the present Bishop of Jerusalem, describes them as very immoral and licentious, priest-ridden and bigoted, but hospitable to travellers, and with no cruelty in their dispositions.

Major Harris says: "Abyssinia, as she now is, presents the most singular compound of vanity, meekness and ferocity—of devotion, superstition, and ignorance. But compared with other nations of Africa she unquestionably holds a high station. She is superior in arts and in agriculture, in laws, religion, and social condition to the benighted children of the sun."

Mansfield Parkyns says that the Abyssinians are of middle stature, averaging about 5 feet 7 inches. He had seldom seen natives above 6 feet, and only one or two who reached 6 feet 2 inches. In colour some of them are perfectly black; but the majority are brown or a very light copper or nut colour. Both men and women are remarkably well formed and in general handsome.

The religion of the Abyssinians tends to increase their isolation from the Mahomedan tribes who occupy the low lands by which Abyssinia is surrounded—the Shohos on the north, the Basé and Hamran Arabs on the west, the Danakils, Tantals, Asubo Gallas, and Wollo Gallas on the east and south.

CURRENCY OF ABYSSINIA.

THE only European coin that is current in Abyssinia is the Marià Theresa dollar of A.D. 1780, and it is necessary that it should have certain distinctive marks, viz., a diadam of pearls, a pearl brooch on the shoulder, and the mint mark S.F. Dr. Beke says that, even if the dollar is of the correct kind, should these marks not be perfectly distinct, he has known the natives to refuse it. Its value is about 4*s.* 3*d.* This dollar is no longer in circulation in Austria, but is still coined at the Government mint for exportation to the Levant, &c.

For smaller monetary transactions the circulating medium consists of blocks of salt 8 inches long by 1½ inches in breadth. These are called in Shoa, according to Dr. Beke, ámolés, and in Tchelga, to the West of Gondar, according to Mr. Dufton, tsho. Their value is given by the former as 2½*d.*, and by the latter as from 2*d.* to 3*d.*; it varies probably according to the distance from the source of supply. These blocks of salt are obtained from a great salt plain situated between Amphilla Bay and Atebidera, which is thus quaintly described by Don Alonzo Mendez, Patriarch of Abyssinia, who traversed it in 1625:—

"The boundary between the kingdoms of Dancali and Tygre

is a plain four days' journey in length, and one in breadth, which they call the country of salt, for there is found all that they use in Ethiopia, instead of money; being bricks, almost a span long, and four fingers thick and broad, and wonderfully white, fine, and hard, and there is never any miss of it, though they carry away never so much; and this quantity is so great that we met a caravan of it, wherein we believed there could be no less than 600 beasts of burden, camels, mules, and asses, of which the camels carry 600 of those bricks, and the asses 140 or 150, and these continually going and coming. They tell many stories concerning this Salt Field, and amongst the rest, that in some part of it, there are houses that look like stone, in which they hear human voices, and of several other creatures, and that they call such as pass that way by their names, and yet nothing can be seen. The Moorish Commander told me, that as he went by there with a Lion, Ras Cella Chistos sent to Moca, three or four of his servants vanished on a sudden, and he could never hear of them afterwards. In one place there is a mount of red salt, which is much used in physic. This is to be passed over by night, because the heat is so violent in the day, that travellers and beasts are stifled, and the very shoes parch up, as if they were laid on burning coals. We entered upon it at three in the afternoon, and it pleased God that the sun clouded, which the renegado Moor attributed to his prayers."

Mr. Dufton says that Abbi Addy in Tembien "is the last place where the salt of Amhara is taken as money. The nearer one gets to the coast where are the salt mines, of course the less is the value of the mineral; and in Tigré cotton cloth takes its place as a medium of exchange."

MILITARY STRENGTH OF ABYSSINIA.

It is very difficult to estimate the military strength of a country like Abyssinia, as, from what has been before stated of the nature of the Government, it will be seen that the forces which a Sovereign of Abyssinia can bring into the field will depend very much upon his own personal character, and upon the number of Chiefs over whom he has any influence. In a Memorandum communicated by the Foreign Office to the Royal Geographical Society, in 1855, there is the following paragraph bearing upon this subject:—

"The immediate troops of the Ras consist of a number of petty chiefs, governing one, two, or more villages, who imitate, as far as they dare, the independence of the greater barons, and who take the field when called on with 500 men, according to their means. Besides these, who are numerous, the Ras has his matchlock men, and four or five bands of rude and disorderly soldiery, his guards. From the low system of government, and the manner of paying these men by quartering them on the country people, with instructions to levy so much grain or other property, it may be supposed that these undisciplined troops, when at a small distance from the camp, are almost equally independent of the Ras, and frequently are simply organized bands of robbers, &c."

The nature of an Abyssinian army may be gathered from the following graphic description, given by Major Harris, of a marauding expedition of the King of Shoa, which he accompanied in 1840:—

He says, "The military system of Shoa being entirely feudal, each governor of the realm is required to furnish his contingent of militia, in proportion to his landed tenure, his peasantry being at all times ready for the foray and expected to purvey horses, arms, and provisions without payment from the State. Four hundred fusiliers, bondsmen of the King, alone receive pay. Little discipline exists in the army thus composed, but considerable tact is evinced in its organization and distribution."

The following was the proclamation by which the King mustered his forces:—"Hear, oh hear! Behold we have foes and would trample upon their necks. Prepare ye, everyone, for war. On the approaching festival of Abba Kinos, whoso faileth to present himself as a good and loyal subject, mounted, armed, and carrying provisions for 21 days, shall be held as a traitor, and shall forfeit his property during 17 years."

On the appointed day they commenced their march. "Immediately in advance of the army, screened beneath a canopy of scarlet broad cloth, were borne, on an ambling mule, the Holy Scriptures and the Ark of the Cathedral of St. Michael. The King rode next upon a richly caparisoned mule, a small space round the Royal person being kept clear by the corps of shield bearers, who were flanked on the right by fusiliers and matchlock men of the Body Guard, and on the left by the band of kettle drums, on donkeys, with trumpets and wind instruments. Numerous governors, judges, monks, priests, and singers followed, and behind them rode a curious accompaniment to a martial expedition. Forty dames and damsels professing the culinary art with elaborately crisped beehive wigs, greased faces, bedaubed with ochre, and arched blue eyebrows, were muffled in crimson-striped robes of cotton—a demure assemblage, rigorously guarded on all sides by austere eunuchs, armed with long, white wands. Beyond, far as the eye could penetrate the canopy of dust which hung over the horizon, every hill and valley swarms with masses of equestrians and pedestrians, henchmen, and camp followers, sumpter horses, asses, and mules, laden with tents, horns of old mead, and bags of provisions. Throngs of women carrying pitchers of beer and hydromel at their backs, and lads with glittering sheaves of spears upon their shoulders, leading gaily caparisoned war steeds,—all mixed and crowded together in the most picturesque disorder and confusion."

"At the termination of the fifteenth mile, the ladies and their eunuchs having hovered about for some time in uncertainty, finally settled down, like a flight of flamingos, in a pretty secluded valley, through which winds the deep, muddy Baroga."

The whole then encamped round the Royal tents covering a space of five miles in diameter.

"Early in the ensuing morning, the Royal drums beat to saddle, and, in half an hour, the army, which had swelled to about 15,000 fighting men, was in motion over a country especially favourable to advance. Some military precautions were now observed. Large brigades of horses serving as flanking parties, and the heights on the right and left being crowned with patrols."

When they arrived on the scene of the proposed campaign, they burst into a valley occupied by a hostile tribe of Gallas, killed every male they could catch, carried the women and children into captivity, burnt the houses, and destroyed the crops.

Theodore's army in his prosperous days was estimated as high as 60,000 men. It probably did not differ much in its constitution from that described by Harris. It is stated that he tried to introduce a better system of discipline in it, but was obliged to give up the attempt, owing to the irritation caused by it among the soldiers. From the accounts of the captives, it does not appear that he can muster now more than 5,000 men. His attempt to have guns and mortars made by his European workmen at Gaffat appears to have been a failure.

Mr. Dufton thus describes a visit to Theodore's army, about 1863 :—"The king's army possibly consisted of some 50,000 warriors, but the number of camp followers is often double that of the army itself, which, of course, adds much to the unmanageableness of the whole; indeed, there is little or no system in Abyssinian warfare. It consists in rushing pall-mall upon the foe, hurling the spear, which is their principal arm, and picking up and re-hurling the spent darts of the enemy. The musket, which is mostly in the hands of the Tigrean soldiers, is even less effective than the spear, and the amount of powder and shot wasted must be enormous. The sword is seldom brought into requisition, as arm to arm combat is unfrequent."

The following notes on the army and fortresses of Abyssinia have been communicated by Captain Webber, R.E., and Captain Hobart, R.A., who received the information from M. Legean, who was French Vice-Consul at Massowah, in 1863 :—

Army.—The battalion is the unit. It consists nominally of 1,000 men, and is commanded by a chief and numerous under officers. The fighting strength only amounts to 250 well-armed men, and about 150 to 200 half-armed followers, the remainder being merely servants. A thousand rations are drawn for each battalion, the number including about 250 women. These details apply only to Theodore's *regular* army, of which he can muster about 60,000,* who are quartered in time of peace on the various districts of the country. Of these, 20,000 are armed with percussion fire-arms; the rest with sword and spear. Owing to the badness of the quality of the fire-arms, they count much more on the latter than on the former. Their powder is chiefly imported. Rigid obedience is exacted to the immediate superior officer, but there is no attempt at formation except for defence, when they form line, the front rank kneeling, and covering themselves with shields of rhinoceros' hides.

There is no attempt to carry artillery in the field, but they have many guns and mortars in the forts, or "ambas."

Monsieur Legean considers that the Abyssinians are brave even to temerity, and that they would not, in the first instance, try to defend the passes, but would rather allow an army to enter the country, and attack them in the open field. He speaks of having witnessed reviews and sham fights.

The irregular army is the feudal following of the great chiefs, and its numbers depend on the willingness of the chiefs to obey the Emperor's summons. They might amount to nearly 100,000 men.

Monsieur Legean considers that it is important that no cause should be given to the population to believe that the war is of a religious character. If Theodore could excite this belief, he thinks that the whole country would join him.

To a Christian army all Tigré would be friendly, and the country people, if they once get confidence, would be glad to furnish all sorts of supplies. They have a character for honesty, and for keeping to agreements.

Forts.—The hill forts, or ambas, occupy the summits of small table mountains where water is to be had. They are scarped on all sides, and have only one means of access—by a winding ascent. It is rarely necessary to fortify the summits, or build a rampart. M. Legean considered them impregnable to assault, and unassailable by mining operations, on account of the basaltic formations. They could generally, however, be taken by stratagem. Their garrisons only consist of 300 or 400 men, and their chief use is as depôts, &c. The greatest number of them being to the south, they are not likely to prove an obstacle early in the campaign.

The following are some of the principal forts:—

Gondar.—Although this is the capital, its capture would not have much effect on Theodore, whose policy is to have no fixed residence, so that it cannot be said, if any one important town is taken, that he has lost its capital. The capture of Gondar would give possession of the richest part of the country.†

Tchelga.—South-west of Gondar. Very strong.

* This evidently refers to an earlier portion of Theodore's career, probably about 1863.

† By the latest accounts it appears that Gondar has been destroyed.

Amba Ras.—South of the Taccazy, in Samen, near the Chaakne* pass.

Amba Gah.—South-east of Gondar. A favourite residence of the Emperor, and a state prison. A very strong natural position.

Selalkulla.—Near Wobo. Very strong.

Magdala.—Said to be very strong, but never seen by M. Legean. (A description of this fort is given by Steudner, page 250.)

Djibella.—Near the Abai river. Very strong, naturally and artificially.

There are some three forts in the country lying between Gondar and Magdala; one near Zengadi, one at Emfras, and one at Mahdera Mariam.

Devra Damo.—North-east of Adowa is a monastery, in a very strong position, overlooking the route. It is also artificially strengthened.

There is a fort near Aoussienne, in the Haramat country, the favourite residence of King Oubi, the great enemy of Theodore.

Between Yaha and Guendepta, north of Adowa, there is a very strong pass.

One of the great obstacles on the route between Adowa and Gondar is the Chaakne Pass, in Wagara. The ascent is a zigzag, and very difficult, occupying a whole day. It is the greatest impediment for guns in the whole route. (This appears to be the Lamalmon Pass described by Ferret and Galinier, see page 56, and by Bruce, see page 242.)

DESCRIPTION OF THEODORE.

THE Emperor Theodore, Mr. Dufton tells us, was born about 1820, in the province of Kwara, which is situated to the west of Lake Tsana, and of which his uncle was Governor. His mother was, according to some, of low extraction, but, according to others, she was of good birth, and could even trace her descent from the Queen of Sheba, the orthodox ancestress of Abyssinian Royalty. Kassa, for such is the Emperor's real name, soon distinguished himself by his bravery and talent for war, and on the death of his uncle he acquired the government of the province of Kwara. He soon enlarged his dominions by the conquest of adjacent provinces, and at length, after having experienced varying fortunes, he defeated Ali Ras of Amhara, whose daughter he had previously married, and effected the conquest of that province. The element of religion, which is so singularly blended in Theodore's character, is shown by the prayer which he publicly offered up after his victory, and which is as follows: "I praise thee, O God, that Thou hast manifested Thy good-

* Apparently the same as the Lamalmon Pass.

ness to a poor sinner like me. Whom Thou humblest is humbled, and whom Thou exaltest is exalted. Thine is the power and glory, for ever and ever."

Being now installed at Gondar as Ras of Amhara, a rank which carried with it the nominal allegiance of the whole of Abyssinia, Kassa sent to claim tribute from Oubie, Prince of Tigré. This being refused, he marched with an army against him, and, having defeated him in the battle of Deraskie, he had himself crowned as Negus Theodorus, or King of Kings of Ethiopia. His assumption of the name of Theodorus appears to have been made in consequence of an ancient prophecy that an Emperor of that name would raise the kingdom of Abyssinia to an unprecedented pitch of greatness. He then attacked the Wollo Gallas, a Mahomedan tribe between Amhara and Shoa, and defeated them in a battle, in which their king, Adara Bille, was slain.

Theodore had thus made himself master of the whole of Abyssinia with the exception of Shoa, which had long been virtually an independent state. Against this kingdom he now directed his arms, and soon succeeded in completely subjugating it, thus reuniting under his sway the whole of the so long disunited provinces of Abyssinia. His next project was to drive the Turks from their possessions on the coast, and thus to acquire for Abyssinia an outlet on the Red Sea, an advantage which that country had not possessed since Massowah was taken by the Turks in the 16th century. This however he was never in a position to attempt. His conquered provinces revolted, and the cruel element in his character which soon developed itself so alienated his subjects, that they gradually fell away from him until now he seems to hold little territory beyond that in the immediate neighbourhood of Debra Tabor which he has made his capital.

Theodore appears to be a man of great talent, courage, and energy, with a singular power of command over others. Mr. Dufton, who saw him in 1863, thus describes him: "His appearance was that of a man of about forty-five, of middling stature and possessed of a well knit but not over powerful frame, conveying more the idea of being tough and wiry than of strong physical development. His complexion is dark, approaching to black, but he has nothing of the negro about him. His features are altogether those of an European. His head is well formed, and his hair is arranged in large plaits extending back from the forehead. His forehead is high and tends to be prominent. His eye is black, full of fire, quick and piercing. His nose has a little of the Roman about it, being slightly arched and pointed. His mouth is perfect, and the smile, which, during the conversation, continually played upon it, was exceedingly agreeable, I may say fascinating. He has very little moustache or beard. His manner was peculiarly pleasant, gracious, and even polite, and his general expression, even when his features were at rest, was one of intelligence and benevolence. On the whole, the physiognomist could find no trace of fierce passion save in the lightning glance of his eyes. I watched for the keen shot of light coming from them at times, and reflected upon what he could be capable of, but they did not strike me as treacherous eyes. I felt that he could act savagely under irritation." He adds in a note, "I here take occasion to remark that, though Theodore consumes a vast quantity of Arracky, he is no drunkard; that is, I have never heard of him being overcome with drink. He always stops at a certain point."

In another place Mr. Dufton gives a curious trait of the Emperor. "All the time that Theodore was speaking of these warlike preparations, he was playing with a little child of M. Bourgaud's, which he had seated alongside of him on the carpet, between himself and M. Legean; and certainly, a stranger who saw him there for the first time, and who knew nothing of his antecedents, would have found it difficult to believe that he was the cruel monster which recent accounts unite in describing him." M. Legean has given in his work a picture of Theodore which agrees very well with Mr. Dufton's description of him.

Since the date of Mr. Dufton's visit, the cruel side of the Emperor's character appears, from the accounts of the captives of Magdala, to have been developed to the utmost. Dr. Blanc, one of the prisoners, writes on the 18th June, 1867, from Magdala.

"A priest who arrived here a few days ago, having accompanied Ras Adilon in his flight, states that 650 Wadela men were executed on the day previous to the desertion of the Yadja people. A follower of the King accused them before His Majesty of intending to run away. The Wadela Chiefs denied the charge; the accuser brought nine witnesses. The case heard, the Wadela people were all conducted into an inclosure prepared beforehand; logs of wood were tied round their necks; their arms and legs bound with leather thongs; helpless and innocent they fell easy victims to the executioner's sword. One of the Chiefs, whilst awaiting his turn, shouted at the top of his voice, so that the King, who was at no great distance, might hear himself the dying words of his former staunch follower; 'Oh, King, you murder me in cold blood; is that the reward of my long and faithful services? I am going to die, but, before long, will meet you before a just judge, where I and my innocent companions will become your accusers. Believe a dying man; you will not survive us more than a few days.'

"It appears that these words made a temporary impression on the superstitious and cowardly mind of the despot. He has since then, it is rumoured, turned up a new leaf, is constantly in prayers, calls himself the 'slave of Christ,' and will probably build a church to obtain absolution from the priests, and wash off, if possible, from his conscience, the innocent blood in which he daily carouses.

"The wife and child of Zalalu fared even worse than the Wadela people. This poor woman and her child were wrapped up in wax clothes and burned like candles. Their cries were heard for many miles around, and increased, if possible, the hatred and desire of vengeance of the already exasperated peasantry.

"In the history of the world there is no parallel case. It is true tyrants and despots have wantonly spilt human blood, great conquerors have overrun large tracts of country: out of 3,000,000 inhabitants he has destroyed more than a third by war, famine, and murder. Nero, Attila, Tamerlane, were lambs when compared to Theodorus. No man was ever so false, so treacherous—no man held friendly or family ties so cheap. His own son, Ras Meshisha, is in chains. Ras Taga, whom he left in charge of his camp during his late expedition, was chained; a reward for his fidelity on the return of his ungrateful master. We are not therefore astonished when we daily hear of fresh desertions."

On the 2nd July, 1867, Mr. Flad, one of the prisoners at Debra Tabor, writes:—

"We are in perpetual fear of our lives. The king during the last six weeks has had 4,000 persons put to death, soldiers and peasants, either burning them, or cutting their throats like beasts, or shooting them. He has caused women and children to be tortured, dishonoured, and starved in an unheard-of manner."

And again on the 13th August:—

"His Majesty enjoys good health now, and is in excellent spirits; he is progressing in subduing his rebellious peasants and deserted soldiers by the sword and by the fire. There is no doubt if he will go on in this manner scarcely a man or female will be left alive. A beautiful island in the Lake Tsana, Mahuska, was lately burnt with its inhabitants, the peasants of both sexes, aged grey-bearded people and little children, were in vast numbers burnt. All the villages about are heaps of ashes, and the churches broken down. Inside the hedge, which surrounds the camp here, His Majesty reigns; outside his name is not known. Peasants hiding themselves in caverns, and deserted soldiers come near to the fence every night, crying out the hardest truths, which are called insults."

There seems little doubt that the subjects of Theodore, incited by these acts of cruelty, are almost universally in open rebellion. The great provinces of Tigré and Shoa have resumed what may be called their chronic state of rebellion. The Wollo Gallas are, as usual, hostile, but their hostility seems to be complicated by the fact of their having two rival Queens.

But the most formidable rebel appears to be the Waagshum Gobazie, who rules over the mountainous and almost inaccessible regions of Waag and Lasta.* His movements, however, seem

* His dominions are shewn on the map at the end of the volume by a green tint.

to be very vacillating. Mr. Rassam writes on the 30th June, 1867:—

"Wakshun Gobazay, the chief rebel of Tigré and Lasta, has sent to the Bishop to say that he was coming to attack this fortress and give him his liberty, and if that be accomplished we shall all be free. The messenger reports that he left his master with an enormous army near Wadala, about five days' journey hence, and that all the rebel districts from Debra Tabor to this have sent deputations to him to join him, and it is premised that, once Wakshun Gobazay approaches this neighbourhood, our royal friend would be done for. There is no doubt that, if Wakshun Gobazay succeeds in capturing this stronghold (according to Abyssinian estimation), he will be proclaimed Emperor by the Bishop, and there is no doubt that, once he gets the heads of the Abyssinian church, the prelate and the chief of the monks (who are both my friends), he will have the greater part of Abyssinia on his side."

But on 24th July he writes:—

"I fear if England does not get us out by force of arms we shall have to spend many an August in this wretched state, and Wakshum Gobaze has done so little since he came up to this vicinity that I have lost all hope of ever leaving this country through the succour of the rebels, for whom I am beginning to have utter contempt. I do not believe there is a more cowardly race than these Abyssinian rebels, who allow themselves to be butchered and plundered and have no pluck to attack their common enemy, against whom they might take a force of not less than 100,000 men (that is to say, if all join), and I am certain with the handful of men His Majesty has now, and the disgust every one has for him, he would not be able to hold his ground one day. The rebel chiefs talk very big, but do very little. We have not heard of the movements of Wakshum Gobaze since I last wrote to you in the beginning of this month; all his big talk about attacking the Emperor and releasing the Bishop and myself and party has dwindled to nothing. One day we heard that he had made friends with the Walloo Gallas upon their promising to pay him tribute, and the next, it is reported that he had to retreat to Lasta for fear of the Gallas, who had gone to Yago to turn him out of it."

Dr. Blanc writing of the same chief on the 18th July, 1867, says:—

"You will have seen by the two small notes enclosed in my two last letters, what the King's two great rivals propose to undertake. I was not wrong when I took it to be simply boasting. From the first nothing more has been heard; the second sent another messenger, who told the Bishop, on the part of his master, 'you know that we cannot take forts, so it would be useless for me to come; but give me your blessing, and I will go and attack my blood enemy.' The required blessing was duly given, but though a fortnight ago, this would-be hero still remains in the Yedjon's country. Some say that he is acknewledged by them, and has appointed the son of Ras Marié (brother of Ras Aali) as Governor of the province; whilst others report that a large section of the population still in arms resist his authority, and have called upon the Queen of the Gallas to render them assistance. It is generally reported that she has gone in that direction, but whether for peace or war is doubtful, though Gallas seldom fight far from their own country; and probably as the rival Queen is to be befriended by Shoa, they will endeavour to form an alliance with Gobazi. Whatever may be his position in Yedjon, acknowledged or not, he will far less attack the King than this mountain; the fallen and humble lion is, even in his weakness, much more dreaded than all the ambas of Abyssinia put together. Gobazi is rising, the King is falling; the first knows well that the King can no more march against him, and must before very long retire here or run to the low country, so Gobazi quietly waits until, by mere necessity, the whole of the Amhara must recognize him. His messages to the Bishop are the natural consequences of his birth, all Abyssinians being story-tellers and boasters, and the greater the man the more he indulges in such freaks. That Gobazi is morally no better, quite as treacherous as the King, a fact that happened a few days ago clearly proves. You remember Rasadiloo, the Yedjon chief about whom I wrote to you some time ago. He went to the Wakshum, was well received and told to remain as a friend and guest, but a few days ago he was seized by his order and put in chains—a simple measure of precaution. This is not the man into whose hands I would like to fall."

Of the provinces more immediately under Theodore's rule, nearly the whole seem now to have revolted. Dr. Blanc, writing on the 18th June, 1867, says:—

"A few months ago the King still had a few provinces left; small remains of his former conquests, now he cannot even call himself monarch of all he surveys, when he gazes on the surrounding plains from the summit of Debra Tabor. The camp and Magdala constitute his kingdom."

PORTUGUESE EXPEDITION INTO ABYSSINIA.

THE only instance, as far as is known, of an invasion of Abyssinia by European troops, was in 1541, when 400 Portuguese were sent to the assistance of the Emperor, whose kingdom was overrun by the "Moors" (apparently the tribes now called Gallas). It has been supposed by some that these troops entered Abyssinia from Zeila, Amphilla Bay, or some other point considerably to the south of Massowah; but there seems little doubt that this was not the case, but that they made their entry from the latter place or from Arkiko, four miles to the south. The event is thus described in "The Travels of the Jesuits in Ethiopia," which was published in 1710:—

"In the year 1541, *Don Stephen de Gama*, then Governor of *India*, entered the Red Sea with a considerable Fleet, and having done much Harm to the infidels on the Coast of *Arabia*, came to an anchor at the Island of *Mazua*, whence he sent his Brother, *Don Christopher de Gama*, with 400 Men to the Assistance of the Emperor of *Ethiopia*. These Men met with extraordinary Difficulties in passing the uncouth Mountains, over which it was almost impossible to draw their canon, but having overcome them, and being met everywhere by the country people, who looked upon them as their Deliverers, they brought down the Empress *Cabelo Oanguel* from the Mountain *Damo* (apparently Debra Damo, between Halai and Adowa), to which she was retired for Safety, there being no way to get up it, but being hoisted in Baskets . . . .

"Intelligence being brought that Five *Portuguese* Vessels were arrived at *Mazua*, *Don Christopher* sent a Captain of his own, with 40 Men to get some supply of Ammunition, and carry Letters for the Viceroy of *India*, and set forward himself, with his forces towards a Country, where a Christian *Abyssine* had been compelled to submit to the *Moors*, and now sent to inform him, that if he would come to him he would find no opposition. He had not gone far before he received an express from the Emperor, desiring he would make haste to join him, because the *Moor Granhe* was advancing towards them, and each apart would be too weak to withstand him. Being come to those Lands, whither the *Ethiopian* commander, above mentioned, had invited him, he was met and presented by him with Eight fine Horses, and informed the Enemy was so near that he could not advance without meeting them. It troubled *Don Christopher* that he could neither join the Emperor, nor stay for those Men he had sent to *Mazua*, however he resolved to Fight, and, encouraging his Men, they all approved of his Resolution. The next Day, the Enemy being at hand, he encamped on a Rising Ground, when *Granhe*, having taken a View of his small Forces, enclosed him with 15,000 foot, armed with Bows and Arrows, Darts, and Bucklers, besides 1,500 horse and 200 *Turkish* Musquetiers, thinking to starve him out. *Don Christopher*, understanding his Design, after some small Skirmishes, drew out all his Men with the Empress in the center, on the 4th of *April*, 1542. The Canon and Muskets made the Infidels keep off, but the *Turks* advanced, and did some harm with their Shot, and *Granhe* himself coming on with 500 horse, the *Portugueses* began to be hard pressed, but that the canon being well played killed many of the Horse and made the rest slacken. Many of the *Portugueses* were now wounded, and *Don Christopher*, shot through a Leg, yet left not the Battle, but encouraged his Men. *Granhe*, on the other side, thinking his Men gave way, came up so close, that he was also shot through the Leg, and his Horse killed under him, whereupon his Men struck their Colours, and carryed him off, the *Portugueses* pursuing them till they were all so spent that it was thought a rashness to go any further, and therefore they returned victorious to their Camp, where they found the Empress and her Women, dressing the wounded Men and binding their Hurts with their own Linnen, for want of other. Of the *Portugueses* Eleven were killed, among the Infidels slain the *Abyssines* knew four of *Granhe's* Commanders of Note, and thirty *Turks*. *Don Christopher* sent that very Night to acquaint the *Portugueses*, who were gone to *Mazua*, with his Success, and hasten them back."

There seems no doubt from this narrative, that Massowah was the starting point and base of the expedition. The Portuguese were subsequently defeated by Granhe, and Don Christopher was killed.

In "Purchas's Pilgrimes" (vol. 2, page 1151), an account of this expedition is also given by Don John Bermudez, who was sent on an embassy to the Emperor of Ethiopia from the King of Portugal, and who accompanied Christopher de Gama and his force. In the account of the preliminary operations, a "Captaine of the King of Zeila" is mixed up in the transactions, which may have originated the idea that the expedition started from Zeila (latitude 11½°, longitude 43½°), but Arquico is afterwards distinctly mentioned as the starting point. In the course of the narration Don John says,—"And beginning to travell, within three dayes we came to *Debarua*. Within a few dayes they vsed Schismaticall and Hereticall Ceremonies, differing from the *Romane*. I satisfied the best I could *Don Christopher* and his men, and the murmuring ceased, and they concluded how to carrie the Ordnance when they iourneyed. They made presently certaine carriages like vnto ours: the which, because in the countrey there was no Iron, they shod them with certain old caleeuers, which brake, because they vould serue for no other vse."

In Rudolph's History of Ethiopia (p. 222), it is said, speaking of the same expedition, "Their commander was *Christopher Gomez*, a Person of great Valour, who, in the month of *July*, in the year 1541, entered the Kingdom with Six small Field-Pieces, and Four hundred and fifty Musqueteers. At first they had a very severe March, for they wanted Horses and Teams; the country being so wasted, that they were forced to carry their Luggage and Conveniences upon their shoulders over most rugged and steep Mountains. Nevertheless, these Souldiers, few in number, but all choice men, and coveting the honor to restore the King of *Habessinia* to his Kingdom and his Liberty, patiently underwent all sorts of hardship. This caused a change of Fortune, so that now the late Victors were everywhere put to flight, astonished at the Execution of the Guns."

There seems no doubt that, until Massowah was taken by the Turks in 1557, it was always the port by which Abyssinia was entered. In 1555 a mission went to the Emperor, and returned by that way. In 1557 the Bishop Don Andrew de Oviedo entered by Arkiko. The capture of Massowah by the

Turks in that year is thus described by the Superior of the Mission in Ethiopia, writing in 1562 to the General of the Society:—"Having mentioned the coming of the *Turks*, I must inform your Reverence that when we came to *Mazua*, an Island on the coast of Ethiopia, and the anchoring place of all ships trading hither from India and Arabia, we there found a *Turkish Bassa* with 500 or more men, designed to conquer Ethiopia, and expected the arrival of our ships; when, seeing those that came could do him no harm, he landed, and that obliged us to depart hastily from *Debaroa;* and though we have been here above five years, we do not know that any letter of ours is past into India, notwithstanding we have tried so many ways, that we feared the men sent by us were killed."

When Massowah was thus closed, as a port of ingress and egress, by the Turks, other routes were tried. In 1588, some Jesuits, not being able to land at Massowah, attempted Zeilah. In 1595 a Jesuit tried to get through Massowah, but, being discovered, had his head cut off. In 1596 a Jesuit effected an entrance by Massowah, disguised as a seaman. In 1603 and 1604 some Jesuits, by making friends with one of the leading Turks, effected an entry by Massowah. In 1607 the Emperor sent an Embassy to Portugal by way of the Nile, "to escape falling into the hands of the Turks at Mazua." In 1620 two Jesuit Fathers entered by way of Suakim. In 1622 four Fathers entered with a pass from the Pasha of Suakim. In 1625 Don Alonzo Mendez and Jerome Lobo entered by Baylur (Amphilla Bay).

Things then seemed to improve. In 1628 five Jesuits, after meeting with some troubles from the Turks, got through Massowah. In 1630 a bishop was suffered to go from the same place to Fremona "without any obstruction."

In 1769 Bruce entered the country by Massowah, and since his time, with the exception of those who entered from the side of the Nile, there seems to have been no instance of any travellers entering by any other route, with the exception of Coffin, whose journey from Amphilla Bay was undertaken from motives of curiosity only. The province of Shoa having been almost always in a state of rebellion or independence, its communication with the sea have been through Tajurrah, as the road to Massowah, besides being much longer, leads through their enemy's country.

ROUTES TO MAGDALA FROM THE NORTH, BY THE EASTERN SIDE OF THE HIGHLANDS.

ANY force advancing from the Northern Highlands of Abyssinia on Gondar and Magdala have two formidable obstacles to encounter. The first is the ascent from the plains to the Highlands, involving a change in the nature of the transport on reaching the latter; the second is the chains of mountains which stretch across the Highlands between the twelfth and fourteenth parallels of north latitude. These mountains attain their greatest elevation in the provinces of Samen and Lasta, famous for their ruggedness, and the difficulty of their passes. The worst portion of these mountains may apparently be avoided by going to the westward by the Lamalmon Pass, which, although in itself a most formidable obstacle, yet does not seem to present that succession of mountains and defiles which are to be found on the Selki road. They may also be avoided by going to the east by Lake Ashangi, by the route followed by Krapf to Antalo. An easterly course has also the advantage of following the watershed of the country, and thereby avoiding the great Rivers Mareb, Tahazze, &c., which, as has been before stated, are formidable military obstacles, and are also very unhealthy.

Any route to the east of Lake Haik would apparently lead through the hot country of the plains, where camel transport is necessary. This evidently would be objectionable, for, from whatever side an expedition approached Magdala or Gondar, they must at some period exchange their camel for mule transport. It is true that the caravans find no difficulty in traversing the low country with camels, and then changing to mules when they reach the highlands; but their requirements are probably limited to 100 or 200 of the latter, which the tribes on the caravan routes have got the habit of supplying. A British force, on the contrary, would require probably over 5,000 mules or horses, and if they advanced for any distance from the coast by a camel road, they would be in one of two predicaments—either, when they reached the Highlands, they would have to wait until the requisite number of animals could be collected, or they would have to bring mules with them across the plains destitute of water and suitable forage, an undertaking probably impracticable. It might, however, be possible, if the route selected along the edge of the Highlands, to supplement the transport by a camel line. As the force advanced, and the line of communications became longer, the number of baggage animals required would be proportionately increased; and if, when opposite Amphilla Bay, for instance, a camel communication were established, the mule traffic might be sensibly lightened. This opportunity would be gained by selecting a route to Ategerat and Antalo, which approaches apparently the nearest to the eastern edge of the Highlands. (See map at the end).

From Antalo there exists, as already stated, a route towards Magdala, described by Krapf as a very favourable one (see extracts, page 137). From Magdala, past Lake Haik, it passes, according to his description, through a beautiful well-watered country, with fertile soil, in which there are many considerable villages, among others Totola, where, when he passed, a market was being held, attended by thousands of persons from Gondar,

Tigré, &c.; portions of this district, however, are only thinly inhabited. He says, "the climate of these plains is beautiful (April), neither too hot nor too cold, the air being always refreshed by the winds blowing from the mountains; there is plenty of water poured out from the veins of the neighbouring mountains." Soon after crossing the twelfth parallel he passes the watershed between the provinces of Angot and Lasta, and here for some ten miles he traversed a wilderness with much want of water. He then again descends to a country well watered and fertile, and, in parts, well inhabited. Soon after passing Lat he traverses a small portion of a rather desolate country, where, he says, "we could scarcely find our way through the thorns and bushes, which caused us many difficulties in advancing towards the river." He then met with a mountain pass, of which he says, "our passage was sometimes extremely difficult and narrow. The banks of the mountain had sometimes the appearance of high walls of rocks, a slip from whence would cause certain death." The country from thence to Antalo was fertile and well watered, but hilly.

There is also another road from Antalo through Sokota, which was followed by Dr. Beke (see extracts, page 172), and which he describes as the main caravan road from Antalo to Begemeder. This road traverses the mountainous districts of Lasta and Waag, and is probably, therefore, a more rugged one than Krapf's route, but better defined, being more frequently travelled over; between Sokota and the River Tzellari there seems to be a want of water. It also crosses the Takazze, but at a point where it does not seem to be a formidable obstacle. This route leads apparently too much to the left, and it is doubtful whether a branch could be found from near Sokota to Magdala, as very high mountains appear to intervene.

From Antalo to the north Krapf followed the road by Atebidera to Addigraht, &c. (see extracts, page 165), but there seems to be the fatal objection on this line of want of water. Krapf, writing in May, says,—"From Chelicut we took our directions to Adigrate. Our road was pretty plain. Sometimes we had to ascend a slight hill; but although we had now a better road, compared with that in Lasta and Waag, yet we were considerably inconvenienced by not having plenty of water, which we had found in abundance in those countries. Besides this, the heat of the valleys of Tigré was an addition to those inconveniences with which our journey through that country had abounded, from the inhospitable reception of the natives, and from the rumours of war, and dissoluteness of the soldier." (When talking of his inhospitable reception, it must be remembered that Krapf was travelling with little or no money.)

There is an intermediate road from Antalo to Addigraht, part of which was travelled by Rüppell (see extracts, page 182), and which appears a favourable one, and not destitute of water. Beke, when at Antalo, went to Adowa (see extracts, page 179), but he says that his reason for this was that the direct road to Massowah, through Agamie, was rendered inaccessible by rebels. He seems to have met with no difficulties on the road.

Salt travelled from Dixan to Chelikot near Antalo, by a route which appears to have been on the whole not unfavourable (see extracts, page 42). He travelled in the dry season of the year, February, and the only portion of the route where he experienced a want of water was in the neighbourhood of Dixan. He says, whilst journeying from the Taranta Pass to that town, "the heats became intense and scorching; vegetation parched, brooks dry. . . . . The country round Dixan at this time of year wore a parched and desolate aspect," but he adds that "large herds of wild goats and kids are brought in (to Dixan) every evening to protect them from wild beasts." After passing Dixan he says that "the whole country had the appearance of being scorched, and we did not reach water until we had passed the high rock of Addicota," apparently about 15 miles from Dixan. After this there seems to have been no want of water, and the country was generally favourable, but with occasional steep and rocky passes.

The country for 15 or 20 miles to the south of Dixan and Halai seems to be subject at the dry season of the year to want of water. Salt's statement has just been quoted. Rüppell in his journey from Adowa to Massowah in June, after passing Granduftofta, about 20 miles from Halai, going north, described the country as an undulating sandstone plain, having very little water, and in his journey from Halai to Addigraht, in May, when at about the same distance from the former, he describes the country as a barren plateau with a few acacias. On this line, however, there appears to be always water near Dogonta at the head waters of the Haddas. Ferret and Galinier, on the other hand, when passing over this tract in November, do not complain of want of water (see page 53), and Combes and Tamisier, in April, describe their first camping ground out of Halai at Marda, as watered by a delicious brook.

It must be borne in mind that there are a great many more roads or paths through these countries than those followed by travellers. The best of the roads, as Mansfield Parkyns says, are merely tracks, and it is probable that, wherever there are villages, communications exist between them. Travellers seem never to have found any difficulty in deviating from their course when political or other reasons compelled them to do so.

From Addigraht there is a road by Sanafé to Dogonta, which was travelled by Krapf (see extracts page 168), and also by Rüppell (see extracts page 18), and which appears a favourable one. From Dogonta there is a pass to the low country, which was followed by Krapf (see extracts page 170), and which he states to be much more favourable than the Taranta pass—this pass communicates with Ansley Bay by the route recommended by Dr. Beke in a very interesting letter to the Athenæum, August 14, 1867, which is reprinted at page 183. There is also a road from near Sanafé to Ansley Bay.*

ROUTES FROM MASSOWAH AND ANSLEY BAY TO THE HIGHLANDS.

THERE seems to be no doubt that the ascent to the Highlands may be most easily effected by Kaya Khor on the west, or by Dogonta and Sanafé on the east; the passes to those places being much more accessible than those leading to Dixan and Halai.

As regards the nature of the different routes from Massowah and Ansley Bay to the Highlands, and the amount of water to be procured whilst traversing the low plains, the following will give some idea:—Col. Merewether says of the route to Ailet, which he traversed in January, 1867 (see extracts, page 82), that the first part is a mere cattle track, traversing dry beds of water-courses, low hills, and bits of level plain; that it is practicable for artillery everywhere except in three places, which could easily be made so. The highest point crossed was 989 feet above the sea. Seventeen miles from Massowah there is a spring of good water always running, but any quantity of water may be obtained by digging a few feet in the sandy beds of the watercourses. Ailet is a fine plain, covered at that time of year with rich verdure, owing to showers falling once in the twenty-four hours. Any quantity of water may be obtained by sinking wells. The distance from Massowah to Ailet is twenty-seven miles.

Of the route to Kaya Khor, M. Münzinger, travelling in the end of January (see page 75), says, that the first twenty-three miles are over open low country, where plenty of water is to be obtained, either in wells or by digging. For the next thirty miles the torrent Alligaddi is followed, where running water is found at intervals; the road is described as good. The last thirteen miles are up the mountains, but the road is not very steep, and presents no obstacles to camels; half-way up running water can be obtained at half-an-hour's distance from the road.

Dr. Beke describes the descent from Kaya Khor to Massowah as a "gradual and easy road, well watered, and occupying two days and a half very easy travelling. This is so much superior to the steep way up the Taranta, that it is very generally chosen by Europeans."

Of a road near to the last one leading to the Agamutta Plateau, Col. Merewether, who explored it in April, 1867 (see

* It appears that Ansley Bay has been selected as the landing place for the expedition, and that the routes by Sanafé and Tohonda will probably be chosen.

p. 79), says, the road traverses plains and low hills for about fourteen miles to Part, where plenty of excellent water was found by scratching about two feet in the sandy bed of a torrent, and in the summer months it is found a little deeper; the vegetation very rich, and cattle abundant; heat by no means excessive. For the next ten miles to Hendrode the road lay along the bed of a torrent, practicable for camels and even artillery; running water found continually; dense wood and high grass. Then commenced a steep ascent, unsuited to draught artillery, easy enough for mules carrying mountain guns, but difficult for laden camels unless having light burdens. The Agamutta Plateau was about nine miles from the last point, and about thirty-three miles from Massowah; it is cultivated, and water can always be obtained by sinking wells; "a finer or richer country than this could not be desired."

The road to the foot of the Taranta and Shumfaito passes has been constantly travelled, being the main caravan line from Massowah to Gondar. From Massowah to Arkiko, a distance of about four miles, the journey is generally made by sea. Here there is a detachment of Egyptian troops. Water is obtained from wells. For about eleven miles after leaving Arkiko the road traverses a low, sandy, sterile plain, with a few stunted mimosas. It then enters the lower spurs of the highlands through the Shilliki pass, and after about nine or ten miles it strikes the River Alligaddi either at Hidale or Woha; distance from Arkiko about twenty-two miles. This portion of the route is for the most part very rugged, and is destitute of water except in the rainy season, but by turning off to the east, wells will always be found at Woha or Sahto. The River Alligaddi is often dry, but water can always be procured in certain parts of its bed by digging, From Hidale the road soon strikes the Haddas stream and follows up its bed to the foot of the passes of the Taranta mountains at Asubo, a distance of about twenty miles; distance from Arkiko about 42 miles. Along this part of the route good water is always procurable at intervals, either by digging in the bed of the stream or from springs, of which there are several, especially at Hamhamo, Tubbo, and Asubo. The road south of Hamhamo is enclosed between high mountains, and is very rocky and rugged, in fact it is little more than the bed of a torrent, which is dry during the dry season, but liable to sudden floods in the wet season. Bruce, when camping at Hamhamo in November, 1769, says, "The river scarcely ran at our passing it, when all of a sudden we heard a noise on the mountains above, louder than the loudest thunder. Our guides, upon this, flew to the baggage, and removed it to the top of the green hill, which was no sooner done, than we saw the river coming down in a stream about the height of a man, and breadth of the whole bed it used to occupy."

Near Asubo the ascent of the Taranta mountains commences. The passes through these mountains are usually generalized under the name of the Taranta Pass, but there appear in reality to be three principal ones; the Sulah Pass, which turns off to the right at Asubo, and leads to Dixan; the Asubo Pass, which leads from near the same place to Halai; and the Shumfaito Pass, which turns off from the valley a mile or two higher up, and also leads to Halai.

These passes are of a very formidable description. Bruce says of the Sulah, "At half-past two o'clock in the afternoon we began to ascend the mountain through a most rocky uneven road, if it can deserve the name, not only from its incredible steepness, but from the large holes and gullies made by the torrents, and the huge monstrous fragments of rocks which, loosened by the water, had been tumbled down into our way. It was with great difficulty we could creep up, each man carrying his knapsack and arms, but it seemed beyond the possibility of human strength to carry our baggage and instruments." Ferret and Galenus say of the same pass, "It took us three hours to climb the Taranta, a frightful path, encumbered sometimes with stones which rolled under our feet, sometimes with enormous rocks, which it was necessary to climb by holding on strongly with feet and hands, led us to the summit." Combes and Tamisier say of the passes to Halai, "Two paths lead from the foot of the Taranta to Halai. One less difficult but broader, has been laid out for beasts of burden; the other, more difficult but shorter, has only been followed by foot travellers. We chose the latter (the Shumfaito one) . . . . . . The sun was burning, the mountain almost perpendicular, and we climbed with difficulty . . . . . . The road was always very steep, strewed with rocks, and often we could only advance by the help of great ladders, which trembled under our feet."

Other travellers, such as Rüppell and Salt, seem to think that the difficulties of these passes have been somewhat exaggerated. The foot of the Sulah pass is about 4,670 feet, and its summit 8,350 feet above the sea.

Instead of turning to the right up the passes of the Taranta mountains, the valley of the Haddas may be followed up to its head near Dogonta,* where the summit of the Highlands is reached. This pass was followed by Krapf in May, 1842 (see page 170), and appears to be of a much more favourable nature than the other. Krapf says of it, "We had a very good and plain road through a woody wilderness. It is much superior to the road of Halai, which leads over the difficult mount of Shumfaito. On the Tekunda road you descend by degrees, and the road might be trodden even by camels, if it could be improved a little by removing some rocks in the way." Near Tekunda there is a spring from which the Haddas stream rises.

The Haddas stream, after flowing down the valley past the foot of the Taranta passes, turns off, two or three miles to the south of Hidale, to the east, and falls into Ansley Bay, near Zulla and the ancient Adulis. This is the route advocated by

* Called sometimes Tekunda or Tohonda.

Doctor Beke (see page 183) as the best means of approach to the Highlands.

Starting from Ansley Bay, it follows the direction of the Haddas to Woha, whence the main road from Massowah can be reached either at Hidale or Hamhamo. The distance of the latter place from Ansley Bay is about twenty-two miles, whilst its distance to Arkiko is about thirty miles, shewing a clear gain of eight miles, or of fourteen miles, if reckoned from Massowah. The road as far as Woha, appears to traverse a tolerably level country, and water can be obtained, according to Dr. Beke, at Zulla, and all along the bed of the stream, by digging. He says, on this head, that when he was at Zulla, in February, 1866, they came to wells about a mile nearer the sea, where numerous horned cattle were being watered; they were small, well shaped, fat beasts, giving a delicious rich milk, the pasture at that time of the year being plentiful almost down to the sea side.

This road is identified by Dr. Beke with the one mentioned in Arrian's *Periplus of the Red Sea*, and which is thus alluded to in Vincent's *Periplus of the Erythræan Sea:* "At twenty stadia (two miles)* from the shore, and opposite to Oriné, lay Aduli, which was a village of no great extent; and three days' journey inland was Koloe (Halai), the first market where ivory could be procured. Two days' journey from Koloe lay Axum, where all the ivory was collected which was brought from the other side of the Nile, through the province called Kuenion, and thence by Axuma to Aduli."

A few miles to the south of Dogonta is Sanafé, formerly a trading town of the Greek merchants, and to which there is a road from Zulla.

Dr. Beke states that there is also a road from Buré, in the south-west corner of Ansley Bay, to Sanafé, which was used in ancient times by the Greeks. He says that this road is as good, if not better, than that from Zulla to Dogonta, the ascent being much more gradual, whilst Sanafé, being 8,400 feet above the sea, has a very temperate climate, and would be well fitted for a depôt. Sanafé would appear also to be in communication with Amphilla Bay, and with the great salt plain which supplies Abyssinia with its present currency, pieces of rock salt, and a military force stationed there would, therefore, have the command of that valuable and essential commodity. Extracts have been given at page 129, from Coffin's journey from Amphilla to Chelicut. He approached the highlands by the pass of Sanafé, but his route is placed in Salt's map as passing forty or fifty miles to the south of the present recognized position of Sanafé, and near a mountain† of the same name.

* Dr. Beke states that the ruins of Adulis are now four miles from the sea, the coast of the Red Sea having risen.

† It appears impossible to reconcile the mention made of Sanafé by travellers otherwise than by supposing that the town and pass of that name are distinct places, as shewn on the map at the end.

He describes, however, the passage of the salt plain, alluded to by Dr. Beke, from which the Abyssinians obtain their currency.

According to Rüppell, there are roads to Amphilla Bay from Barakit, Omfeito, and Ategerat.

EXTRACTS FROM THE WORKS OF TRAVELLERS.

The following extracts are taken from the works of various travellers, and have been selected as giving information on the different routes in Abyssinia.

The routes which have been referred to are as follows:—

1. "Bruce's Travels to discover the Source of the Nile, 1768 to 1773." Bruce was a Scotch gentleman, who made a journey to Abyssinia to discover the sources of the Nile.
2. "Salt's Voyage to Abyssinia," 1809–10. Mr. Salt was Secretary to Lord Valentia, nephew to the Governor-General of India, and was sent by him on a mission to the Emperor of Abyssinia.
3. "Ferret et Galinier. Voyage en Abyssinie," 1839 to 1843. These gentlemen were Captains in the French Etat-Major, and were sent by their Government to make explorations in Abyssinia. They published some very good maps of the country.
4. "Voyage en Abyssine, exécuté pendant les années, 1839, 1840, 1841, 1842, 1843, par une commission scientifique composée de MM. Théophile Lefebvre, Lieutenant de Vaesseau, &c., A. Petit et Quartin-Dillon, Docteurs mèdecins, &c. Vignaud Dessinateur." This commission, of which Lefebvre was the head, was sent out by the French Government. The work is accompanied by beautiful illustrations of the scenery, natural history, &c.
5. "Highlands of Ethiopia," by Major Harris, 1841–2. He was sent by the Indian Government on an Embassy to the Governor, or *de facto*, King of Shoa.
6. "Royaume de Shoa:" D'Héricourt, 1842. M. D'Héricourt was a French traveller, who made two journeys into Shoa, the second one of which was under the auspices of the Royal Academy of Sciences at Paris.
7. "Isenburg and Krapf. Journey to Shoa from Tajurra in 1839." Published in the "Journal of the Royal Geographical Society." They were German Missionaries.
8. "Isenburg and Krapf's Mission," 1839 to 1842.
9. "Voyage en Abyssinie par Combes et Tamisier," 1835 to 1837.
10. "Reise in Abyssinien:" Rüppell.
11. "Life in Abyssinia:" by Mansfield Parkyns, 1843.
12. "Hamilton's Sinai, the Hedjaz and Soudan," 1854.
13. Colonel Merewether's description of the Routes from

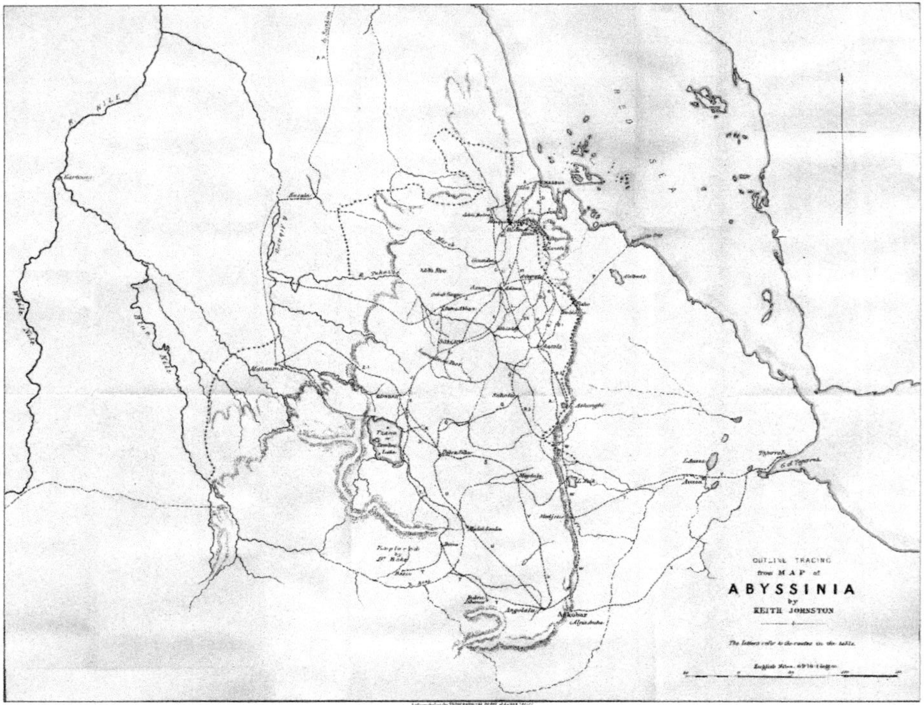
OUTLINE TRACING
from MAP of
ABYSSINIA
by
KEITH JOHNSTON
The letters refer to the routes in the table.
RED SEA
R. NILE
White Nile
Blue Nile
Kartoum
Metamma
Gondar
Magdala
Ankobar
Angolala
Tajurrah

Massowah to the Table Land, 1867. Given in the Blue Book on Abyssinia.

14. M. Münzinger's description of the Routes from Massowah to the Table Land, 1867. Given in the Blue Book on Abyssinia.

15. "Reise nach Abyssinien, &c.:" von M. Ch. von Heuglin, 1861 and 1862.

16. "Travels of the Jesuits in Ethiopia in the 16th and 17th Centuries.

17. "Purchas's Pilgrimes."

18. "Rudolph's History of Ethiopia," 1864.

19. Dr. Beke's Works on Abyssinia.

In order to facilitate reference to the various extracts which have been made, a table is added, which shows the different parts of the country to which they relate. The letters refer to those on the accompanying sketch map.

Route.	Traveller.	Date.	Page.
Massowah to Ailet (A).	Ferret and Galinier	August, 1842	62
	Mansfield Parkyn	Summer, 1843	62
	Merewether	January, 1867	82
Massowah to Kaya Khor (B). (Kiaquor).	Münzinger	January, 1867	75
	Merewether (H.)	January, 1867	83
	Beke	. .	2
	Lefebvre's Routes, 1	. .	208
Massowah to Agametta Plateau (C).	Merewether	April, 1867	79
Massowah to Dixan (D) and Halai.	Ferret and Galinier	November, 1840.	47
	Bruce	November, 1769.	221
	Salt	March, 1810	42
	Combes and Tamisier.	April, 1835	189
	Lefebvre's Routes, 2	. .	208
Massowah to Dogonta (Tekunda) (E).	Krapf	April, 1842	170
	Beke	. .	183
Ansley Bay	Beke	. .	183
Routes from Massowah to Taranta Passes (F).		. .	36
Ailet to Adowa, by Addi Bahro (G).	Ferret and Galinier	August, 1842	58

Route.	Traveller.	Date.	Page.
Ailet and Kaya Khor to Adowa (H).	Mansfield Parkyns .	Summer, 1843 .	65
	Merewether . .	January, 1867 .	83
	Lefebvre's Routes, 1, 8	. .	208
Dixan or Halai to Adowa (I).	Ferret and Galinier .	October, 1840 .	52
	Bruce . . .	November, 1769.	225
	Combes and Tamisier.	April, 1835 .	189
	Lefebvre . . .	June, 1839 .	200
	Lefebvre's Routes, 2 .	. .	208
Adowa to Gondar, by Lamalmon Pass (J).	Ferret and Galinier .	May, 1842. .	56
	Bruce . . .	November, 1769.	233
	Mansfield Parkyns* .	July, 1845. .	73
	Lefebvre's Routes, 15, 19 . . .	. .	208
Adowa to Devra Tabor, by Selki Pass (K).	Combes and Tamisier.	June, 1835 .	191
	General description .	. .	4
Adowa to Antalo (L).	Beke . . .	April, 1842 .	179
	Lefebvre's Routes, 22	. .	208
Dixan to Antalo (M).	Salt . . .	March, 1810 .	43
Halai to Adegrat (N).	Rüppell . . .	May, 1832. .	181
	Krapf . . .	April, 1842 .	168
	Lefebvre's Routes, 3 .	. .	208
Adegrat to Antalo (O).	1. Krapf. . .	April, 1842 .	165
	2. Rüppell† . .	June, 1832 .	182
	3. Ferret and Galinier‡ . . .	November, 1841.	
	Lefebvre . . .	October, 1840 .	204
	Lefebvre's Routes, 3, 4, 17 . . .	. .	208
Amphilla Bay to Atsbie or Atebidera (P).	Coffin . . .	January, 1810 .	129
	Don Alonzo Mendez .	May, 1825. .	194
	Lefebvre . . .	June, 1841 .	198
	Merewether . .	1867. . .	9
	Lefebvre's Routes, 9 .	. .	208

* As far as the Taccazy only.
† To Tackeraggiro or Takirakira only.
‡ As far as Atsbi or Atebidera only.

Route.	Traveller.	Date.	Page.
Antalo to Debra Tabor (Q).	Beke . . .	March, 1852 .	172
Antalo to Lake Haik and Magdala (R).	1. Krapf. . . . 2. Pearce* . . . Lefebvre† . . . Lefebvre's Routes, 15, 18, 22 . . .	March, 1842 . September, 1807.	137 45 208
Gondar to Magdala (S).	Steudner and Heuglin	February, 1862 .	245
Gondar to Angolala and Baso (T).	Lefebvre . . . Combes and Tamisier. Lefebvre's Routes, 20, 21 . . .	May, 1843. Dec. and Jany. . . .	 208
Debra Tabor to Angolala (U).	Combes and Tamisier.	October . .	
Magdala to Angolala (V).	Krapf . . .	1842 . .	
Antalo to Angolala (W).	Lefebvre . . . Lefebvre's Routes, 22	January, 1843 . . .	 208
Tajurrah to Ankobar (X).	Harris . . . Isenburg and Krapf . D'Héricourt . .	June, 1841 . April, 1839 . September, 1842.	83 120 118
Tajurrah to Aussa and Shoa (Y).	General description . Lefebvre . . . Lefebvre's Routes, 27, 28, 29 . . .	. . . .	8 208
Gondar to Metemmah (Z).	1. Bruce . . . 2. Krapf. . . .	December, 1771. May, 1855. .	134 11
Suakim to Kasala (AA).	Hamilton. . . .	March, 1854 .	130

* Pearce did not visit Magdala, but returned by Socota to Antalo.
† Lefebvre did not visit Magdala, but continued south to Angolala.

MR. SALT'S JOURNEY FROM MASSSOWAH TO CHELICUT NEAR ANTALO IN MARCH, 1810 (marked D, M, on map page 39), AND MR. PEARCE'S FROM ANTALO TO LAKE ASHANGI AND BACK, IN SEPTEMBER, 1807 (marked R 2 on map page 39).

THE plain we had to cross, extended in a gradual ascent from Arkeeko to the first ridge of mountains, and was occasionally covered with a species of mimosa, called Girá. We saw great numbers of camels, sheep, asses, and goats in the course of the day, and passed two villages, Dukona and Dábi, round which were several enclosures of kush-kush or juwarry; at sunset reached a station on a rising ground at the bottom of the first line of hills, called Shillokee, and encamped.

On the 26th February travelled over a rugged ridge of low hills (granite over a bed of micaceous earth) to Wéah, course, S. by W. 8', found pits of rain water.

27th. Left Wéah, went S.W., through a forest of girá trees, towards a break in the mountains, leaving a high hill on the left; in two hours we began to enter among the mountains, where the road became intersected with deep gullies formed by the rains, and soon came to a small pass, appearing as if cut through a rock of iron stone, beyond which commences the Samhar country. In half an hour entered a ravine between two ranges of mountains, rising almost perpendicularly on both sides, up which is a circuitous road to Taranta; a little further, passed two encampments of Hazorta, who came with cattle from the upper country, bought three cows for fifteen dollars; and in half an hour halted at Hamhammo, a small circular spot in a nook of the mountains, distant a few hundred yards only from the stream; course, about nine miles, S.W.

28th. Left Hamhammo. The pass seldom exceeding one hundred yards in width, the ground forming an irregular ascent, causing the stream occasionally to be lost under ground; but it seldom ran any distance without again making its appearance; in two hours we halted at Sadoon, a small verdant spot, under the shelter of some bushy trees, an abundance of partridges and other game. At 1 P.M., set out, and shortly reached Tubbo; here the cliffs, and rugged precipices were covered with vegetation. At 3 P.M., arrived at Leila, and camped.

March 1. Left Leila at 5.45 A.M., and soon reached Assaba; a little beyond on the left, a pass or gully in the mountain, opens into the road, considered the most dangerous spot; a wild set of Bedowee residing there, who plunder the cafilas travelling to and from Massowa. In 1809 they plundered a party of native soldiers of two thousand goats, a serious loss to people depending on them for support. We halted a little beyond this point, on a steep jutting rock, commanding the ravine and road. When our party had passed we proceeded in its rear to Taranta, and camped. The station is furnished with water from a bason formed in a rock, a short distance up the northernmost ravine; in the rainy season a tremendous torrent occasionally rushes. Rocks, reddish granite. A spring which rises about a mile higher, affords a supply of water throughout the year.

Passage of the Mountains.—About half-way up, the road divides into two tracks, one leading to Dixan, the other to Halai.

March 3rd. Marched at 5.50 A.M., ascending the mountain of Taranta. For about a mile the ascent is gradual, and much encumbered with loose stones and fragments of rock. Passing this in a W. by S. direction, we arrive at a steep and rugged part of the mountain, thickly covered with kolqual. This continued for two miles, when we reached a very precipitous ascent, and shortly reached Mijdevella, where there is a spring. From here the road runs S.W., in some parts extremely steep, compelling most of the party to dismount. To walk, or rather to climb, required no trifling effort for people so long unaccustomed to exertions of this nature. We soon reached a point where a road branches off to the left, leading to Halai. A little beyond is a high rock, called Gorezo. The abyss below is frightful to behold. Above this part of the mountain the vegetation changes its character, and clumps of trees are found called Wàra, of a moderate height, with leaves like a willow, and branches covered with lichens. Further on, the road appeared to have been cut through a bed of chalk-stone, and where this prevailed, an extensive grove of a hardy kind of cedars called Tud grew.

From this point we had a considerable descent to make before we again mounted, when in half-an-hour we reached one of the summits of the mountain, near a station bordering on a small pool of water called Turabo. Two hours and a-half had been occupied in the ascent. Thermometer 61°.

On descending the southern side of Taranta, the steepness of the path rendered riding unsafe. An hour's walk carried us down to the worst part of the road. We remounted, and went through a wild rocky district, along a winding pathway towards Dixan. The change of climate became very apparent. The heat became intense and scorching, vegetation parched, brooks dry, and the cattle had been driven across the mountain in search of pasture. At one we arrived at Dixan, seven hours ten minutes after leaving our camp this morning. The country round Dixan at this time of the year wore a scorched and desolate aspect. The only cattle were milch goats and kids. Large herds are brought in every evening to protect them from hyenas and wild beasts.

March 5th. Left Dixan at 6 A.M., proceeded westward, and in an hour reached a lofty hill on which is the village of Hadehadid. From hence we proceeded south across the plain of Zarai, which at this time looked very bare of verdure, the stream passing through it being completely dried up. The whole country had the appearance of being scorched, and we did not meet with water until we had passed the high rock of Addicota. At no great distance beyond, we came to a large daro, standing in the bed of a torrent, where we found some pits of water. Here we rested, and in the afternoon went on to Ambakauko, two miles distant on our right.

March 6th. Our course lay to the south, and after passing Ascoriah, we descended a steep declivity that brought us into the eastern end of the fine plain of Serawé, which is thickly interspersed with tombo-trees, and seems to extend westward, on a low flat, to Hamazen. This plain may be considered as part of the western boundary of the mountains of Taranta, the country through which we had hitherto descended, constituting only the lower ridge of that extensive range. Soon afterwards we arrived at Abha.

March 7th. At 5 A.M., after proceeding south one mile, the hill of Cashaat bore east; we turned a little to the west, and travelled eight miles through a wild forest, until we reached an agreeable station by the side of a river called Seremai, which flows through the bottom of a small valley, surrounded by steep and rugged hills: in a nook of which, about a mile to the eastward, lay a large town called Logo, whence the surrounding district takes its name; rested for a time, and then proceeded S.S.W., through a wild, uncultivated country, crossed a stream called Mai Belessan, left the high hill of Amba Anvas on our right, and after mounting a steep ascent, reached the village of Legoté, which, in appearance, somewhat resembled Dixan, eight miles from our last station.

March 8th. Descended from Legoté, crossed an extensive and well-cultivated plain, to the left of which, as we proceeded southward, lay the mountain of Devra Damo, which formerly served as a place of confinement for the younger branches of the king's family. This mountain is completely scarped on every side, and very difficult of access, having only one path leading up to it, and resembles the hill forts in India. After travelling a few miles, came to a pass in the mountains, called Kella, taking its name from the castellated appearance of the rocks in its neighbourhood; about a mile further we came to a beautiful glen, where a large daro tree stood by the side of a winding stream, with banks richly covered with verdure. Here we stopped, and, I think, must have reached a high altitude, for though journeying south, and the sun proceeding to the north, we found every day the climate more temperate, and vegetation more backward.

At 3 P.M. proceeded, and after a considerable descent, came to the river Angueah, which runs through a bed of granite in a north-west direction till it joins the Maley. Beyond this, we had several steep and rugged precipices to mount to the house of Ayto Nobilis, where we remained.

March 9th. Started at 3 P.M., and proceeded across a fertile valley towards a range of hills to the south, leaving the mountains of Adowa about twelve miles on our right, then traversed a steep pass leading to a fertile valley, and afterwards to a lofty hill, on which stood the mansion of Ozoro. After remaining the night, proceeded south down a highly cultivated valley, through which a stream runs named Mai Feras. The land is highly productive; the first crop of peas had not been gathered in, though the second crop of wheat and barley was making a rapid progress to an abundant harvest. This productiveness is owing to the skill in irrigating the land by digging small channels from the higher part of the stream, and conducting the water across the plain in square compartments. We proceeded to a village on the top of a lofty hill, and halted for the night.

March 10th. Started at daylight. Travelled three hours through a rugged and mountainous district, where the path was often so steep as to compel us to dismount from our mules. This path brought us to an open country, exhibiting an extensive plain stretching to the hills of Agamé and Haramat (20m on our left) in a westerly direction to the River Tacazze, through the rich districts of Gullibudda and Temben. This plain divides the mountainous district of Tigré from the no less elevated districts of Giralta and Enderta. The soil of the former being in general sandy, the rocks rising in perpendicular strata of slate over schistus and granite—in the two latter the strata are more horizontal: the surface of the valleys consists of a rich black loam, well calculated for the cultivation of barley.

After crossing the plain, we came to a steep pass, leading to the same country as the one of Atbara, this brought us into the district of Giralta. Halted.

Monday 11th. Left Mugga at break of day. The vale through which the first part of the road conducted us, wore a beautiful aspect, and was interspersed with groves of trees. In two hours we arrived at a point where another road turns off, towards the pass of Atbara. The route we had taken by Mugga saves this very difficult ascent. At noon we rested at the village of Aderaza, and after partaking refreshment proceeded. At four we arrived at a very steep declivity, down which our mules were led with considerable difficulty; this brought us into a deep ravine which extends in a direct line to the valley of Gibba. A broad stream runs through the middle of the ravine, with groves of flowering shrubs and trees on each side, scarcely admitting a passage. This country abounds with game, and is frequented, as it is said, by lions and other wild beasts, who resort at night to the river for water. When we emerged from this ravine, we shortly arrived at Gibba, situated in a small secluded valley, surrounded by woody hills, almost encircled by a stream, abundantly supplied with fish and wild fowl. Here are the Galla oxen, celebrated for the remarkable size of their horns. The country from Gibba is very hilly, and the road for a considerable distance lay along the edge of a steep precipice. The descent led to the rich and fertile country of Gambéla, and on the left stood the hill and town of Moculla. At ten o'clock we gained the summit of a hill overhanging the vale of Chelicut. Descended into the valley and reached the town of Chelicut.

MR. PEARCE'S JOURNEY FROM ANTALO (marked R 2 on Map, page 39).

September 28th., 1807. Having crossed the narrow and mountainous district of Wojjerat, Mr. Pearce arrived in eight hours at an extensive and uncultivated plain, inhabited by tribes of negroes called Doba, which are interspersed through all the regions of Africa.

September 29th. After seven hours' travelling, he reached a district called Jyah, held by a tribe of Galla; the country they inhabit is one continued forest, abounding with deer and guinea fowl.

September 30th. He left Jyah and proceeded to Mocurra. This .own is situated about a mile from a fresh-water lake, named Ashangee, said to be nearly as large as Lake Tzana, in Dembea. To the south of this lake extends the mountainous district of Lasta.

October 1st. Mr. Pearce left Mocurra, traversed the eastern side of the lake, and passed through the district of Wófila. The same night, after leaving a small lake called Guala Ashangee on his left, he reached Dufat, a village situated on one of the high mountains of Lasta. Here the cold was intense, and an hoar frost lay upon the ground.

October 2nd. Continued the journey to Senaré.

October 3rd. After seven hours' march, Mr. Pearce slept under a tree at the top of a high mountain; the weather extremely cold; and on the following day

October 4th. Descended into the plains of Maizella to a village near the sources of the Tacazze. This river, which may be considered one of the larger branches of the Nile, rises from three small springs named Ain Tecazze (or the eye of the Tacazze), emptying themselves into a reservoir, whence the waters issue into a collective stream. In this journey from Antálo no river of importance was passed until the Tacazze was reached, with the exception of a small one running northward through Wojjerat.

October 5th. Course nearly due north, following the windings of the Tacazze for eight hours to Mukkine, where, from the accession of smaller streams, the river swells into some importance, as it is thirty feet wide.

October 6th. In five hours reached Sela-Gerré, a lofty hill, eight miles from the Tacazze, and from this place on

October 7th, he proceeded six hours N. by E. to Socóta, the reputed capital of Lasta. This province is extremely mountainous throughout, and forms an almost impenetrable barrier between the two great divisions of Abyssinia, generally comprehended under the names of Amhara and Tigré, two passes only existing through the mountains, which are easily commanded by a small number of troops.

The Lasta soldiers are remarkable for their horsemanship, a quality not common among mountaineers; but attributable to the connection subsisting between this province and Begemder, whose natives pride themselves on their breed of horses, and are also distinguished by the skill with which they train them for service. They speak Amharic, resemble the Galla, are great boasters, but not deficient in courage.

The town of Socóta is ten miles from the Tacazze, is larger and more populous than Antálo; these towns are six days journey apart.

Soon after leaving Socóta, he arrived in the district of Waág, and thence leaving Bora and Salowá on his right, he proceeded three days northward along the banks of the Tacazze, through Gualiu, the country of the Argows, until he came within thirty miles of Maisada. During this march he did not meet any river of consequence running into the Tacazze, though he had crossed particularly about Mukkiné, a great number of small streams and rivulets.

The Agows have a peculiar prejudice against furnishing water to a stranger; Mr. Pearce was readily supplied with milk and bread; but never with water, which was not difficult to procure; the aversion may arise from some ancient superstition, or veneration of the waters, connected with the history of the Nile, an idea strongly confirmed by the circumstance of this people always selecting the banks of the great branches of this river for their residence.

October 9th. He crossed the Tacazze at a ford, where the river is nearly 300 yards broad, into the province of Samen, whence, after travelling about four miles up a steep ascent, he arrived at the village of Guinsa; while here, presents were made of goats, honey, milk and other articles of which the party stood in need.

14th. He had now gained about two-thirds of the ascent of one of the highest mountains of Samen, along a path leading up a deep gully formed by torrents. The landscape presented lofty trees growing among the rocks, beyond which was a boundless extent of country.

15th. The evening of the 15th brought him to Segonet, one of the principal towns of the province, situated on the east side of Amba-Hai.

17th. He got to Mishekka from this; the road is rugged and difficult of ascent; snow and ice in the hollows made it piercingly cold.

18th. Passed over the summit of Amba-Hai, which was tremendously difficult of ascent, and they had a heavy snow-storm; not coming down with violence, but quietly descending in large flakes like feathers.

In the evening after a gradual descent of five hours arrived at Inchetkaub.

24th. He reached Mishekka. Another fall of snow so thick that he made way through it with difficulty.

25th. He arrived at Segonet.

26th. He descended the mountains, and at night reached an Agow village, eight miles from Tacazze.

27th. The river being swollen, was difficult to cross. In the evening arrived at Maisada, and the 28th brought them to Asgevva.

29th. Arrived in the neighbourhood of Antálo.

FERRET ET GALINIER.—JOURNEY FROM MASSOWAH TO ADOWA IN NOVEMBER, 1840 (marked D and I on Map, page 39).

Au sortir de chez le gouverneur, nous visitâmes Messawah. Cette île, située près de la côte d'Afrique, par 15° 36' de latitude septentrionale et par environ 37° 15' de longitude orientale, n'a que 1,000 mètres en longueur, de l'est à l'ouest, et 400 seulement en largeur, du sud au nord. Elle est formée tout entière par un banc de corail poussé un jour à la surface de l'eau par l'effet de ce soulèvement que l'on remarque sur tout le littoral du golfe Arabique. Le point le plus élevé ne dépasse guère de 4 mètres le niveau moyen de la mer. Là, pas une source, pas un arbre, pas même un brin d'herbe; partout la roche qui se montre à nu et la pierre stérile. Pour avoir de l'eau potable, il faut creuser des citernes et y recueillir précieusement les pluies, si rares d'ailleurs, qui tombent de loin en loin depuis Octobre jusqu'en Mars; faible ressource, puisqu'elle suffit à peine aux besoins du gouverneur et de ses amis. L'eau que boit le peuple vient des sources saumâtres du continent, d'Arkiko et de Moukoullou. La moitié de l'île appartient aux morts, l'autre aux vivants. A l'est les tombes, à l'ouest la ville. Les citernes se trouvent parmi les tombeaux.

Quel triste et malheureux aspect que celui de Messawah!

* * * * * *

Messawah est un des points les plus ardents du globe. En novembre, le thermomètre centigrade, placé à l'ombre, donnait à neuf heures 31 (88° Fahr.) degrés, à midi 34 (93° Fahr.), à trois heures 33 (91½° Fahr.), à neuf heures du soir 30 (86° Fahr.). En été, l'atmosphère est une haleine de feu. Dans le mois de Juillet 1842, nous avons eu jusqu'à 53 (127° Fahr.) degrés de chaleur! Les hommes du Nord ne sauraient s'habituer à cette température excessive.

* * * * * *

Pour comble de malheur, tant que se prolonge la saison d'été, l'île est très-malsaine. La partie de la grève que les eaux basses laissent à découvert couve des miasmes pestilentiels. L'air se corrompt, et de cruelles maladies, la dyssenterie, la fièvre, la plaie de l'Yémen, la petite vérole enfin, exercent dans l'île d'effroyables ravages. Néanmoins, le croirait-on? sur ce roc que la nature semble avoir maudit, on voit des exemples de longévité tout à fait extraordinaires.

* * * * * *

Stérilité absolue, climat brûlant, maladies nombreuses et terribles, tout concourt à faire de Messawah l'endroit le plus inhospitalier du globe. Son port seul offre quelques-uns des avantages dont le littoral de l'Afrique, dans ces parages du moins, est totalement dépourvu. Formé par l'île au sud, par la côte du golfe au nord, c'est un des plus sûrs de la mer Rouge: il peut contenir une soixantaine de ces petits navires de 50 à 180 tonneaux, dont se servent les Arabes. L'entrée, dont la largeur est d'environ 400 mètres, se trouve au nord-est de l'île. C'est à la sûreté de son port et à sa position à l'entrée de la seule route qui conduit de la mer Rouge dans l'Abyssinie septentrionale que la ville de Messawah doit son existence. On y compte 4,000 habitants, suivant tous la loi du prophète; et, sur ce nombre, les artisans, dont la plupart est employée à la construction des barques, viennent tous de l'étranger. Nous avons passé en revue les professions diverses ; nous avons trouvé dans l'île un armurier, un orfèvre, un cordonnier, un tailleur et enfin deux barbiers. Les deux barbiers exercent la médecine, cela va sans dire. A eux seuls est confié le soin de combattre l'excessive rigueur du climat meurtrier de Messawah ; c'est pour cela que l'on y voit tant de tombeaux.

* * * * * *

Quant au caractère général de la population, qu'il nous suffise de dire que les Messawanis sont pétris de tous les vices. Ils sont intéressés, fanatiques, insolents, voleurs et ingrats jusqu'à l'impudeur.

* * * * * *

Les sequins de Venise et les thalaris d'Autriche à l'effigie de Marie-Thérèse sont les seules monnaies reçues à Messawah. Pour les menus achats on se sert de petits grains de verre, appelés là-bas *bourchokos*, et connus dans le commerce européen sous le nom de conterie de Hollande. Trois *bourchokos* valent un *kebir*, quarante *kebirs* un *harf*, trente-deux *harfs* un *thalaris* ou 5fr. 25c. Ainsi un thalaris vaut 3,840 *bourchokos*.

* * * * * *

Les poids en usage Messawah sont :—	
Le *rotoli*, qui vaut en kilogrammes	0·445
Le *mimes*	1·000
Le *farasselli*	8·900
L'oghia pour peser l'ivoire	19·130
L'oghia pour l'or et le musc	0·064
Les mesures de capacité sont :—	
Le *hemlé* qui vaut en hectolitres . . .	2·088
L'*ardep* est le quart du hemlé. Il se devise en 52 *kelés*, et le kelé 2 *roubs*	
Le *gobeh* est une mesure de capacité qui sert pour le beurre et le miel; son évaluation est ordinairement calculée sur 8 rotolis	3·560
Les mesures de longueur sont :—	
Le *pic endasi*, pour draps et soieries . . .	0·72
Le *pic beledi*, pour tout autre tissu . . .	0·62
Le *madah* vaut 4 pics endasi	2·48

After experiencing some opposition from the Nayb, or chief of the neighbouring tribe, they commence their journey.

* * * * * *

C'était le 10 novembre : malgré l'heure avancée du jour, nous ne voulûmes pas attendre le lendemain pour nous mettre en route ; aussi bien le Nayb pouvait-il nous susciter quelque nouvel obstacle, et nous le tenions suspect désormais. En vain voulut-on nous représenter que la nuit allait nous surprendre, et que nous serions obligés de coucher sous le ciel et dans la plaine, exposés aux bêtes féroces qui infestent le pays : bêtes féroces pour bêtes féroces, notre choix était fait, et nous partîmes immédiatement. Du reste, nous n'allâmes pas loin. Le soleil disparut bientôt de l'horizon, et nous dûmes nous arrêter aux environs d'Adde-Habib, hameau situé seulement à deux lieues au sud d'Arkiko.

* * * * * *

Nous partîmes le lendemain un peu avant six heures. Nous cheminions encore dans la même plaine que la veille ; seulement elle change de nom et prend ici celui de Zabararaguay, tandis qu'elle porte le nom de Catra entre Arkiko et Adde-Habib ; c'est une plaine basse, sablonneuse, stérile. On n'y voit que quelques mimosas rabougris, rongés par la poussière et dévorés par le soleil. Après deux heures de marche, toujours dans la direction du sud, nous atteignîmes le premier contre-fort des montagnes. Une gorge fraîche, d'une montée facile, on l'appelle Chilleké, nous amena sur un fertile plateau dominé en avant par une chaîne plus haute et plus compacte que celle que nous venions de franchir. Notre chemin était de traverser ce plateau en marchant vers le sud-ouest; mais, comme nous manquions d'eau, nos guides nous firent dévier un peu de notre route pour nous conduire vers le sud à Ouéah, où passe l'Oligadé, qui se jette, dit-on, dans la mer Rouge à Zoulla, l'ancienne Adulis. Nous arrivâmes à onze heures sur le bord de l'Oligadé. Fâcheux mécompte! le lit du ruisseau était à sec, et du torrent il ne restait qu'une mare d'eau stagnante; mais, après tout, l'ombre d'un bois d'acacias nous flottait sur la tête et nous garantissait des rayons brûlants du soleil; il n'en fallait pas plus pour que ce lieu nous parût un jardin de délices. Des arbres et de l'eau, peut-on désirer autre chose après une longue et pénible étape sous le ciel des tropiques? Nous étions bien décidés à passer le reste du jour dans cette halte, sous cette belle tente de feuillage, mais nos guides trouvaient le site moins enchanteur; ils redoutaient la visite des lions qui viennent se désaltérer la nuit dans le ravin. Nous le sûmes depuis, nos guides ne nous trompaient pas: il n'est pas rare que quelque voyageur y soit dévoré. C'est égal, la fatigue d'abord, et puis le désir d'admirer le roi des animaux dans toute sa liberté, nous eussent fait oublier toute prudence, si nous avions pu trouver le moindre gîte; mais le manque d'abri rendait le danger trop sérieux, et il fallut nous éloigner en donnant raison à nos guides.

D'après nos observations barométriques, Ouéah est élevé de 190 mètres au-dessus de la mer. Le thermomètre, placé à l'ombre, marquait à midi 33 degrés centigrades ($91\frac{1}{2}$° Fahrenheit).

Au bout d'une heure, en suivant le lit desséché de l'Oligadé, dans la direction de l'ouest, nous atteignîmes Hédali, hameau abandonné qui se trouve à l'extrémité occidentale du plateau et au pied des montagnes. Là, rompus de lassitude, nous prîmes à la hâte un modeste repas, et, nargue du lion! sans y mettre autrement de bravade, sans y avoir songé, sans avoir pu s'en défendre, chacun de nous se laissa surprendre par le sommeil. Quand le soleil parut, il se trouva que personne n'avait veillé, et la nuit s'était passée sans accident.

Le 12, de bon matin, nous reprenions notre route, parfaitement remis des fatigues de la veille. De hautes montagnes se dressaient devant nous, et semblaient barrer notre marche. Déjà nous cherchions, non sans inquiétude, sur leurs flancs escarpés et nus, la trace de quelque sentier, lorsque, entre le mont Adodah et le mont Oligadé, nous aperçûmes un défilé étroit qui présentait un passage sûr et facile. A la bonne heure, au moins! le contentement nous revint au cœur. C'était la vallée de Haddas qui, par une montée presque insensible, mène jusqu'au pied du Tarenta, dernier échelon pour arriver sur le plateau de l'Abyssinie.

A vol d'oiseau, cette vallée n'a pas plus de douze lieues de longueur. Cependant les nombreux détours du chemin, et des débris de rochers écroulés çà et là sur le sol, en rendent le parcours long et quelquefois difficile. Toutefois, c'est la route la plus directe et la plus pittoresque qui conduise de la mer Rouge dans l'Abyssinie septentrionale. Malheureusement, elle n'est point practicable toute l'année. En été, on y trouve rarement de l'eau; pendant la saison des pluies, au contraire, elle sert de lit à un torrent impétueux qui grossit dès qu'éclate l'orage, et alors elle est inondée. Mais au commencement ou à la fin des pluies, à l'époque enfin où nous l'avons parcourue, quel admirable contraste avec les contrées que nous laissions derrière nous ! Ce n'étaient plus ces paysages monotones, ces plages basses et stériles qui fatiguaient nos regards depuis notre sortie d'Égypte; tout d'abord, la vallée étale devant nos yeux étonnés une nature sauvage, pleine d'accidents et de variété. Les montagnes qui nous entourent ont un aspect effrayant : elles s'élèvent verticalement, et quelques-unes des roches, qui les composent, minées par le temps, menacent de s'écrouler et de nous écraser dans leur chute; mais le fond de la vallée offre un paysage bien différent : on dirait le plus riant jardin. Près de nous coule un ruisseau limpide ; l'air s'embaume des parfums des plantes et des fleurs que nous foulons sous nos pas, et une végétation puissante projette sur nous de verdoyants parasols et ombrage notre route. Cette première partie de la vallée de Haddas s'appelle Hammamo ; c'est un délicieux vallon que nous parcourûmes sans fatigue et presque à notre insu, tant notre pensée se perdait hors de nous dans le charme d'un tel spectacle. Devant nos pas se croisaient des perdrix d'une grosseur extraordinaire, des gazelles et des lièvres que notre approche n'effrayait pas. Sans pitié pour ces gracieuses créatures, si douces et si confiantes, nous les tirions presque à bout portant, et nous voyions, il faut l'avouer, avec une joie barbare, notre table depuis longtemps frugale, devenir tout à coup abondante et splendide. Le bruit même de nos armes nous égayait ; nous le jetions comme un défi aux mille échos de la vallée, et nous nous amusions à épouvanter des myriades de singes qui s'enfuyaient avec des glapissements désespérés.

Après Hammamo, le paysage change ; le ruisseau disparaît, et nous voilà tout-à-coup dans un lieu affreusement stérile ; le chemin est encombré de pierres ; tout est morne, silencieux ; malgré nous, la gaieté fait place à la tristesse, le bonheur à l'ennui. Mais au détour de la vallée, une nouvelle oasis apparaît, et nous nous retrouvons de nouveau dans un site délicieux. C'est ainsi qu'on chemine dans ce sombre vallon, traversant tantôt de charmants paysages, et tantôt des lieux sauvages, arides, rocailleux et désolés.

Cependant, à mesure que nous avançons, la vallée prend un caractère de beauté plus grandiose ; l'aspect des montagnes devient plus sévère et plus majestueux. Cette double muraille qui nous environne semble se resserrer par moments pour nous clore le passage, et nous n'apercevons plus le ciel qu'en regardant au-dessus de nos têtes. Mais qu'avons-nous besoin de regarder ailleurs qu'autour de nous ? Nous voici à Tobbo : partout de la verdure, partout des fleurs ; c'est un jardin enchanté, un paradis de câpriers, de tamarins, de sycomores, entre des montagnes gigantesques. Sur la cime des montagnes s'amassent les nuages sombres ; à leur pied, la gazelle bondit, l'écureuil saute, et des essaims d'oiseaux de toutes les couleurs s'envolent du milieu des feuillages comme des fleurs ailées. Séduits par la beauté du site, nous nous reposons délicieusement sous ses mystérieux ombrages, et nous y passons la nuit.

Le lendemain, 14, nous arrivons au pied du Tarenta. Là, campés sous un vaste sycomore, nous fîmes immédiatement nos observations barométriques. Nous étions en ce moment à 1,425m. au-dessus de la mer Rouge. Le thermomètre centigrade marquait, à midi, 26°80 (89° Fahrenheit), environ 7 degrés de moins qu'à Arkiko.

Selon nos conventions avec le Nayb, nous renvoyâmes les chameaux ; ces animaux si utiles, si précieux dans les plaines de sable, ne sont pas faits pour les pays de montagnes. C'est avec l'aide des Chohos que nous devions maintenant transporter nos bagages jusqu'à Dixah.

Les Chohos habitent les plateaux qui dominent cette vallée de Haddas que nous venons de parcourir. Leur pays s'appelle *Samhar*, et ce nom s'applique aussi quelquefois à la vallée. Une de leurs tribus, la plus nombreuse et la plus puissante, est celle des Hassaortas. Quelques-uns de ces Hassaortas descendent de temps en temps dans le vallon pour y faire paître leurs bestiaux. Les flancs de la montagne creusés en cavernes, des huttes de branchages, leur servent de demeures. C'est parce qu'on est exposé à rencontrer ces sauvages pasteurs, qu'il est impossible de parcourir ces lieux sans la protection du Nayb, leur chef et leur maître. Féroces et voleurs, ils vous dépouilleraient inévitablement si vous n'étiez accompagné par des hommes de leur tribu, qui répondent de vous sur leur tête. La tribu des Hassaortas, comme toutes les autres tribus des Chohos, a des troupeaux immenses, et ne se nourrit guère que de laitage; elle ne cultive pas, ou cultive à peine ses champs; le grain nécessaire, elle l'achète aux habitants du Tigré, qui en récoltent au delà de leurs besoins. Là se bornent les relations des Hassaortas avec les Abyssins, qu'ils méprisent. C'est avec Messawah, dont les habitants sont musulmans comme eux, qu'ils ont des rapports plus fréquents: ils approvisionnent cette île de beurre, de moutons, de chèvres, de lait; et Messawah leur fournit, en échange, des toiles, du tabac, et une foule d'autres objets.

Avertis par un de nos guides, une dizaine d'Hassaortas vinrent le lendemain, avant le jour, amener leurs bœufs pour remplacer les chameaux: c'est le seul moyen de transport usité pour l'ascension du Tarenta; faute d'autre, nous l'acceptâmes; mais, dès les premières paroles, nous reconnûmes le Choho effronté et rapace. A en croire ces larrons, il ne fallait pas moins de douze bœufs, tandis que cinq pouvaient suffire. Grâce à l'intervention de Zaraïl, il fut convenu que nous en prendrions sept pour partager à peu près le différend. Lorsque, après d'interminables démêlés, nous fûmes d'accord sur le nombre, les Chohos, querelleurs et paresseux, examinèrent de nouveau les paquets; ils les trouvèrent trop incommodes et trop lourds. Force fut donc de les refaire, et de les refaire en leur présence. Si nous ne fûmes pas volés, nous en rendons grâce à Dieu comme d'un miracle; mais, à peine l'opération des paquets terminée, voici un pêle-mêle épouvantable: chacun de ces lâches coquins se jetait sur le paquet qui lui semblait le plus léger, et on s'arrachait, ou plutôt on arrachait nos bagages: injures, menaces, il n'y manquait rien, rien que les coups. Pour dédommagement, nous espérions voir ces furieux s'infliger l'un à l'autre le châtiment qu'ils méritaient si bien, mais ce plaisir ne nous fut pas donné. Zaraïl leur proposa de tirer à la courte-paille; ils se soumirent au jugement du sort, et oublièrent aussitôt leurs injures réciproques.

Nous partîmes à 7 heures. Le Tarenta était couvert de végétation; il nous montra tout d'abord un arbre que nous n'avions vu nulle part, et qui ne se trouve pas ailleurs en effet, si ce n'est en Abyssinie; c'est le *qolqual*, semblable à un immense candélabre; ses branches ne portent jamais de feuilles, elles donnent seulement, à leur extrémité, un bouquet de fleurs, dont le coloris change et passe, suivant la saison, du jaune vif au cramoisi. Plus haut que les qolquals, nous vîmes des oliviers sauvages, et de grands arbres appelés *ouarras* par les Abyssins. Leur bois est dur et propre aux constructions. Tous ces arbres prennent racine dans les fentes des rochers, car il y a peu ou point de terre végétale. Presque partout la roche se montre à nu; aussi est-il facile d'étudier la composition géologique de la montagne. On y voit, à partir de la base, la syénite rouge ou rose verdâtre, le gneiss, qui est souvent grenatifère et traversé de filons de quartz blanc laiteux; la protogine traversée par des filons et typhons de diorète, enfin la protogine qui reparaît et recouvre toutes les autres roches.

Nous mîmes trois heures pour franchir le Tarenta. Un sentier affreux, encombré tantôt de pierres qui roulaient sous nos pas, et embarrassé tantôt par d'énormes rochers qu'il fallait gravir en se cramponnant fortement des pieds et des mains, nous mena à son sommet. Là, nous nous arrêtâmes, exténués de fatigue, en un lieu qui s'appelle *Ouady Saasseh*. Quel magnifique spectacle se fit alors devant nos yeux! Du côté de la mer, nous dominions toutes les chaînes de montagnes que nous avions précédemment traversées; elles se pressaient, elles s'entrecoupaient au dessous, et nous les voyions si petites, que nous les comparions aux vagues de la mer. Du côté de l'Abyssinie, nos regards s'étendaient au loin sur les montagnes du Tigré, dont les sommets légers et gracieux se confondaient avec l'azur du ciel. C'est vers ces montagnes que nous nous tournions de préférence; cette vaste contrée que nous venions visiter de si loin, ce royaume vers lequel nous avions marché si longtemps, cette région à peine connue que nous allions explorer pour en dire quelque chose à la curiosité de l'Europe savante, l'Abyssinie était là devant nous; nous la voyions, nous la touchions; ce fut pour nous un moment de bonheur et de douce espérance.

Nous ne connaissions pas encore la hauteur du Tarenta; nous fîmes aussitôt nos observations, et le baromètre nous apprit que nous nous étions élevés à 2,543 mètres au-dessus du niveau de la mer. La base de la montagne ayant une altitude de 1,425 mètres, nous avions donc franchi, en trois heures, une hauteur de 1,118 mètres. Cette différence d'élévation entre la station de départ et celle d'arrivée se traduisait d'une manière bien sensible dans l'abaissement de la température: à 10 heures du matin, le thermomètre ne marquait que 14 degrés (57° Fahrenheit), et il ne s'éleva qu'à 16 (61° Fahrenheit) à midi. Habitués aux chaleurs excessives des parages de la mer Rouge, une température de 14 degrés nous parut glaciale. Aussi, en attendant les bagages, que nous avions devancés, nous nous assîmes avec plaisir en plein soleil, au lieu de chercher l'ombre comme nous le faisions depuis notre départ d'Egypte.

Dès que notre caravane fut réunie dans l'Ouady Saasseh, nous pressâmes pour que l'on se remît en route; nous voulions atteindre *Dixah* avant la nuit; mais Zarail ne se souicait pas d'arriver chez lui vers la fin du jour: il lui fallait du temps pour nous trouver un logement convenable; sa fierté l'empêcha de nous en faire l'aveu, et il retarda notre départ sous prétexte que les bêtes de somme étaient fatiguées. Nous couchâmes à *Maiharassat*, à environ une lieue de Dixah, et, le lendemain enfin, nous fîmes notre entrée dans la résidence du *Baharnagass*, du roi de la mer, où notre ami nous faisait espérer depuis longtemps une hospitalité généreuse.

* * * * * *

Un groupe de mauvaises baraques échelonnées au hasard sur le sommet d'une montagne nue et décharnée, un village misérable contenant une population tracassière d'environ 1,500 âmes, chrétiens et musulmans réunis, voilà *Dixah*. Il est situé par 14° 59′ de latitude et par environ 37° 8′ de longitude; changez la route qui vient de la mer Rouge à Gondar, et Dixah deviendra un village obscur et ignoré; car sa position seule fait son importance.

* * * * * *

La caravane d'Agaouderés, de Fanta et de Gouchou, était arrivée à un mille de Dixah. Ces bons négociants apprennent que nous sommes encore chez le Baharnagass, et ils s'empressent de nous rendre visite.

* * * * * *

Arrivés au campement de la caravane, les négociants nous offrirent à dîner.

* * * * * *

C'était le 20 Octobre : dès que le soleil parut, on se prépara pour le départ. Plus impatients que le reste de la caravane, nous fûmes prêts avant tout le monde ; nous avions hâte de quitter Dixah, dont le séjour avait failli nous être si funeste.

Pour faire la route d'une manière plus commode et plus convenable, nous venions de nous monter (on a mule), comme l'on dit. En Abyssinie, il ne faut pas voyager à pied si l'on veut jouir de quelque considération ; encore moins faudrait-il voyager sur un âne.

* * * * * *

Le chemin que nous suivons est un des plus fréquentés de l'Abyssinie ; c'est par là que les caravanes se rendent à la mer et qu'elles en reviennent. Ne vous y trompez pas cependant, ceci ne ressemble guère à un de nos grands chemins, à ces routes commerciales qui traversent notre France ; un sentier tout simplement, rien qu'un sentier. Le pied des voyageurs l'a tracé lentement, et il serpente au hasard sur les flancs des collines, au milieu des prairies ; çà et là il côtoie quelques arbres épineux ; prenez garde alors, vous courez risque d'y déchirer vos vêtements et vos jambes. La campagne qu'il traverse est partout accidentée, et nous parut assez bien cultivée. A une lieue de Dixah, on passe près de *Adde-Haddit*, situé sur le sommet d'une montagne ; une heure après nous atteignîmes un autre village nommé *Adde-Goutta ;* puis enfin nous entrâmes dans le district de *Degoundouga.*

Nous cherchâmes d'abord un lieu désert afin d'y faire halte ; débarrassés ainsi des importunités des indigènes, nous passâmes le reste de la journée dans une charmante quiétude. Après les tracas de la veille, les angoisses de la nuit et les fatigues du voyage, nous avions le plus grand besoin de repos. Nous jouîmes de ce plaisir des plaisirs, qui se nomme le *far niente.*

Le lendemain, lestes et dispos, nous reprenions notre route à sept heures ; à neuf heures, nous étions sur la crête des collines qui dominent le district de *Gounzobo*, et voici ce que nous avions devant les yeux : près de nous, sous nos pieds, le village de *Saïdah* dans une gorge étroite ; plus avant, un immense amphithéâtre entouré de murailles prodigieuses et terminé au loin par des édifices croulants, flanqués de tours colossales ; cet amphithéâtre de géants, c'étaient les montagnes arides, bizarres de Gounzobo, dont la stérilité contraste d'une manière étrange avec la fertilité de la plaine qu'elles enserrent. Nous campions à leurs pieds, vers une heure, dans un lieu sans habitations appelé *Maourray ;* là se tient un marché hebdomadaire où se réunit la population du district de Gounzobo et des environs. Mal servis par la circonstance, nous arrivâmes précisément le jour du marché, au milieu d'une foule considérable. On nous avait dit d'avance : gens de Gounzobo, gens de mauvaise foi, turbulents et voleurs. Ils ne nous ont pas fourni l'occasion de démentir la renommée.

A Maourray, la caravane qui passe paye un droit de douane, c'est la coutume ; mais comme la coutume a omis de fixer une taxe régulière, la perception, quelque peu capricieuse, amène nécessairement de longs débats.

* * * * * *

Le 22, nous partîmes à sept heures. Après être sortis des montagnes de Gounzobo, nous entrâmes dans une grande et belle plaine malheureusement inculte, couverte d'herbes et de mimosas; çà et là paissaient de nombreux troupeaux. Les bergers, pour un peu de tabac ou de poivre noir, nous régalaient d'un lait tiède et pur qu'ils s'empressaient de traire sur notre demande. Deux rivières, le *Tséréna* et le *Bélessa*, arrosent cette plaine; elles coulent vers le nord-ouest et se jettent dans le *Mareb* après s'être réunies, nous dit-on, à environ trois lieues à l'ouest de notre route. En ce moment, les eaux étaient très-basses; mais, d'après le dire des gens qui nous entouraient et d'après les traces des dernières crues, il paraît qu'à la saison des pluies ces rivières roulent des masses d'eau considérables. Les communications se trouvent alors interrompues. Les Abyssins ignorent l'art de construire des ponts, et se servent rarement de radeaux pour passer les rivières. A une heure de l'après-midi, nous nous arrêtâmes dans le district de *Neggot*, à une lieue au sud du Bélessa.

Nous approchions donc d'*Adoua* et du camp d'Oubié. Le pays était sûr; il était tranquille, chose peu ordinaire en Abyssinie. Charmés de cette sécurité précieuse, les négociants de la caravane voulurent en jouir sans doute, et les voici cheminant du pas dont on se promène, prolongeant d'ailleurs les haltes au gré de leur nonchalante nature, si bien que la caravane mit trois jours pour aller de Neggot à *Guendepta*. Trois longs jours pour franchir une distance de onze lieues!

* * * * * *

Nous quittâmes Guendepta le lendemain à huit heures. Après une descente rapide, nous entrâmes dans la vallée de *Robber-Aeïni*. Cette vallée riante et fertile, ceinte de montagnes élevées, couvertes d'arbres de toute espèce, nous mena jusqu'à *Mariam-Chaouïtou*, église vénérée, entourée de sabines, d'oliviers sauvages, de mimosas odorants qui la couvrent de leur mystérieux feuillage. Là, nous quittâmes MM. Rouget et Bell, qui devaient suivre la caravane jusqu'à Adoua, et, accompagnés d'Agaouderés, nous nous dirigeâmes vers le camp d'Oubié, qui se trouvait à une lieue environ sur notre droite. Nous y arrivâmes à onze heures.

Le camp était situé sur le sommet aplani d'une haute montagne dont les flancs en précipice n'offraient une route accessible que par un seul côté: c'était une espèce de ville bâtie au hasard, comme presque toutes les villes d'Abyssinie. Notre arrivée fit beaucoup de sensation parmi les soldats: c'était à qui nous verrait le premier et nous adresserait le premier la parole. Par malheur, le roi Oubié ne se piqua pas du même empressement. On nous dit de sa part qu'il avait consacré ce jour aux prières et qu'il ne pourrait pas nous recevoir avant le lendemain. Ce retard ne nous paraissait pas de bon augure. Quel accueil nous réservait Oubié? Notre incertitude devenait plus inquiète; la nuit nous parut longue et se passa sans sommeil: l'impatience nous empêchait de fermer les yeux, et la vermine bien plus encore.

* * * * * *

La capitale du Tigré, bâtie à la fois dans la plaine et sur le penchant d'une colline, n'a rien de remarquable; elle ne fait pas meilleure figure que la plupart de nos villages de France; les rues sont étroites, irrégulières, encombrées de pierres, de fumier, et bordées de murs de clôture ou de maisons d'une apparence misérable. La population, chrétiens et musulmans réunis, ne dépasse pas quatre mille âmes. Les églises, entourées de grands arbres qui les couvrent d'une ombre épaisse, sont au nombre de quatre; mais la vénération du pays distingue et honore entre toutes l'eglise de *Médani-Alem*, située au milieu de la ville, et dédiée, comme son nom l'indique, au Sauveur du monde.

Excepté vers le sud-ouest, Adoua est environné de hautes montagnes composées principalement de schistes redressés. Du côté du nord, on distingue le mont Chelloda, remarquable par sa grande élévation et la forme de son sommet, que les Abyssins comparent, avec assez d'exactitude, au dos d'un cheval. Au milieu de ces montagnes coule une petite rivière nommée *Assam* qui arrose la ville. Elle prend sa source non loin d'Adoua, se dirige vers le sud, se grossit de deux ou trois petits cours d'eau, et se jette ensuite dans le Warié, l'un des affluents de la rive droite du Taccazé.

D'après nos observations astronomiques, consistant dans neuf séries de hauteurs circumméridiennes tant du Soleil que de la Chèvre et de Jupiter, Adoua est situé par 14° 9′ 34″ de latitude septentrionale.

Sa longitude, prise au chronomètre, par rapport aux longitudes d'Axoum et d'Intetchaou, que nous avons déterminées un peu plus tard, est de 36° 35′ 9″ à l'est du méridien de Paris.

Nous avons trouvé, au moyen de nos observations barométriques, que la ville, près de Médani-Alem, s'élève à 1,900 mètres au-dessus du niveau moyen des mers.

* * * * * *

La ville d'Axoum est située au pied d'une montagne, partie dans la plaine, partie dans une gorge étroite où elle oublie, à l'ombre des sycomores et des oliviers, les jours de sa splendeur passée. Autrefois son nom était célèbre dans toute l'Ethiopie. De nos jours, trois ou quatre cents masures, les unes groupées sans ordre, le plus grand nombre disséminées au hasard, c'est là toute la ville d' Axoum. L'église, qui passe pour la plus belle de l'Abyssinie, ne présente rien de remarquable. Imaginez un immense dé à jour, un cube de 35 mètres de long sur 16 de large et 14 de hauteur, avec un rang de piédestaux rompus, alignés devant la façade principale. On pénètre dans le temple par un grand escalier de vingt marches, dont les douze degrés inférieurs n'ont pas moins de 50 à 55 mètres de long et sont bien conservés. A une petite distance de l'église, sur une grande place, on voit un obélisque magnifique qui, depuis plus de vingt siècles, se dresse fièrement vers le ciel. Ce superbe monolithe a 80 pieds de haut ; à considérer son volume et la hardiesse de son élévation, l'œil s'étonne, on se sent frappé d'admiration, et l'on admire l'invention de l'homme aux prises avec les forces de la nature. Comment les anciens ont-ils pu manier une pareille masse ? C'est là, sans doute, une question curieuse, un problème difficile à résoudre. Les Abyssins se le sont proposé, et l'expliquent d'une façon commode : en faisant intervenir la puissance du diable, qui voulait, disent-ils, construire une grande tour pour escalader le ciel.

A côté de cet obélisque, il s'en trouve deux autres d'un travail également remarquable ; mais ceux-ci sont brisés en plusieurs morceaux, et gisent sur le sol rongés par le temps et la poussière. Tout près de là, passe un petit ruisseau, dont les eaux, toujours vives, alimentent un petit réservoir construit dans le milieu du quinzième siècle par l'*abouna* Samuel. Ce réservoir est placé à peu près au centre des ruines qui furent l'ancienne ville. A deux kilomètres, au sud-est d'Axoum, cherchez encore d'autres restes de l'antique capitale des Axoumites ; vous ne trouverez plus qu'une masse de décombres, dominée par un monticule grisâtre ; ce monticule passe pour être le tombeau de Ménilek, fils de Salomon et de la reine de Saba. Quant aux décombres, insignifiant amas de briques ou de pierres taillées, ils méritent à peine qu'on fasse une demi-lieue de chemin pour les visiter.

En revanche, vers l'est de la ville, ce qui a survécu aux temps anciens offre un plus grand intérêt. De ce côté, à douze cents mètres environ sur la route d'Adoua, vous rencontrez plusieurs autels en pierre, ainsi qu'un petit obélisque encore debout sur sa base. Cette aiguille a vingt pieds de hauteur et ses proportions sont parfaites. On nous a montré également dans cet endroit la pierre sur laquelle est gravée l'inscription grecque dont Salt a donné le premier la traduction au monde savant, et qui a fait la fortune de son voyage.

FERRET ET GALINIER. RETURN JOURNEY FROM GONDAR TO ADOWA, MAY, 1842, AND FROM ADOWA TO MASSOWAH BY THE WESTERN ROUTE, JULY (marked J and G on Map, page 39).

Les troubles et la guerre nous fermaient la route des provinces de l'Abyssinie qu'il nous retait à visiter. D'un autre côté, nos forces s'étaient usées à la fatigue, notre santé était sérieusement atteinte, et la saison des pluies, qui était déjà commencée, pouvait la compromettre davantage. Pour éviter ce danger, nous résolûmes de quitter l'Ethiopie et de retourner en France.

C'était le 15 mai. Dès le matin nous prîmes congé de l'Impératrice et de l'*Etchequié*. Nous serrâmes la main à M. Arnaud d'Abbadie, dont nous ne reçûmes pas les adieux sans regret, et nous sortîmes ensuite de Gondar.

Notre petite caravane se composait de 18 à 20 personnes. Nous marchions vers le nord-est.

* * * * * *

Chemin faisant, nous atteignîmes la frontière méridionale de la province du Waggara. Arrivés dans le district de Massal Danghia, nous laissons à notre gauche quelques huttes de branchage, et nous entrons ensuite dans une grande plaine semée de mamelons arides. Pas un habitant, pas une masure. Le soleil labourait péniblement le ciel à travers d'épais nuages. Le vent du sud-est s'était élevé, courait à grandes rafales, et courbait violemment les branches des arbres. De quelque côté que l'on se tournât, les yeux étaient aveuglés par des tourbillons de poussière. Il n'y avait pas à se faire illusion, l'orage se formait dans le trésor des nuées, et la tempête allait nous assaillir des quatre points de l'horizon. En effet, un moment de plus, et l'obscurité du ciel se déchira dans tous les sens; les éclairs se nouèrent et se dénouèrent comme des couleuvres de feu; le tonnerre ébranla la profondeur de la voûte céleste; il y eut un instant de silence, et la pluie tomba à torrents.

* * * * * *

L'orage dura six heures, et pendant six heures, nous reçûmes sur le corps une pluie torrentielle, tandis que au-dessous de nous, comme au-dessous d'un pont, coulaient de petits ruisseaux qui se heurtaient, en se courrouçant, contre notre lit de pierre, et allaient ensuite se perdre dans une flaque d'eau voisine.

* * * * * *

A droite, à gauche, partout, la plaine qui s'étendait autour de nous offrait l'aspect d'un vaste désert. La guerre avait passé par là, les armées de Marso, d'Oubié et de Ras-Ali avaient traversé cette partie du Waggara, pillant et détruisant, brûlant ce qu'elles ne pouvaient pas emporter. Les habitants s'étaient enfuis avec leurs troupeaux dans les montagnes.

* * * * * *

A une heure enfin, nous trouvâmes un village debout et de bon accueil. Aussi était-ce un *guédam*, le guédam de Feras-Saber. Ombrages délicieux, prairies verdoyantes, terre aimée des cieux; on aurait dit une riante oasis au milieu du désert.

De Feras-Saber, c'était le 17, la caravane s'achemina dans la direction du nord-ouest. Nous traversâmes quelques prairies où paissaient de gros troupeaux de bœufs et de moutons. Puis, après deux heures de marche sans avoir eu à monter d'une façon bien sensible, nous nous trouvâmes sur le Lamelmon, dont le sommet, situé à 2,750 mètres au-dessus de la mer, se confond avec la plaine du Waggara. Tout-à-coup un gouffre immense se creusa sous nos pieds. Nous mîmes pied à terre, et conduisant nos mules par la bride, nous descendîmes la montagne par un sentier à pic, encombré de fragments de roche mobiles qui menaçaient à chaque pas de nous précipiter dans l'abîme creusé à notre gauche. Arrivés au village de Debbe-Bahar, nous faisons décharger les bagages.

* * * * * *

En ce moment passaient deux soldats du Godjam qui venaient de prélever quelques contributions sur un village voisin, et s'en allaient rejoindre l'armée d'Oubié, campée alors dans les environs d'Inchetcab.

* * * * * *

Le lendemain, lorsque notre caravane se remit en route, le soleil se levait au milieu d'un ciel vide, et nous promettait une journée des plus chaudes; le thermomètre accusait une température de 22° cent. (71½° Fahrenheit). Le terrain était raboteux, accidenté, d'un accès très-difficile. Sur la droite les montagnes gigantesques du Samen dressaient leurs crêtes à une hauteur qui défiait le regard, et montraient à leur sommet des prismes, des pyramides, des colonnades de la forme la plus irrégulière, comme pour rappeler au voyageur que ce n'était pas une main d'homme, mais la main de Dieu qui avait pu jouer avec ces masses.

Bientôt nous rencontrons un petit ruisseau qui creuse sa vallée d'arbustes, de sable et de broussailles dans un terrain aride et desséché par le soleil. Ce ruisseau se dirige du sud au nord. Notre caravane le traverse huit fois, cheminant tantôt sur la rive gauche, tantôt sur la rive droite; à midi, elle s'arrête enfin sur les bords de la Zarima.

La Zarima est une rivière qui descend du versant occidental de la chaîne du Samen. En ce moment elle roulait un volume d'eau peu considérable; mais ses rives ombragées par des arbres magnifiques étaient couvertes d'un gazon abondant, tapis naturel charmant à voir, plus doux encore à fouler. Nous y établîmes notre campement.

* * * * * *

La crainte d'une attaque nous décida, le 19, à pousser en avant.

Une heure après notre départ, la caravane laisse à gauche le couvent du Waldouba, nous franchissons successivement les vallées de l'Enzo et l'Ancéa, et après avoir traversé le district d'Adderké, nous arrivons sur les bords de l'Yama. A gauche la campagne semble aride et brûlée par le soleil, sur la droite, au contraire, en se rapprochant des montagnes, elle est parsemée de hameaux adossés à des bosquets d'oliviers, de wanzas et de sycomores, dont le feuillage varié nuance harmonieusement le paysage. Ici, plus encore que dans le reste de l'Abyssinie, tout ce qui vient de l'homme est chétif et fragile. Parcs à brebis formés de branchages, huttes en pierres sèches, portes de roseaux, villages pauvres et mal bâtis. L'anarchie qui règne habituellement dans cette malheureuse contrée a sans doute averti les habitants du peu de durée et de la vanité de constructions plus solides.

* * * * * *

Nous fîmes près de trois lieues au milieu d'un brouillard vivant. Notre caravane arriva ensuite sur les bords du Boéa, et s'arrêta à quatre heures du soir devant le hameau de Maï-Tsaberi, où nous devions passer la nuit.

Le lendemain en route au lever du soleil. Le sentier que nous suivions était tracé dans une plaine aride qui s'étendait vers l'est à perte de vue. Peu de temps après notre départ, nous atteignîmes un petit cours d'eau qui s'appelle Sourencia; nous traversâmes ce joli torrent sans perdre une minute; une demi-lieue plus loin nous nous arrêtions au pied de quelques palmiers qui couvrent de leurs verdoyants parasols la source délicieuse du Maï-Aeni. C'est là le rendez-vous des caravanes qui fréquentent les routes du Lamelmon et du Walkaït.

Un grand nombre de marchands se reposaient au bord de la source. Nous laissâmes passer les ardeurs de midi en nous entretenant avec eux des intérêts de l'Abyssinie; après quoi nos domestiques sellèrent nos mules, les marchands chargèrent leurs paquets sur leurs épaules, et nous nous dirigeâmes tous ensemble vers le Taccazzé.

Nous formions une caravane de plus de cent cinquante personnes. Il était six heures lorsque nous arrivâmes sur les bords du fleuve.

* * * * * *

Le lendemain, dès la pointe du jour, la caravane se hâta de quitter cette vallée inhospitalière.

* * * * * *

Arrivés à Diga nous dîmes adieu à nos compagnons de voyage, et pressant le pas de nos mules, nous allâmes coucher le soir même sous les ombrages de Bélés, à côté d'un groupe d'esclaves que l'on allait vendre à la mer.

* * * * * *

Le 24, nous quittâmes Bélés à la pointe du jour, dirigeant notre marche vers le nord-est. Nous traversâmes successivement les districts le Seleklaga, Wogro et Maï-Brasio. Sur toute cette route, un an auparavant, nous avions vu la campagne riche et bien cultivée; nous la revîmes pauvre et inculte. Des bandes de guerillas battaient la plaine, pillaient les villages, rançonnaient les paysans et interceptaient les communications. Près d'Addi-Hosso, un chef de bande, il s'appelait Guidié, arrêta notre petite caravane et exigea un droit de péage. La résistance était inutile. Nous lui donnâmes tout ce que nous possédions, c'est-à-dire un mauvais foulard, le seul object qui nous restât de tous les effets que nous avions apportés d'Europe. Maigre était le cadeau. Guidié s'en contenta toutefois. Il fit plus. L'un de nous l'avertit que nous étions des amis d'Aréa, aussitôt il nous fit donner de la bierre et nous offrit un soldat pour nous accompagner jusqu'à Axoum.

Halte de vingt minutes au pied des murs de la ville sainte, pour laisser reposer nos mules. A trois heures, nous poursuivons notre route dans la direction de l'est. Nous franchissons rapidement les ravins et les montagnes qui nous séparent d'Adoua. Enfin, après d'incroyables efforts de diligence, après quinze lieues faites dans la journée, nous saluons à l'entrée de la nuit la capitale du Tigré, le bon gîte et le bon repos.

* * * * * *

Adowa to Massowah.

Nous prîmes deux routes différentes qui devaient nous ramener à Messawah, dernier terme de nos explorations dans les contrées éthiopiennes : la route de Dixah, dont nous avons déjà parlé, et la route de l'Hamacen que nous allons décrire.*

Le 29 juillet était le jour fixé pour notre départ. Ce jour-là, de bonne heure, les nombreux amis que nous avions à Adoua se réunirent chez nous afin de nous dire un dernier adieu.

* * * * * *

Il était neuf heures. Notre petite troupe traversa l'Assam et arriva presque aussitôt dans un pays très-accidenté. Là passe la ligne de partage des eaux du Mareb et des eaux du Taccazzé, ligne formée par la crête de montagnes nues, déchiquetées, dont les sommets légers et grêles se détachent en formes capricieuses sur l'azur des cieux.

Vers quatre heures nous atteignons Nagah. C'est un pauvre village à quatre lieues d'Adoua. Mais nos gens se mouraient de peur ; ils s'éloignaient à regret de la capitale du Tigré ; impossible de leur faire hâter le pas. Nous avons donc mis tout un jour pour franchir une distance de quatre lieues.

* * * * * *

Le lendemain, au lever du soleil, nous nous hâtames de quitter ce lieu inhospitalier, et nous vînmes coucher à Addissi-Addi. De Nagah à Addissi-Addi, sol inégal, schisteux, mamelonné, à peine boisé. Çà et là un peu de culture, mais généralement le pays est pauvre, la terre végétale y manque, et la roche se montre à nu sur de grandes surfaces.

Addissi-Addi n'est ni moins triste que Nagah, ni moins inhospitalier. De la curiosité, nous y en trouvâmes davantage. Plus éloignés d'Adoua, les habitants du lieu n'avaient guère vu d'hommes blancs, peut-être même n'en avaient-ils jamais vus.

* * * * * *

Addissi-Addi domine la vallée du Mareb. Nous descendîmes pendant une heure le flanc escarpé des montagnes, et nous arrivâmes enfin au fond de la vallée. Là le Mareb coule dans une plaine sablonneuse, ombragée de mimosas de toutes sortes. Cette plaine a environ cinq mille mètres de largeur. Dans la saison d'été, à peine y distingue-t-on le lit de la rivière ; car, à cette époque, il est presque toujours à sec ; mais, au retour des pluies, cette rivière devient considérable, elle emplit la longueur de la plaine et présente alors une nappe d'eau véritablement imposante.

Après un moment de repos, nous gravîmes péniblement les montagnes qui forment la ceinture septentrionale de la vallée, et nous arrivâmes à Goundet, mouillés et trempés jusqu'aux os, car la pluie tombait à torrent depuis près de trois quarts d'heure.

Goundet n'est pas une ville. C'est un hameau construit sur une montagne, et qui donne son nom au pays environnant. Nulle part, en Abyssinie, nous n'avons vu un canton plus peuplé. Non pas que le pays soit productif en céréales, il est hérissé de pierres et manque de terres végétales, mais il est entouré de vallons où paissent d'innombrables troupeaux, et ces troupeaux constituent pour les habitants une véritable richesse.

* * * * * *

Nous partîmes de Goundet le 10 août, à huit heures du matin. Ce

* Quoique nous ayons voyagé séparément, pour ne pas changer la forme du discours, nous avons continué à employer, dans ce chapitre comme dans le reste de l'ouvrage, la première personne au pluriel. Celui de nous qui passa par Dixah fut pillé sur la Tarenta et perdit dans cette conjoncture les observations de longitude faites à Gondar, quelques itinéraires, trois ou quatre bocaux remplis d'insectes et un gros paquet de plantes.

hameau, nous l'avons déjà dit, est bâti sur le sommet d'une haute montagne. Nous voici donc descendant une pente extrêmement rapide, qui nous mène, ou plutôt qui nous jette dans le fond d'une vallée tributaire du Mareb. Cette vallée se nomme Chakné. Elle se dirige vers l'ouest, son sol est une syenite rouge, limitée au sud par le granit, au nord par le basalte. De toutes parts les flancs de la vallée sont taillés en précipices. Ils se couronnent au sommet d'un immense plateau formé de basalte, qui a une étendue d'environ vingt lieues. Ce plateau est parsemé de collines boisées, et bien que la roche y perce la terre en maint endroit, on y voit pourtant des localités assez fertiles. Vers le milieu de la journée, nous passâmes à côté d'Add'Ougala. Nous franchîmes ensuite quelques ravins, et pressant le pas de nos mules pour éviter l'orage suspendu au-dessus de nos têtes, notre caravane s'arrêta devant le hameau d'Add'Eganah pour y passer la nuit.

Le lendemain, nous reprenons notre route avec la matinée.

* * * * * *

Le pays était toujours très-peuplé, les villages succédaient aux villages. Ces villages sont ordinairement bâtis sur les flancs des collines qui dominent la route. Nous passons entre Ounouayela et Addi-Agoga. Ounouayela à notre droite, Addi-Agoga à notre gauche. Une demi-heure après nous arrivons à un endroit où le chemin se bifurque pour réunir ses deux branches à Mailehous. Si l'on prend à droite, on passe par Koudde-Falassi, district fertile où se tient un marché hebdomadaire. Nous prenons à gauche, nous laissons derrière nous Zebanhona, Addi-Mengounti, villages peu importants, nous arrivons enfin à Mailehous, où nous nous arrêtons jusqu'au lendemain matin.

* * * * * *

Le lendemain, à sept heures du matin, malgré une pluie fine qui tombait depuis une demi-heure, nous nous remettons en route. Le chemin s'en va toujours à travers la plaine.

Le sol est fertile et couvert d'abondant pâturages où paissent de nombreux troupeaux. Nous laissons Torabné à notre droite. Nous passons par Addi-Takhita, Tsallem-Ebni, Adde-Guebray; enfin, après quatre heures de marche, notre guide nous montre, à la distance d'une lieue, et à l'orient de la route, un groupe de cabanes qu'on appelait Debaroua.

Le nom de Debaroua revient souvent dans les annales éthiopiennes. Poncet, qui visita l'Abyssinie en 1700, fait de cette ville une magnifique description. Suivant son récit, elle avait deux lieues de circonférence, et toutes les maisons construites en pierres, au lieu de toits en chaume, supportaient des terrasses. S'il n'y a pas là imagination ou méprise, assurément il y a exagération. Que cette ville ait eu des jours meilleurs, qu'elle ait été jadis beaucoup plus considérable, l'histoire l'atteste, mais que Debaroua ait été une cité opulente, nous en doutons encore, puisque aucune ruine ne porte témoignage à son ancienne splendeur.

Il est quatre heures. Nous atteignons Adde-Bahro. Adde-Bahro est situé au pied d'une chaîne de montagnes qui ferme au nord la plaine du Serawé. C'est un village assez considérable. Mais ses maisons basses et ses toits plats ne permettent pas qu'on le distingue au loin. Le pays est riche en céréales ainsi qu'en bestiaux, et l'hospitalité s'y exerce avec magnificence.

D'apres les renseignements que nous avions recueillis, nous ne devions pas être éloignés des sources du Mareb. Où se trouvent ces sources? Ce fut notre première question; les gens du lieu nous répondirent qu'elles se trouvaient à peu de distance du village, sur la route de

Messawah. Ainsi pour indiquer le cours du Mareb sur notre carte, et le placer le plus exactement qu'il se pouvait, nous commençâmes par déterminer la latitude de ce village est de 15° 9'. Cette opération terminée, nous voulions reprendre notre chemin, mais le mauvais temps nous retint plusieurs jours à Adde-Bahro. Nos domestiques continuèrent à débiter leur fable, et nous, sans le savoir, nous passâmes encore pour les frères de l'abouna, ce qui veut dire qu'on nous accueillit avec la distinction la moins méritée.

Rien ne nous fit faute. Chacun s'empressait de nous envoyer des provisions de toute sorte. On nous visitait avec dévotion, et nous nous laissions admirer.

* * * * * *

C'était le 6 août. A peine sortis d'Add-Bahro, notre caravane s'engage dans un pays très-accidenté. Nous franchissons rapidement les montagnes qui s'étendent vers le nord-ouest, et après une marche de deux lieues, nous arrivons à la source du Mareb.

La source du Mareb est donc suffisamment bien placée sur notre carte, puisque nous avions commencé par prendre la position d'Add-Bahro. Pourquoi donc sur les cartes de ceux qui nous ont précédés, la source de cette rivière n'est-elle pas indiquée comme sur la nôtre? Parce que ces voyageurs n'ont pas visité le pays que nous parcourons en ce moment; parce qu'ils ont parlé de cette source d'après des renseignements recueillis de trop loin. Ces renseignements ont trompé Bruce et Salt; de là vient qu'ils ont donné à la rivière de Mareb un cours très-différent de son cours véritable. Comme nous, MM. Combes et Tamisier ont visité l'Hamacen, aussi sont-ils les premiers qui ont connu la direction du Mareb et la position de sa source. Seulement, ils ne faisaient pas d'observations astronomiques pour fixer la place exacte des lieux qu'ils visitaient, et dans leur estimation peu rigoureuse, ils ont indiqué la source du Mareb beaucoup trop vers le nord.

Nous passâmes la nuit à Adde-Questan, et le lendemain avant midi nous faisions halte à Asmara, dernier village abyssin situé à l'extrémité du plateau de l'Hamecen. Encore quelques pas, et nous allions franchir les limites de l'Abyssinie; mais nous avions besoin d'un guide pour traverser le pays des Taltals, et il nous fallait demeurer à Asmara le reste de la journée, parce que la moindre affaire ne demande guère moins d'un jour en Abyssinie. Dieu est grand! d'ailleurs on nous reçut assez bien, et, grâce à cet accueil, nous prîmes facilement en patience l'éternelle lenteur des indigènes.

Le choum nous chercha lui-même un guide, nous l'amena, et le lendemain, à la pointe du jour, nous reprîmes notre route vers la mer. Peu de temps après nous atteignîmes l'extrémité du plateau éthiopien. Du point où nous étions, notre vue s'étendait au loin par-dessus les montagnes des Taltals qui moutonnaient à nos pieds. On nous fit chercher des yeux un pic assez élevé d'un accès très-difficile, et l'on nous dit que là se trouvait le monastère de Bisan. Ce monastère célèbre renfermait jadis un très grand nombre de moines. Quoique déchu, il n'en reste pas moins un lieu saint et vénéré où les Abyssins vont souvent accomplir de pieux pélérinages.

Nous nous engageons sans perdre de temps dans les montagnes. D'ici jusqu'à la mer le terrain s'abaisse considérablement, il ne nous reste donc plus qu'à descendre en traversant le pays des Taltals, pays de triste renommée. Les voyageurs le redoutent, et ce n'est pas sans raison, lorsqu'ils arrivent par la mer pour entrer dans l'Abyssinie. Quant à nous, ces lieux mal famés ne nous effraient nullement; les nomades habitants de la contrée reconnaissent l'autorité du Naïb d'Arkiko, et s'il nous arrivait d'être mal traités par eux, nous savions à qui recourir pour obtenir justice, non pas que l'équité du Naïb ait toujours paru fort rassurante aux voyageurs, peut-être se souvient-on qu'elle ne nous avait pas rassurés nous-mêmes quand nous entrâmes en Abyssinie, mais depuis notre passage sur son domaine, le roitelet de ces contrées sauvages avait dû rabattre beaucoup de son insolence à l'égard de nos compatriotes. La France avait envoyé un agent à Messawah. Protégés par ce représentant de la mère-patrie, nous étions sûrs que nos réclamations, si l'on nous donnait lieu de réclamer, seraient écoutées avec une juste bienveillance.

Le premier endroit habité que nous rencontrâmes fut Guendah, où nous arrivâmes après six heures de marche. Une tribu de pasteurs s'y était établie sous la tente, dans le voisinage d'une source qui suffisait à peine aux besoins des hommes et des bestiaux. Les tentes, et elles étaient fort misérables, se dressaient de manière à former un immense quadrilataire. Dans ce quadrilataire entraient, avec la nuit, tous les troupeaux de la tribu.

Le cheik nous reçut très-amicalement et nous offrit un gîte à côté de lui. Son hospitalité fut douce à notre cœur comme son lait fut doux à nos lèvres.

Le lendemain nous vînmes coucher à Haylet. C'est un village où réside un chef assez puissant. Le chef tenait à gagner les bonnes grâces de l'agent français. Son accueil le disait. Excellent homme d'ailleurs, et dont nous n'aurions pas eu à nous plaindre sans doute, lors même qu'il n'aurait pas eu cet intérêt à nous bien traiter. Haylet ne s'élève pas à plus de 180 mètres au-dessus du niveau de la mer. L'altitude d'Asmara est d'environ 2,500 mètres, nous étions donc descendus en deux jours de 2,320 mètres. C'était un changement considérable dans la température : ici la chaleur devenait accablante. Nous l'éprouvâmes surtout dans la vallée de Sahati, entre Haylet et Messawah. Là nous fîmes halte pour déjeûner. Il y avait une source et un mimosa rabougri. Faute de feuillage nous jetâmes nos taubes sur le mimosa pour obtenir un peu d'ombre. Nous portions le costume abyssin. Le taube enlevé, il ne nous restait plus qu'un mince caleçon en toile de coton. La chaleur fit rage. Nous ne tardâmes pas à sentir des picotements dans tout le corps. Le soir, comme nous arrivions à Moukoullou, où s'était établi M. Dégoutin, agent consulaire de la France, nous étions rouges, gonflés de la tête aux pieds, harassés de fatigue. Pendant plusieurs jours tout contact nous causa une douleur. Nous souffrions à nous coucher, nous souffrions à porter le plus léger vêtement.

Moukoullou n'est plus qu'à une lieue et demie de Messawah. M. Dégoutin, qui nous accueillit en frères et nous prodigua durant huit à dix jours les soins de la plus douce hospitalité, voulut bien se charger de nous procurer une barque, et le 22 août nous partîmes de Messawah.

MR. MANSFIELD PARKYNS' JOURNEY FROM MASSOWAH BY AILET TO ADOWA AND SOLEKLEKHA, DURING THE SUMMER OF 1843 (marked A, H, on Map, page 39).

In a conversation about the comparative heat of different places, an officer of the Indian navy remarked that he believed Pondicherry to be the hottest place in India, but still that it was nothing to Aden, while again Aden was a trifle to Massàwa. He compared the climate of the first to a hot bath; that of the second to a furnace; while the third, he said, could be equalled in temperature by nothing but ——, a place which he had never visited, and which it is to be hoped neither he nor any of us will. Towards the latter end of the month of May I have known the thermometer rise to about 120º Fahrenheit in the shade, and in July and August it ranges much higher. Such a climate is of course most unhealthy,—especially so during the summer months, when a number of dangerous diseases prevail, such as dysentery and the usual fevers of the tropical countries. The island is a mere rock of coral, without a vestige of vegetation to enliven its bare face. There are cisterns for collecting the rain-water (no spring existing), but most of these have been allowed to fall into disuse, and the inhabitants of the island are obliged to trust to Arkiko, a village on the mainland, distant some three or four miles, for their supply. This water, moreover, is rather brackish. The extreme heat of the place would not appear extraordinary to any one acquainted with its position. Massàwa is open on the one side to the sea, while the other is shut in by an amphitheatre of distant hills,—sufficiently near, however, to prevent its receiving a breath of air from that direction, but, on the contrary, to collect, as it were, the rays of the sun into the narrow slip of land they enclose.

The village is situated on the western extremity of the island, which is scarcely a mile long, measured from east to west, by not quite half that breadth. The eastern portion of it is occupied by the burial-ground, a guard-house, and the cisterns I have just noticed.

I had made up my mind to pay a short visit to Ailat,—a hot mineral spring a day's journey from Massàwa. Accordingly on the following day I set out about three o'clock, crossing in a ferry-boat from the island to a point on the mainland which serves as a pier or landing-place. Here I observed lying on the ground a stone capital and some fragments of an ancient column, which, on inquiry, I was informed were brought from one of the neighbouring islands (I believe "Dhalac"), where many similar ones are to be found. The first part of my road lay through a flat sandy country, partially overgrown, in some places, with stunted shrubs, many of which appeared to me to be very curious; but not being a botanist, I could form no decided opinion of their merits. Among them was a shrub which bears a round orange-coloured fruit, in shape and appearance much resembling the colocynth, and which, when dried, is in this country used for making snuff-boxes, the seeds having first been carefully extracted. The whole air was alive with insects of every variety both in species and hue, many of them most brilliantly coloured; and as I advanced farther inland, I observed two or three different varieties of sun-birds,—one kind of a dark-brown colour, excepting his throat, which is scarlet, and his head, which is changing-green and purple; another, almost all changing-green, with a bright canary-coloured breast, and two long feathers in his tail. Wandering on, I came to a place where the sea runs in like a creek, and, seeing a copse of fine bay-trees overhanging the water's edge, and so completely surrounding and shading a little corner as entirely to screen any one who might bathe in it both from the view of passers-by and the more trying glance of the sun above, I took the opportunity of refreshing myself and paying to the salt water my last visit for many years. The water was about five feet deep, with a smooth sand bottom. Nothing could be more delicious,—far preferable to the finest marble swimming-bath in Europe. Having bathed, I proceeded on my way, and soon saw the man with the camels descending to meet us at the point where we landed: telling him to follow with my servant as fast as possible, and inquiring if I was in the right road, to which he replied "All right," I continued my march. Excited to the highest pitch by the workings of my fertile imagination, which induced me to expect every moment to tumble into a Happy Valley, I almost ran along, bearing such a load of castles in my head as would have puzzled Hercules to carry, had they been constructed of any other material than air. Thus I trudged on, full of everything I saw, till, on arriving at three roads, I found myself at a loss which to pursue, as they all appeared to take nearly the same direction: so remembering the old adage, "medio tutissimus ibis," I choose the centre one, which seemed to be the most trodden; or rather, like Don Quixote's good old hack, I took the first that came, and followed it, thinking of nothing in my happy state of mind till I suddenly observed that the sun was gone down; and as in these countries there is no twilight to speak of, it struck me that I had no time to lose if I did not wish to be caught in the dark. I therefore quickened my step till, half an hour afterwards, finding no Moncullou, nor any sound or sign of humanity to warrant me in the supposition that I was near it, I concluded that I had overshot my mark, for at Massàwa I had been informed that it was only an hour's journey, whereas I had been walking fast for double that time. I therefore decided on employing the last remnants of light in preparing myself some corner wherein to sleep, and was just poking about with this intention when I heard voices approaching me, and running towards them was met by five little slave-girls returning to the village with wood they had been collecting. I accompanied them, and in a short time arrived at the house of Hussein Effendi, where, having been provided with milk and other refreshments, I was told that the French Consul's lady had sent to request me to pass the night at her house, whither my beasts had preceded me.

(They started again the next day.) The country through which we passed during the early part of the day is rough, wild, and, in some parts, rocky and mountainous. Large trees are rarely met with; nothing in fact but shrubs and some of the different species of the mimosa tribe, the tallest of which seldom exceed twenty feet. To the sportsman I could say more in favour of the country. Before nine o'clock we had shot several guinea-fowl and some large birds of the partridge or grouse kind, and had seen several gazelle, but these were too wild to allow us to get within shot of them. There are also numbers of jackals, which might afford sport to the fox-hunter, were not the country too difficult to ride over.

Towards nine o'clock we descended into a thickly wooded valley, bordered on each side by rocky hills. Our road for some distance lay along the bed of a former torrent, of which, as the dry season was now long set in, there remained but a small rippling stream, which, here trickling on a little, there losing itself entirely in the sand, still contained sufficient water to attract to its edge several sorts of wild ducks and geese, —some of beautiful plumage. No description can possibly convey an idea of the sensations of one who for the first time feels himself really in a tropical country, with a tropical vegetation;—the burning sun, the orange sand, the bright rich green of the foliage, bordered by a sky of the deepest blue without a cloud;—everything so different from our own chilly clime. We were glad to be told by the guide that there was a spot close by, convenient for its shade and the vicinity of a spring, where we might take our breakfast, and rest during the heat of the day. So we alighted, collected fuel, which was plentiful in the jungle, kindled a fire, and prepared to cook the game we had shot in the morning.

Our game proved excellent, and we remained quiet to digest it till between three and four o'clock in the afternoon. We then remounted, and continued our journey without any adventure till nightfall, when we arrived at our destination.

The village of Ailat, which is composed of many scattered huts built of a framework of wood filled in with branches of trees, straw, &c., and thatched, is situated on the edge of a large sandy plain, covered with bushes, and surrounded by hills of no great size. No country in the world could be better adapted for covering game, and none could be better stocked than it is. One cannot go a hundred yards from the houses without seeing something. In the morning one is awakened by the distant cry of the guinea-fowl as it leaves its perch on the trees. Grouse, partridges, wild boar, gazelle, and antelopes of every size and description, abound in the immediate neighbourhood; while elephants, rhinoceros, ostriches, and sometimes giraffe, are in the proper season found a little further off, and beasts of prey are everywhere to be met with. The hot spring, which is situated at some distance from the village, is considered to be a favourite haunt of the lion.

The inhabitants of Ailat are Bedouins of the Bellaw tribe, which occupies all the tract of country lying about Arkeeko, and thence to the neighbourhood of Ailat.

I remained some weeks longer at Ailat, shooting and collecting specimens of natural history, till a letter from Mr. Plowden (whom I imagined to be at Adoua) reached me from Kiaquor, a village about three days' journey from Ailat, where, as he informed me, he lay in a state of great weakness from the effects of a severe fever, which both he and his companion, Mr. Bell, had contracted during their stay at Massâwa.

Having little preparation to make, we were afoot the next morning long before the sun was up, and when he arose we were some way advanced on our road, with our backs turned to him. The plain which we had to cross before arriving at the hills literally teemed with guinea-fowl, which at that early hour appeared unwilling to quit their roosting-places on the trees; and when, as we approached them, they did condescend to budge, they collected on the ground in coveys of some hundreds each. The road, as we advanced, became more and more rough and difficult, till at last we found ourselves ascending and descending almost perpendicular hills, covered with large, round, loose pebbles, and well garnished with the usual proportion of thorny trees, neither of which, as may be imagined, contributed to the comfort of a barefooted pedestrian in one of the hottest climates of the world.

We proceeded in this way for nearly two hours, when we arrived at the top of the hill. The country, it must be admitted, had its redeeming qualities; for the scenery, though rough and wild in the extreme, was not devoid of interest. Here the guide gave us the welcome news that water (the first we had met with that was drinkable since our departure from Ailat) was to be found in the valley below us.

We set off at once for the water, which we reached in about a quarter of an hour, and were agreeably surprised on finding a magnificent stream dashing down between two cliffs, which, overhanging it at a few yards only one from the other, shaded its course as it fell from rock to rock in cascades, each of which had, by centuries of perseverance, hollowed out of the hard stone a basin for its waters to repose in. Some of these cavities were very large, and of considerable depth;

and some had pretty water-plants growing from their edges, and now in full flower. Could any thing be more refreshing than such a sight to a hot and wearied traveller?

Thus refreshed, we continued our march through the same style of country as before. Antelopes, gazelles, baboons, monkeys, and wild boars passed close to us; but, fatigued as we all were, I let them go, rather than add unnecessarily to the people's load. I always hoped we should again meet with some fowl; but at half-past one o'clock, the time of our arrival at the place where we were to halt and dine, we had found none, and we then regretted the gazelle. On being informed that we had arrived at our halting-place, I made the inquiry most natural to a thirsty man—Where is the water? To which our guide replied by scraping a hole with his hands in the sand, which soon became half filled of a dingy, suspicious-looking aqueous matter, which, however, he assured me would (like many young men in Europe) become more respectable when settled.

Having slept nearly an hour, the guide awoke us, and we continued our journey. The road, instead of improving, appeared to grow worse as we advanced; there was, in fact, no regular road, and our guide did not appear over clever in his calling, for, after frequently climbing a mountain, we found, on a careful inspection of the country that we had taken a wrong direction, and were obliged to return by the way we had come, and seek another. This was not a little perplexing and vexatious, and my companions expressed loudly their discontent at the guide's want of ability; and words increasing, I was obliged to interpose my authority to prevent a serious quarrel, especially as we were altogether at the "Shoho's" mercy, who was in the country of his own people, while we were all strangers; and he might at any time have taken us out of our road, and at night decamped, and left us to fish for ourselves. However, about an hour after sunset we had the pleasure of descending into a little plain among the hills, and of hearing voices of men and the lowing of cattle, and shortly after we arrived in sight of the village fires. It was a "Saho" or "Shoho" camp; for though these people build themselves huts instead of tents, they in other respects follow the customs of all nomadic tribes, only remaining in one spot as long as there is good pasture for their cattle, and, when this is eaten up, seeking another.

The villages are composed of huts, formed of straw and boughs of trees, neatly enough fashioned, and thatched; they are placed so as to form a circle, with one or two spaces left as entrances, in which the cattle are penned for the night, the entrances being closed by bushes strewed before them. The people are Mohammedans; their language is altogether different from that of any of their neighbours, resembling neither the Abyssinian, nor the language of Massàwa and the coast, nor yet the Arabic. In some respects it resembles the language of the Galla tribes, especially in the numerals, many of which are nearly the same. This is somewhat astonishing, as between the Gallas and the Shohos there lies a very large tract of country, among whose several dialects no trace of a link can be found. But as on these points of language I can by no means call myself an authority, never having made it an object of research, I cannot do better than refer such of my readers as may be desirous of further information on the subject to a work by my esteemed friend Monsieur Antoine d'Abbadie.

We were hospitably received by these people, who lent us skins for beds, and provided us with fire-wood, as we preferred the society of the cows outside to that of their masters' parasites within the huts. Shortly after, the cows being milked, we were supplied with a large bowl of milk for our supper, and, having made our homely repast, were soon all sound asleep. Next morning, having carefully wrapped up the skins on which we had slept, we started before either the sun or our good hosts had risen. The events of this day were in most respects similar to those of the day preceding, even to our being received in like manner at its close by the inhabitants of another Shoho village. On the following day, after crossing a vast plain similar to that of Ailat, and which I was told abounds with elephants at certain seasons of the year, we arrived at about two o'clock in the afternoon at Kiaquor, which is the first village in this direction belonging to Oubi, nominally Viceroy, but in fact the absolute monarch of Tigrè.

We here leave the temporary hut of the migratory Shoho for the more solid but equally rude cabin of the Abyssinian, which is usually built with stones and mud, with a thatched roof, and sometimes plastered inside with mud. The difference of costume is also observable. The tressed hair of the Abyssinian Christian contrasts strangely with the bushy wig of the Shoho, who arranges his woolly hair into two large tufts, one of which is on the top of the head, and the other behind.

After one night's rest, we prepared for our departure. This journey promised to be more agreeable than the last,—with the serious drawback, however, of poor Plowden's ill health. I was lucky enough to find that he had an extra mule, so that I could spare my legs. We had a good many servants, about eight porters for luggage, and the little variety of provisions which the country could furnish: moreover, we were about to travel on a beaten track through a populous district, in the villages of which we could always renew our supplies, should they fail us. The greater part of our road lay through the fine province of Hamasayn, a vast table land, varied with beautiful hill and valley scenery. The most careless observer, in passing through this country, cannot fail to mark the extreme richness of the soil, and the great capabilities of the land were it properly cultivated. To us it presented itself under the most disadvantageous circumstances. Civil war, the perpetual scourge of Abyssinia, and the principal cause of its remaining in its present state of poverty and barbarism, had passed over this fair land, and reduced it to such a state that wherever you turned you saw nothing but devastation and ruin.

Our first halt was at noon, when we rested for a short time under the shade of a large sycomore, near a ruined village; and, having refreshed our animals and reinvigorated ourselves with a little bread, honey and Cayenne pepper, we proceeded for an hour, when the rain coming down heavily we were obliged to take refuge in a house at the village of Addy Killàwita, a small hamlet very prettily situated on rising ground, and surrounded by remarkably picturesque scenery. Here, for the first time, I observed the quolquol, a species of Euphorbia, which grows like a cactus, the leaves and branches being both of a fleshy substance, and containing a large quantity of milky sap, which flows out plentifully on a sprig being wounded or broken. This milk is poisonous, and is used by the natives for intoxicating the fish in the small rivulets, which being dammed above and below the holes where the fish are known to lie, a quantity of the quolquol juice is put into the water, and in a short time the fish are seen to float insensible on its surface.

So soon as my companion had recovered a little from this fit of ague, we continued our journey as far as a village called Maiya, about six miles distant, in the hope of finding better accommodation. But it was a vain hope! At first we found none at all; and it was not till after a vast deal of persuasion and great promises that we induced the good people of the village to consent to our occupying a dwelling for the night; and when they did so, that which they offered was so bad, so very far inferior even to the last, that, rather than be stifled in a hut, we preferred lying in the open air, covered with hides as a protection from the rain, which kept pouring for several hours. Plowden's continued illness compelled us to remain here two days and nights, during which time we amused ourselves as well as we could, contriving tents and huts among the rocks near the inhospitable village. But soon tiring of this sort of life, we again started, carrying the patient in a litter made of boughs; and after crossing the river Mareb, which here flows in nearly a southerly direction, we arrived at the village of Shaha, where, our lodging being of the same wretched description, we remained the night only, and continued our journey the next morning, hoping to arrive the same day at Kouddofelassy, where we thought of staying a few days, having heard that it was a market town, and that all sorts of supplies necessary for my sick friend could be obtained there. We approached it with a feeling that our troubles were about to end, at any rate for a time. Great, however, was our disappointment and vexation on arriving; for on inquiring in every direction for a lodging, we were absolutely refused one, either for love or money. In vain we pleaded the sickness of our companion, offering a handsome payment for what ought to have been gratuitously provided as common hospitality. All was useless; so we were for a time obliged to put up with the partial shade of a small tree as the only protection against a broiling sun.

We remained in this town for five days, during the whole of which time we could procure no supplies beyond a little honey and a few miserable fowls. The morning of our departure advanced us but little on our way; for almost immediately on starting we were caught by a pelting shower, which compelled us to seek refuge in a village some distance from the road, and situated on a little hill, on which there is a church dedicated to St. Mary, whence the place takes its name Beyt Mariam.

Leaving Enda Mariam next morning, we arrived a little after midday at a large village called Addy Hai Hai. Here we were better received than at any place since we left Kiaquor, for we had been seated only a short time under the shade of a tree when we were invited into the house of a petty chief of the place.

Next day we arrived at Goundet, situated on the hills which rise from the eastern bank of the Mareb. On the following day we crossed this stream, which makes a turn here, and flows nearly north. It is of considerable breadth, and where we forded it is up to our waists, it being the rainy season, during which period it is very variable—sometimes, after a heavy fall, rising so high in half an hour as entirely to obstruct the road, and falling again as rapidly as it rose; while in the dry season there is barely water enough to wet the ankles.

Next day we slept at Baysa, and the following noon, in a heavy shower, arrived in sight of Adoua, the capital of Tigrè.

When we arrived in sight of Adoua, I galloped on ahead of the party, anxious to obtain shelter as soon as possible; but being mounted on a weak and tired mule, and the road being of a stiff and greasy clay, and in many places very steep, I gained but little by my haste; for the mule slid down all the hills, and stumbled or tumbled over all the inequalities of the plain. My attention being thus occupied, and the rain driving in my face, I had not leisure to enjoy a distant view of the city we were approaching; nor could I, till within a short distance of it, see enough to enable me to determine whether Adoua was built in the Grecian or Moorish taste. I own I rather expected to see columns or obelisks, if not an acropolis on some of the neighbouring hills. Judge then of my astonishment when, on arriving at this great city, the capital of one of the most powerful kingdoms of Æthiopia, I found nothing but a large straggling village of huts, some flat-roofed, but mostly thatched with straw, and the walls of all of them built of rough stones, laid together with mud, in the rudest possible manner. Being wet, moreover, with the rain, the place presented the most miserably dirty appearance. Before entering the town we had to cross a brook, and to scramble up a steep bank.

We started, a few days after, in the direction of the camp, then at Howzayn. Our first day's journey did not advance us far on our way, for we had scarcely been an hour on the road, when, while halting in a shady place to rest Plowden, (who was still a great sufferer,) I was suddenly seized with a fainting fit, and was in consequence carried into a neighbouring house.

Early on the following morning we again started, and, after descending the precipitous rock which forms the natural boundary of the province called Dabba Garema, we passed the village of Guddiba, and finally entered the district called Assa. Here we were obliged to halt in the middle of the road, Plowden's fever having returned very severely, and there being no house within three or four miles; but in a few hours he felt so far better as to enable us to continue our journey, and we succeeded in carrying him to the summit of a hill, on which is situated a village called Addy Nefas (the Village of Wind, so called from its elevated position), where an uncle of one of our servants resided. The road up to it was exceedingly difficult, both from its roughness and steepness; but the fatigue we endured in the ascent was amply repaid by the kindness and hospitality with which we were received on our arrival. Honey, milk, eggs, and various other good things were speedily offered us, and we gladly consented to remain there the night, although we might have gone several miles farther, the day not being nearly closed. Towards the evening of the following day we reached a village of Ha Haily, called Devra Berbery, the people of which, having already suffered great annoyance from the frequent stragglers to and from the camp, were much inclined to treat us inhospitably.

Next morning the elders of the village, having, it would appear, formed a better opinion of us than they did on our first arrival, sent two boys to show us the road, as not far onward was a deep gap or ravine, which might be crossed by a foot passenger in a few minutes, while the mules could only arrive by a long détour. The path, in fact, which we followed was so steep and slippery down the face of the rock, that we were obliged in some places to slide down in a sitting posture.

After a short time we arrived at another small hamlet belonging to the same district as the village where we had slept the previous night.

Our next day's journey brought us to a small village called Addy Argoud. We passed a very uncomfortable night at Addy Argoud. The pouring rain obliged us to sleep in the hut, which we seldom did when the weather permitted us to remain outside.

On the following day we passed the mountain of Haramat, one

of the strongest fortresses in Tigrè, now occupied by a rebel and brigand of the name of Iskyas. A former viceroy (I believe Ras Welda Selassy) is said to have laid siege to the mountain, and, unable to take it by storm, blockaded it for seven years. Mountains almost impregnable by nature are common in this country. Many are in the hands of priests, who have on their summits a monastery and sanctuary, such as Devra Damo, and many others; and to these the people of the neighbouring provinces send their property for safety in times of war or other disturbances. Almost every great chieftain has likewise his mountain, to which he retires in a moment of need. Cisterns, either natural or artificially hollowed, are on the summit of each, and large supplies of provisions are generally kept ready for any emergency. Many of these rocks cannot be ascended except by the aid of cords or rope ladders, which are let down and drawn up at pleasure. Numerous amusing anecdotes are related of the stratagems employed by some of the more powerful chiefs to get possession of some of these mountain fastnesses.

Towards the afternoon we arrived, in a heavy shower of rain, at the camp of Howzayn, and proceeded immediately to the dwelling of Bejerundy Càfty, the "Ikkabeyt" or steward of the Prince's household, who was appointed by his Highness as Bell's "balderàbba," or introducer, when he visited this country on a former occasion.

The appearance of an Abyssinian permanent camp is singular, but by no means unpleasing. The diversity of tents—some bell-shaped, some square, like an English marquee; some white, and others of the black woollen stuff made principally in the southern provinces of Tigrè; huts of all sizes and colours, and their inmates scattered about in groups, with their horses, mules, &c., form altogether a picturesque and very lively scene. In the centre is the dwelling of Oubi, which consists of three or four large thatched wigwams and a tent, enclosed by a double fence of thorns, at the entrances through which guards are stationed, the space between them being divided into courts, in which the soldiers or other persons craving an audience of the King await his pleasure. Close around this is the encampment of the "Ikkabeyt," or steward, and his "Chiffra," or followers, of whom he has a large body, used as porters in case of the Prince's changing quarters, and as soldiers in time of war. Around these again encamp the "Zeveynia," or guards. In front of these come the "Nefteynia," or bearers of fire-arms, with the "Negarit," or great drums, while "Fit-Owraris," or generals of advance guard, occupy the front position. [I don't know the derivation of "Fit-Owraris." May it not be from "Fit," face or front, and "Owrari," Rhinoceros, alluding to the offensive weapons of that animal, which are so prominent in front of his face?]

Behind the Prince's tent is the camp of the "Sheff Zagry," or swordbearers, while the "Dejjin," or rear guard, occupies the hindmost position. On each side of the royal abode are the great men, or chiefs of provinces who may have joined their master with their forces. Every corps of about fifty soldiers has an officer called a "Hallika." His hut is rather larger than those of his followers, and is built in the centre, while they encamp in a circle around him. The "Hallika" is generally a favourite servant, whether he be in the employment of a Prince or that of any other chieftain; and when his master is levying fresh soldiers, every volunteer for service demanding a "balderabba," a favourite servant is named for this office, and in this way his "Chiffra" or company is formed, he becoming "Hallika" to those volunteers to whom he is thus appointed "balderabba." As "Hallika" he receives and distributes the pay and allowances of his "Chiffra." The only power he has of exercising his superiority over them lies in his right to deduct a small sum from their allowances. Thus in every point the relations of the Abyssinian "Hallika" to his "Chiffra" are much the same as those of the "Boulouk Bashy" of the Turkish irregular infantry to his "Boulouk." This officer is elected by choice of the "Sanjak" (a chief of four hundred), and deducts a small sum from the pay of his soldiers, with part of which he is expected to give them one meal per diem. The troops in Abyssinia are for the most part collected from among the worst of the people, who prefer idleness in peace and plundering their neighbours in war to the more honest but less exciting occupation of agriculture. They have neither tactics nor discipline, and their dress is the same as the ordinary costume of the country, but usually cut in a somewhat smarter manner.

I started for Addy Abo, towards the end of September, 1843, accompanied only by a few native servants. On leaving Adoua, the westward bound traveller, after half an hour's ride, passes the little church of St. John (Beyt Yohannes), a mere hut, perched on a small pyramidical hill, or heap of stones, on whose barren sides grow a few scattered bushes, principally of the quolquol, of which I have already spoken. An undulating road, abounding in picturesque scenery, especially from those points which command distant views of the hills beyond Adoua, leads to the church dedicated to the Saviour (Enda Yessous). This building, little superior in architectural beauty to that last mentioned, may be considered as half-way between the ancient and modern capitals of this part of Abyssinia, Axum being anciently considered the capital, while Adoua rose to importance from a mere village of huts so lately as the reign of Ras Michael (about sixty years ago). The princes who succeeded him increased its size, and built for themselves a house, a sort of palace compared to the ordinary huts; but Oubi, from the situation not agreeing with his health, or from fear of poison, never resides there, preferring his camp, and the house has in consequence been allowed to fall into ruin. Enda Yessous is built on a small but well-wooded hill, on the verge of the splendid plain of Hatzabo, which extends nearly all the way to Axum, a distance of several miles. It is famed for its fertility, producing remarkably fine white teff, the species of corn most esteemed in this country. Near the church, but at some distance from the road, is a spring of delicious water.

From the plain may be seen some of the mountains of Simyen, which, though at a great distance, form a pleasing boundary, thus relieving the eye from the continued flatness of the foreground. As you approach Axum, however, a range of small hills rises on the right hand abruptly from the road. On the rocky summit of one of these, at a short hour's distance from the town, is the church of St. Pantaloon (Abouna Mentellin), a saint formerly held in great esteem by the people, and therefore much attended and rich.

The road skirts the foot of the hills for a considerable distance, till at last a small plain obelisk, on the right hand, and farther on, to the left, a large stone tablet inscribed in Greek characters, proclaim to the traveller his near approach to the city of Axum. From the tablet a sharp turn to the right brings him in view of half the town, which, being situated in an amphitheatre of hills, and possessing a tolerably well-built square church, probably of Portuguese construction, forms altogether a rather agreeable *coup d'œil*.

The church is prettily situated among large trees, and surrounded by rustic but neatly-built huts. From the tablet, however, to the church there is a distance of several hundred yards, along which lie scattered, every here and there, unfinished or broken columns, pedestals, and other remnants of the civilization of former ages.

To the east of the column and town is a large reservoir, supplied by a stream or torrent which pours down from the hills during the rainy season, and for some time after, till it has drained them of the superfluous water they may have collected during the three wet months; but when this is finished and the dry season commences, it discontinues its supply. The tank, however, which, with some wells on the other side of the town, furnishes the inhabitants with water, holds out for nearly the whole year. I would often have gladly plunged into it when coming fatigued and hot from the road, but, as it is set apart for drinking, a penalty is very properly inflicted on any one washing clothes or bathing in it.

The grape in this country is very little cultivated, although, from the nature of the climate and soil, it might succeed admirably. Here and there a few detached plants produce just enough to satisfy an observer as to the capability of the land; but only at Axum, in Tigrè, and at a village in Dembea, are they grown in sufficient quantities for making wine. The vintage of Axum altogether would not amount to the quantity made by the poorest peasant in the south of France, as only one or two persons attempt it.

We were obliged to remain at Axum a few days, having met with some difficulty in procuring provisions for the journey. We left it on the 23rd of September, early in the morning. The road from the part of the village where we lodged passes close to the church, and we were told that custom and respect for the sanctuary, which is one of the most reverenced in Abyssinia, required all persons to dismount and walk till they had altogether passed its precincts.

For some distance after passing the Church, we continued in the great Gondar road. This appellation may give an idea of macademizing, with footpaths alongside, milestones, fences, &c.; but here the high road is only a track worn by use, and a little larger than the sheep-paths, from the fact of more feet passing over it. The utmost labour bestowed on any road in this country is when some traveller, vexed with a thorn that may happen to scrath his face, draws his sword and cuts off the spray. Even this is rarely done; and I have been astonished at seeing many highways, and even some of the most used, rendered almost impassable by the number of thorns which are allowed to remain spread across them. An Abyssinian's maxim is, "I may not pass this way for a year again; why should I give myself trouble for other people's convenience?" The road, however, here, as in many parts of Tigrè, is abundantly watered, not only by those torrents which, though they dry up shortly after the cessation of the rains, leave a supply of water in the holes for many months, but also by several tolerably copious streams, which flow all the year round. These are most useful to the numerous merchants who pass constantly between Gondar, Adoua, and the Red Sea, with large caravans of laden animals, offering not only ready means for watering their cattle, but often green food for them near the banks, when all the rest of the country is parched up and dry, and a cool grassy bed for their own weary limbs to repose on.

The Gova-Dirra and Mai-Shut together fall into the Mai-Tchaou and Werry, and thence into the Taccazy, whither also flow the Mai-Shum and Dabba Bourrou, after having united their waters: these are good-sized rivulets, and all crossed our road.

As for the appearance of the country, it is in general hilly and tolerably well wooded, but much varied in feature. Sometimes you are climbing or descending a hillock, and at others pursuing your way down a shady valley or along the level summit of some table-land. We once lost the track, to the great annoyance of the poor inhabitants of a nearly deserted village, among whose corn-fields we wandered for nearly two hours, till some one coming to order us off was obliged at least to show us which way we were to go in order to obey him. Following his directions we regained the road, not much more than a mile from the place we had left it. A little by-path leading from it up the hill side indicated a village in that direction, and, as evening was near, we ascended till it brought us to a hamlet, so snugly placed among the hills as to be almost out of sight of the road. It was called Addy Nefas, or the Village of Wind, from its elevated position, and belongs to the district or parish of Aghabseràì, in the province of Màitowàro.

Shiré, once the richest and most productive province of Tigrè, and still capable of becoming so under a prudent government, might indeed have laid golden eggs for a wise ruler. Now the land is nearly deserted. Where once were populous villages, with their markets and a happy and thriving people, the traveller now sees but a few wretched huts, vast tracts of fertile land lying uncultivated, and, of the few inhabitants that remain, many that were formerly owners of several yoke of oxen each are now to be found clubbing together to cultivate just enough corn to pay their taxes and keep themselves and their families from starvation.

Next morning, having parted with Temmenou, who had insisted on accompanying us some distance to set us right on our way, we resumed our route.

Near the village I found a great quantity of large round pebbles, which, on being broken, were found to be hollow, and lined with an incrustation of beautiful amethyst-coloured crystals, some nearly an inch long.

After leaving the lovely plain of Mai-Towàrs, the road passes for some distance through a hilly and rocky tract of country, winding through woods of acacias and other shrubs, and at one part running along the brink of an almost precipitous ravine. A few torrents,—at this season of the year still containing a little water, but soon to become dry courses, till the next rains replenish them—and a hamlet or two were passed before we arrived at the stream of Tambukh, which borders the plain of Solekhlekha. What a different style of scenery was that now before us from that which we had just left!—the one a vast plain, apparently fertile, but altogether uncultivated; the other a wild, barren mass of forest and rocks. The change was, as it were, from one country to another, quite differing from it in every feature, and yet without any gradation. One foot might have been in the Alps while the other was in a gentleman's park in England.

At this point we halted for a short time to rest the porters. We were about to leave the road to Gondar, which runs west, while that which we were to follow takes a northerly direction.

MR. MANSFIELD PARKYNS' JOURNEY FROM ADOWA TO THE TAKAZZE, JULY 1845 (marked J on Map, page 39).

THE rains were just setting in; still I felt that I must make up my mind for a start. Accordingly I set about my preparations instanter.

These are soon made when a man travels as I do, with next to nothing in the baggage department; but what between paying the necessary visits of adieu to my many friends, and a decided tussle between duty and inclination, I did not get away from Adoua till late in June (I will not be sure if it was not the first week in July). We reached successively Axum and Maitowaro without any occurrence save an attack of ophthalmia, which kept me a day in the latter village. Our road branched off from Axum to Addy Abo, just after entering the plain of Solekhlekha. Three more days ride, over a fertile and well-watered table-land, brought us to Devra Abbai (the great monastery), after passing the villages of Belliss, Addy Giddad, and Adega Sheikha. This part of our journey was somewhat devoid of interest; we had a few words at Belliss with a party of soldiers, who stopped our porters, pretending that they wanted Customs duty, but on my coming up the matter was soon set to rights, as the leader of the party happened to be an old acquaintance of mine.

As for the scenery by the road, I heard say that it was tolerably monotonous, with the exception of occasional glimpses of distant mountains; but I saw nothing, for my eyes were bandaged from the effects of my late attack of ophthalmia. About half way across the plain we passed a ravine and stream, called Gammalo, where may still be seen scattered bones of some of the fugitives from the great battle of Mai Islamai, the field of which is about two days' journey further on. This serves to show with what cruel determination the Gallas pursued and slaughtered their vanquished foes, even to this distance.

At Adega Sheikha we left the Gondar caravan route, which turns in a south-westerly direction, while ours continued westward. We found the village of Devra Abbai built in a deep hollow or chasm, and so nearly concealed, that, when approaching it from some directions, you would scarcely imagine yourself to be near habitations, seeing nothing but a wide tract of table-land before you.

We started early, in order to effect our passage of the Taccazy as soon after noon as possible. Every moment was precious to us; the rains had already so much swollen the river that no one had attempted the upper ford (on the ordinary Gondar road) for several days past. We procured a guide, whose business it was to assist us in crossing the torrents, and to show us the way over the wild, uninhabited district that lies between this part of the country and Walkait. He told us that we should perhaps have to retrace our steps, if we found the river too deep and strong for us; but that, as the ford to which he was about to conduct us was very broad, and consequently shallow, we might possibly get over.

Never did I feel in better spirits than that morning. We rode for some hours over a very wild, picturesque country, varied with table-land, valleys, and hills of all shapes and sizes, passed near the scene of the battle of Mai Islamai, and about noon began the actual descent towards the river. For an hour or two we were buried in deep ravines, with rock and trees overhanging us, till at length we emerged into a broad and woody flat, through the trees of which the reflection of the afternoon sun on its waters showed us the Taccazy, now swollen to a majestic river, at a distance of about half a mile. Most of our party set off at a run, eager to get a nearer view of it. I, for my part, had seen nothing like a river since I left the Nile; for the Mareb is, as I have said, but a rivulet in the dry season. Some of our people had never before seen a river of any sort, and looked upon it with awe and wonder. Indeed, it was a noble stream, in many places nearly, if not quite, as broad as the Thames at Greenwich; but in its rapid, boisterous descent, more like the Rhone as it leaves the Lake of Geneva. On the opposite shore appeared a belt of forest, similar to that we had just crossed, though neither so wide nor so flat, and in rear of this rose a dark mass of abrupt rocks. We ascended the stream for a considerable distance before we arrived at the ford where we were to cross. As the river did not appear so high as the guide had feared, he recommended a short halt before we entered the water; and, in the mean while, the baggage was made up into convenient parcels, and perishable articles packed in skins, so as to protect them as much as possible from a wetting. After sitting a few minutes we began to strip, and tied up our clothes in bundles, which we were to carry, each man his own, turban-like, on his head. I was proceeding very leisurely in my preparations, finishing a pipe, and waiting to be summoned, when I heard one of the Abyssinians call out, "Come back, come back!" A black who was with us answered him, "Oh, never fear, he's a child of the sea!" I looked up, and saw Yakoub wading out in about two feet of water, and occasionally taking a duck under, as if to cool himself. Aware that he was ignorant of the language, I called to him, telling him that he had better not go alone, but wait till some one, acquainted with the peculiarities of the river, should guide him. He answered, laughing, that he was not going much farther, and that he could swim. I did not think there could be any danger if he remained where he was, the water not being more than a yard deep, and he had told me before that he was an extremely good swimmer; but the guides had cautioned me of the danger of the whirlpools, currents, and mud, which they said rendered it impossible for anything, even a fish, to live in some parts of the torrents; so when on looking up I saw him moving about, I again called to him, begging of him with much earnestness to return. He answered something that made me laugh, at the same time swinging his arms about like the sails of a windmill, so as to splash the water all round him. He might have been thirty yards from the shore, and a little lower down the stream than where I sat. Still talking with him, I looked at what I was doing for a single instant, and then raising my eyes, saw him as if trying to swim on his back, and beating the water with his hands, but in a manner so different from his former playful splashing, that, without knowing why, I called to him to ask what was the matter. He made no answer, but seemed as if moving a little down the stream for a yard or two, and then quicker and quicker. I was up in an instant, and ran down shouting to the people to help him, though at the same time I thought that he was playing us a trick to frighten us. A thick mass of canes and bushes, under the shade of which most of the servants had been sitting, overhung the river for several yards' distance, just below where I was. Having to pass behind these, I lost sight of him, and before I reached the other end of them, the horrible death-howl of the Abyssinians warned me that he had sunk to rise no more.

MR. MUNZINGER'S ROUTE FROM MASSOWAH TO KAYAKHOR, 1867 (marked B on Map, page 39).

THERE are only two roads leading to Northern Abyssinia, which are worth close examination, because they alone are practicable to camels:—

One is by Tokonda, following the well-known torrent of Hadas, to the foot of the Taranta Mountain, where it turns to the left, then along another torrent to the ascent to Tokonda, which could easily be made practicable for any carriage.

The other is to Kiagnor.

This last was the object of my excursion from the 28th to the 30th January. I went up by the way of Dasuas, and returned by that of Aly Gady. The former is somewhat the shortest, crossing the mountains by two steep ascents. The direction is almost south-west till longitude 39° 5' and latitude 15° 10!' where it joins the other road of Aly Gady. It has been followed by several travellers, but from its comparative difficulty need not be described here.

The latter, by which I returned, is the best road, and is as follows:

It starts from Arkeko, and is divided into three parts—

I. The open low country.

II. Following the torrent Aly Gady.

III. The ascent and plain to Kiagnor.

The distances are indicated in hours.

Part I.—From Massowah.

No. of Stations.	Name of Station.	Direction.	Distance in Hours.*	Remarks.
			H. M.	
1	Arkeko	S.	2 0	Road over hard open plain. Large village. Detachment of Egyptian troops here. Water plentiful.
2	Entrance of Shilliki Pass	S.	3 20	Road over plain. Flat open country. Many gum trees (Babul), Acacia Arabica. Water obtained by digging a few feet in bed of torrent.
3	Pass of Shilliki ..	S. 10 W.	1 25	This pass is a large smooth dry bed of a torrent; in rainy season carries water off to the sea. Water obtained by digging.
4	Hill Hadaley ..	S. 20 W.	1 25	Road level at first, but hilly towards Hadaley. Water obtainable by digging. There is water in wells at Saheto, 30 m. East; and plenty at Wooja, 1 h. E. 10 m. S.; and at Zoola, 4 h. E. At Wooja the Ali Gady and Hadas torrents unite, and form one to Zoola. The road to Halai and Tokonda is up the Hadas.
5	In the torrent Ali Gady..	W.	1 0	Road open and level; rather woody; small hills about.
	Total of I ..	..	9 10	

* Each hour about 2½ miles average.

Part II.—Road along the Ali Gady Torrent.

No. of Stations.	Name of Station.	Direction.	Distance in Hours.	Remarks.
			H. M.	
1	First halting place	W.	1 20	Torrent broad and tolerably level; partly wooded (Babul); low hills on either side.
2	Second halting place	W. S. W.	1 0	Torrent broad; running water.
3	Off Eygerrey ..	S. W.	4 25	The torrent winds much, and has some narrow rocky places, but without difficulty. The narrowest places are after 1 hour, where is running water; after 2 h. 40 m., and close to Eygerrey, the hills on both sides get nearer and higher.
4	Marsa Hamsa ..	W. S. W.	2 0	Winds much, but no narrower.
5	Haufur	W.	1 0	One narrow road, more difficult from boulders, but can be easily made good.
6	Gannalce	W.	0 45	Road more open.
7	Augal	N. W.	0 50	Road good. A large torrent joins here from the left, bringing down the waters of all the country between Halai and Kiaguor, especially those of the torrent Tshoat, which offers a very fine smooth but longer road to Kiaguor. I had no time to examine it.
8	Torrent changes here, becomes Aydereso	N. W.	1 0	Here joins the first-mentioned road from Arkeko. Water above and below. Many Babul trees. Torrent large and country more open.
9	.. ..	..	0 15	To the upper water and to the foot of ascent where the torrent is left.
	Total of II ..	..	12 35	

Part III.

No. of Stations.	Name of Station.	Direction.	Distance in Hours.	Remarks.
			H. M.	
1	Ascent	S. 30 W.	1 0	Not very steep; no obstacle for camels; round stones in road, but no narrow passage; road could easily be made very good.
2	From top of ascent to hill 1 h. from Kiaguor	S. 20 W.	3 20	For 1 h. 50 m. road is open, very bare, dry, woodless, and strewn with round stones. The last 1 h. 30 m. pass through a large forest to the torrent Tshoat. Half-an-hour from the road there is running water. Hills on right and left.
3	From the hill to Kiaguor	S. 30 W.	1 0	Plain open country; plenty of cultivation. Kiaguor surrounded by a hill. I did not go beyond the Hill Station No. 2.
	Total of III ..	..	5 20	

G

From Kiaguor to Gotafalasee is eight' hours journey (twenty-two miles). Behind Kiaguor the road ascends the hill, not practicable at present for camels, but can be easily made so. Once on the top the road passes over level plateau to the village of Mya (four hours), there descends, crosses the Mareb, and ascends again the other side, no difficulty whatever, to the village of Shaha (two hours), and on to Gotafalasee in two hours more.

The Mareb is passable at all times.

Gotafalasee is the great market-place for Northern Abyssinia, and lies in the midst of the fertile plain of Saramey. It is only about nine hours' journey from Tsazaga, the residence of Djaj Hailoo, the King's Governor of Hamasen.

General Observations.

As will be seen, there is no want of running water anywhere except in the first part of the journey, but even there it may be get by digging; and at Wooja, one hour east from Hadaley, there is plenty of good water.

From Arkeko to Hadaley is six hours ten minutes march, less than twenty miles.

From the commencement of Part II., from the torrent the whole way up there is plenty of water.

These are styled torrents, but they are dry watercourses becoming torrents in the heavy rains.

Wood is to be found in all places.

Few or no supplies in the shape of grain, excepting at Kiaguor and Gotafalasee. Beef and mutton plentiful everywhere. Grass to be found at all the places from October to May.

The tropical rain line is at Eygerrey.

The country till the Ady Gady torrent is reached belongs to any one.

On the left of the torrent upwards to Hadas belongs to the Tora, nomades nominally Egyptian subjects, Mussulmans, 700 or 800 spears.

The country on the right of the Aly Gady belongs to the Catholic tribe of Zana-Fagley, which has fixed settlements on the highlands near Halai. Eight small villages, of which the largest are Akrom and Saganaytee. These people have extensive land, but only muster 800 spears.

The Tora and the Zana-Fagley in the winter months bring their flocks and herds down to the low lands near the coast to graze. During the tropical rains from June till September they remain on the highlands to cultivate. On the right and left of the Aly Gady torrent are fine plateaus, 4,000 to 5,000 feet above the level of the sea, large plains, well cultivated, water plentiful, The one named Agamatta is only about eighteen miles from Massowah, and would make an excellent encamping place.

From Aydereso upwards the ground belongs to the tribe Angana, having five villages, one of which is Kiaguor.

The Angana are subject to a larger tribe, the Aggala Goora, having seventeen villages. The chief of these two tribes is the son of Anda Mikael, who does homage to Waksheim Gobazye, the rebel leader of Lasta and Tigre.

On the right of Kiaguor are the tribes Laber and Wagerthee, having eight and four villages respectively. They are subject to Dejaj Hailoo, as is all the land beyond the Mareb.

The heights are given from observation of the boiling point.

(Signed) W. MUNZINGER, *Acting Consul*

COLONEL MEREWETHER'S REPORT ON THE AGAMETTA PLATEAU. (Marked C on Map, page 39.)

MY LORD, *Massowah, April* 6, 1867.

I HAVE the honour to report that I left this on the morning of the 1st instant at half-past seven o'clock for the Agametta plateau. The route followed passed Moncooloo, and by the Eylat road for one and a-half mile from that place, then turned off abruptly to the left in a south-westerly direction through low hills for five and a-half miles, when it rounded the point of a small range, and went due south by the bed of a mountain torrent named Khooloo, to Part, three miles, where excellent water was found in plenty by scratching about two feet in the sandy bed. From the moment of entering this torrent the country commenced to show the luxuriant vegetation which was found everywhere afterwards until again nearing the coast. The valleys and hill-sides were covered with rich grass, wild flowers, and in some places of most dense wood, consisting chiefly of Babul (Acacia Arabia) bush, other thorny bushes, and wild fruit trees with whose names I am unacquainted, and to which the native denominations gave no clue. Large numbers of very good-looking cattle were collected at this watering-place belonging to the Tsanatigli tribe. Water is to be found here by digging at all times of the year. This was made the first halting-place. The march took from 7·30 A.M. until 12·15. Though it was the month of April, and the sun powerful, the heat was found by no means excessive.

The journey was resumed at 7·15 the following morning, the 2nd instant. The road lay the whole distance [in the bed of the ?] Ukbuloo torrent, and though quite practicable for camels, and even artillery, was difficult in some places. The obstructions could either be removed or turned. Running water was found continually. The hills on either side increased in height, and, as well as the narrow valley, were dotted with dense wood and high grass. The general course was south-south-west, but as the torrent had to be followed, the windings were very great, the distance marched ten miles, time taken four and a-half hours, moving slow. The camp was made at a pool of water called Henrote. The breadth of the torrent was very narrow at that part, not above fifty feet. It had varied the whole way up, opening out in some parts to a moderate-sized plain, in others narrowing, as at Henrote.

Started at 6·35 A.M. of the 3rd instant, and after proceeding three-quarters of a mile commenced a steep ascent, unsuited to draught artillery, easy enough for mules carrying mountain-guns, but difficult for laden camels, unless having light burdens. Those with the party surmounted it, but some of them did not reach the next encamping ground till the following morning. The highest point the road took was 1,300 feet above Henrote. A higher peak called Fathak was ascended, which afforded a capital bird's-eye view of the surrounding country. This was

1,476 feet above Henrote, and 3,222 above the sea-level. To the north was seen the country traversed by the road to Bogos; to the north-east and east that which had been passed over; Massowa and the sea beyond, dotted with islands of the Dunbar group; to the south-east and south the high mountain of Gudan, Annesley Bay, and the road to Kiaguor by the Agadey watercourse; on the west the high mountains which bound Abyssinia Proper. Immediately before us to the west, and about five miles distant, were the patches of comparatively level ground styled the Agametta plateau, the point to be reached. Hills densely clothed with verdure were scattered about around in the most fantastic manner. Large boulders of rock, blackened by exposure, were lying on the hill-sides as if some violent explosion had taken place. In a direct line Massowa appeared to be certainly not more than 20 miles bearing north 39° east. Passing this peak the road descended abruptly several hundred feet, but then rose gradually again to the plateau of Agametta, which was found to be twenty feet below Fathak, or 3,202 feet above the sea. From Fathak the road wound a good deal to the plateau through the same rich country, with grass and wild flowers waist-high in most places, or dense bushes with grass underneath. On the plateau were patches of towaree (millet), cultivation belonging to the Tsanatigli tribe. The encamping ground was about two miles short of this cultivation, on the edge of a torrent, in which running water was met. Half-a-mile lower this trickled over bare sheet rock (granite) into a deep, densely-wooded ravine below. When the torrent is flooded by heavy rains, there must be a beautiful cascade here, and judging by the appearance of the rocks, the discharge of water must sometimes be very considerable. The distance from Henrote to the camping-ground by the road is not more than five miles, but it took two hours and a-half to traverse. The best site for a permanent residence would be on the plateau itself, where the cultivation is, where it is open to the breeze, and where water could always be obtained by sinking wells. The banyan tree (ficus indica) was found here; also the guava tree. The Babul had almost entirely disappeared, and the fir-trees which were seen were in poor condition.

A finer or richer country than this could not be desired. The scenery and general appearance very much resemble the "ghats," on the western coast of India, and the verdure is equal to what is seen in those favoured parts immediately after the monsoon. The reason of this is that the cool season is the rainy one in the portion of the country lying between the Abyssinian highlands and the sea. There is a north-east monsoon; in fact, from October to the end of March there is some rain every day, but never very heavy, nothing approaching the Indian south-west monsoon. The climate is very similar to that of Mattemma. An April sun during the day was warm, but the heat did not make being out in it unpleasant, and in the open plain or on the hills there was always a delicious cool breeze blowing from the sea. At night there was heavy dew consequent on the recent rains, and it was so cold that blankets were necessary sleeping inside a tent. Water is obtainable almost everywhere, and wells could easily be sunk. The ascent from Henrote to Fathak, with subsequent descent, is, as it at present exists, undoubtedly difficult, but the track followed went straight up the hill. A zigzag might be made at no very great cost, up which laden camels, and even carriages, might pass. The country is full of game in the cool months. Recent traces of elephants were seen, but the guides said these animals had passed on to higher grounds to escape from the flies, which are very numerous at this season of the year. Tracks of lions, leopards, wild hog, and deer, were noticed daily: and of birds there were abundance of guinea fowl, spur fowl, and quail.

The great want in the country is population. After leaving Part not a soul was seen. The cultivation at Agametta, though nearly ripe, was unwatched. The land had been ploughed, and the grain sown in October, and the crop left to mature as it might. The splendid grazing-grounds were undisturbed by cattle. From inquiries made it appears that the whole of this extensive and fertile tract of country between the mountains and the sea, from the Agadey torrent to nearly twenty miles north, giving an area of some 400 square miles, belongs to one tribe, the Tsanatigli Christians, mustering not more than 800 adults. They also hold lands in Abyssinia, where they reside in the summer months, migrating with their cattle, of which they have large herds, to the lowlands nearer the sea, from October to the end of April. As they pass down they cultivate patches here and there, reaping the produce as they return. This dearth of population I conclude to be owing to the absence of any one fixed good government as before mentioned. About Eylat the soil is considered to belong to Abyssinia. While the Egyptian Government holds the country, it is nominally supposed to protect the people. The latter have therefore to pay dues to both.

In returning the same route was followed as far as Henrote and for half-a-mile beyond. Then the Ukbuloo torrent was left, a turn being taken to the right, and the road passed over a hill about 250 feet high, descending into another torrent, which four miles further on joined a larger one, and this was continued until the sea was reached at the village of Arkeeko. Three miles down the larger torrent running water was found, and was met with at intervals for the remainder of the way.

The nature of the country was the same as on the upward journey, but this was a much more direct route to the sea, though not to Massowah, shorter in consequence, and with the exception of the hill just below Henrote, which was bad for camels, a better road. Three marches were made going up, only one coming down: the distance by the former twenty-nine miles, by the latter to Arkeeko twenty-two miles. From Arkeeko we went by boat to Massowah, three-and-a-half miles; had we marched by land seven miles, the distance would have been identical, twenty-nine miles. The camping ground near the plateau was left at 6.30 A.M., and Arkeeko reached at 3.25 P.M., deducting two hours thirty minutes halt for breakfast. The actual time in marching was six hours thirty minutes. The last eight miles from Galata being a good road, was done more quickly than the first part. The whole was on mules, excepting down the hill from Fathak to Henrote, and down the hill beyond Henrote, which was walked.

The heights were taken by readings from two aneroids, by the boiling point of water. A tabular route is annexed; also a rough sketch by Captain Merriman. It should be mentioned that this is no regular road into Abyssinia. It is nothing more than a track by the Tsanatagli tribe in their annual migrations. I availed myself of the services of Captain C. J. Merriman, Royal Engineers, Executive Engineer, to observe the country, and of Surgeon James, civil surgeon, Aden, to inspect the state of the work-people colony at Moncooloo, taking these officers with me on duty.

I have,

W. H. Merewether.

LIEUT.-COLONEL MEREWETHER'S REPORT ON THE ROADS FROM MASSOWAH TO ADOWA, BY AILET.

(Marked A and H on Map, page 39.)

Massowah, January 15, 1867.

MY LORD,

WHILE waiting for intelligence I employed the time in making excursions in the neighbourhood to see places and obtain information about the country.

The first spot visited was Ailat, twenty-seven miles nearly due west of Massowah. The road, a mere cattle track, is chiefly along the dry beds of water-courses, or over low hills, occasionally on bits of level plain; with the exception of three places, where gorges have to be ascended or descended, there is no part of the road over which artillery would not drive with ease, and the three exceptions might very soon be made passable by sappers. The first is short, only 200 yards, but steep, having a direct rise of eighty feet in that distance; it could be zigzagged. The second is much longer, being one mile and a half, but has only a rise of about 400 feet, and were it not for its narrowness, and being clogged with boulders, would be easy enough. There is plenty of earth and bushes on either side to afford material for a road which would stand well until heavy rain washed it away, and a heavy fall is not of frequent occurrence. The highest point crossed was found to be only 989 feet above the level of the sea; there is a gradual rise the whole way. From this highest point, the road descends 200 feet on to the plain of Ailat. At seventeen miles from Massowah, thirteen from Moncooloo, there is a spring of good water always running; but any quantity of water may be obtained by digging a few feet in the sandy beds of the watercourses.

Ailat is a fine plain, at the foot of the mountains bordering Abyssinia, twelve miles long from north to south, by five broad, covered at this season of the year with rich verdure, owing to showers falling once in the twenty-four hours. It gets some also of the summer rains, so that it is well off in this respect; any quantity of water may be obtained by sinking wells. The whole plain might be cultivated to great advantage with cotton, wheat, maize, jowaree, &c., were there a good Government to arrange for the cultivators, but the place is unfortunate in this respect; it pays a small tribute to the Egyptian Government for the nominal protection afforded by ten soldiers, and the justice they are supposed to have administered to them by the Governor of Massowah, while they also pay a grazing fee to the Emperor Theodorus, because the ground is considered Abyssinian. The result is, the valley is poorly populated by people who live almost entirely by their cattle, cultivating only a very little jowaree. The village, a scattered one, of about thirty grass and mat huts, is in the centre of the valley. The heat is said to be very great in summer.

This would be an excellent place to locate a large body of cavalry on first landing, there being plenty of wood, water, forage, and meat; grain would be scarce at first, but it would soon come in from the countries north and north-west. Besides the water to be obtained by digging, there is a perennial spring on the west side of the valley, two and a half miles from the village. Here also are the celebrated hot springs used by the invalids of the country.

There is a road direct from Ailat into Abyssinia, but not a good one. The best in this direction for laden camels passes about eight miles south of Ailat, from which it can be joined. It starts from the same place, Moncooloo, and through the same sort of country as the road to Ailat; but is less difficult, both in the first part and after entering the mountains. Water is more plentiful on it, and is found at the following places, eight to ten miles apart, Zagu, Yamgous, Amba Tokan, Bamba, Ayderess, Kiaguor; Kiaguor is only sixty miles from Moncooloo, and once there, you are on the plateau of Hamozeyn, in Abyssinia. From Kiaguor to Gootofellassee (Koodoofelassy in the map) is only thirty miles of fair road. Gootofellassee is a very healthy place, dry, and not feverish; it is where several roads converge, supplies naturally come to, and could be easily reached from the coast, in nine marches, by an advanced force. From Gootofellassee to Adowa is under fifty miles; the Mareb has to be crossed about midway, but there is no difficulty whatever in the dry, cold season, the depth of the water then being only a few inches. From Adowa any road could be taken as required; but though it is a larger place than Gootofellassee and the capital of Tigré, the latter is the best place for first rendezvous, being most healthy.

I proceeded down Annesley Bay to the bottom, and, landing at different places, inquired quietly, through M. Munzinger, about the different roads from that neighbourhood. There are two, one to Halai, taken by Dr. Beke last year, and which is not a good one for laden camels, the other the old Greek caravan road, from Adoolis to Tenafee; this last has not been much frequented of late years, and is overgrown with jungle, but it was much used when Adoolis was a Greek colony, and I cannot help thinking will, on further inquiry, prove as good as any, at any rate well worth examination hereafter should entrance into Abyssinia become necessary. The country on the western side of Annesley Bay is richly green now, in consequence of the constant rain, and the plains are covered with herds of cattle from all parts. Wild elephants were found close to the sea-shore at the bottom of the bay, feeding quietly in the plain. There is a party of 200 Egyptian troops here, protecting the Customs levied on the salt brought from an extensive salt plain, about twenty miles to the south.

There are three other routes into Abyssinia, which formerly were regularly used by caravans, but lately have not been much so: one from Amphilla Bay, direct to Adowa; another from Ead to Sokota; and the third from Tagourra, by Lake Haik, to Begemeder. This last is the most direct to where the Emperor now is confined to; but, from all I have been able to learn, there are great difficulties on the score of water, none but very brackish being procurable.

MAJOR HARRIS'S JOURNEY FROM TAJURREH TO ANKOBAR IN JUNE, 1841. (Marked X on Map, page 39).

From this eventful epoch each sultry day did indeed bring a numerical accession to the beasts of burthen collected in the town; but they were owned of many and self-willed proprietors; were, generally speaking, of the most feeble description, melancholy contrasts to the gigantic and herculean dromedary of Egypt and Arabia; and no trifling delay was still in store through their arrival from different pastures bare-backed, which involved the necessity of making up new furniture for the march. The Dankáli saddle is fortunately a simple contrivance; a mat composed of plaited date leaves thrown over the hump, supporting four sticks lashed together in couples, and kept clear of the spinal process by means of two rollers as pads, having been proved by centuries of experience to be not more light than efficient. Accoutrements completed, and camels ready for the march, other provoking excuses for delay were not wanting, to fill, even to overflowing, the measure of annoyance. At length however no further pretext could be devised, and nine loads being actually in motion towards Ambábo, the first halting ground on the road to the kingdom of Shoa, the schooner "Constance," getting under weigh, stood up the bay of Tajúra, and cast anchor off the incipient camp, of which the position was denoted by a tall cluster of palms.

The almost insurmountable difficulties thus experienced in obtaining carriage, but now happily overcome, had so far delayed the advance of the Embassy, as to oblige it to cross the Tehâma during the height of the fiery and unwholesome blast which, during the months of June and July, sweeps over that waterless tract from the south-west; and had moreover rendered it impossible to reach Abyssinia before the setting in of the annual heavy rains, when the river Hawásh becomes impassable for weeks together. Independently of the natural apathy of the camel-owning population, the fact of the season of all intercourse with the interior, by Káfilah, having already passed away, rendered every one averse, under any consideration of gain, to so hazardous a journey. Grain was to be carried for the consumption of horses and mules during the passage of arid regions, where, during the hot season, neither vegetation nor water exists; and the wells and pools having notoriously failed in every part of the road, during three consecutive seasons of unusual drought, it was necessary to entertain a large proportion of transport for a supply of water sufficient to last both man and beast for two or three days at a time; whilst, neither grass nor green food remaining near the sea-shore, the hundred and seventy camels now forming the caravan, had been individually assembled from various grazing grounds, many miles distant in the interior.

A sufficient number of water skins had fortunately been purchased at exorbitant prices to complete the equipment, together with mules for the conveyance of the European escort and artillery; and the greedy Sultán, besides receiving the lion's share of the profits on all, had sold his own riding beast for three times its worth in solid silver. But the forage brought over from Aden being long since consumed, the whole were fed upon dates, and to the latest moment the greatest difficulty continued to exist in regard to followers. The services of neither Dankáli, Bedouin, nor Somauli, were obtainable at whatever wages; and the whole of the long train of live stock was consequently to be attended by a few worthless horsekeepers, enlisted at Aden, aided by a very limited number of volunteers from the shipping, whose indifferent characters gave ample promise of their subsequent misdeeds.

With a feeling of pleasure akin to that experienced by Gil Blas, when he escaped from the robbers' cave, the party at length bade adieu to Tajúra. Of all the various classes and denominations of men who inhabit the terrestrial globe, the half-civilized savages peopling this seaport, are perhaps the most thoroughly odious and detestable.

The tall masts of the schooner of war, raking above the belt of dwarf jungle that skirts the tortuous coast, served as a beacon to the new camp, the distance of which from the town of Tajúra was less than four miles. A narrow footpath wound along the burning sands, across numerous water-courses from the impending mountain range of trachyte and porphyry, whose wooded base, thickly clothed with mimosa and *euphorbia antiquorum*, harboured swine, pigmy antelope, and guinea-fowl in abundance. Many large trees, uprooted by the wintry torrent, had been swept far out to sea, where in derision of the waves that buffet their dilapidated, stag-horn looking arms, they will long ride safely at anchor. The pelican of the wilderness sailed through the tossing surf, and files of Bedouin damsels, in greasy leathern petticoats, bending beneath a load of fuel from the adjacent hamlets, traversed the sultry strand; whilst a long train of wretched children, with streaming elf-like locks, who had been kidnapped in the unexplored interior, wended their weary way with a slave caravan, towards the sea-port, whence they were to be sold into foreign bondage.

An avenue through the trees presently revealed the white tent, occupying a sequestered nook on the course of a mountain stream near its junction with the shore. Here horses and mules were doing their utmost, by diligently cropping the scanty tufts of sunburnt grass, to repair their recent long abstinence from forage, whilst the abbreviated tails of those which had been improved by mutilation, formed the jest of a group of grinning savages. Clumps of lofty fan palms, and date trees loaded with ripe orange-coloured fruit, still screened from view the village of Ambábo, the straggling Gothic roofed wigwams composing which have the same waggon-like appearance as the huts of Tajúra,—a similar style of architecture extending even to the unostentatious mosque, alone distinguishable from the surrounding edifices, by uncarved minarets of wood.

Three hours after midnight, the galloper gun, fired within the limits of the British camp as a summons to the drowsy camel-drivers to be up and doing, was echoed, according to previous agreement, by the long stern chaser of the "Constance,"—a signal to the "Euphrates," still anchored off Tajúra, to thunder a farewell salute as the day dawned. The work of loading was merrily commenced—the tent went down—and camel after camel moved off towards Dullool; when, on the departure of the last string, it was observed with dismay that the ground was still strewed with baggage, for which carriage had unquestionably been paid and entertained, but for which none was forthcoming.

Izhák's absent camels being now brought in, the ground was speedily cleared of the remaining baggage; and satisfied with the specious assurance of the Ras el Káfilah, that he would on no account tarry beyond nightfall of the following day, the party, relieved from their anxiety, mounted after five o'clock, and galloped seven miles along the sea-beach to the camp at Dullool,—the loose sand being so perforated and undermined in every part by the hermit crab, as to render the sieve-like road truly treacherous and unpleasant.

The grassy nook occupied by the tent was situated at the abutment of a spur from the wooded Jebel Goodah, evidently of volcanic origin, which gradually diminishes in height, until it terminates, one hundred yards from the shore, in a thick jungle of tamarisk and acacia, the former covered with salt crystals. Hornblende, in blocks, was scattered along the beach, and, wherever decomposed, it yielded fine glittering black sand, so heated under the noontide sun as to burn the naked foot. The movable camp of a horde of roving Bedouin shepherds, who, with very slender habitations, possess no fixed abode, was erected near the wells; and a quarrel with the followers, respecting the precious element, having already led to the drawing of creeses, silver was again in requisition to allay the impending storm.

The heat on the 2nd of June was almost insupportable; but the sultry day proved one of greater quiet than had fallen to the lot of the Embassy since its first landing.

At gun-fire the next morning, the arrival of the whole party being reported, orders were issued to strike the tent.

The schooner had meanwhile fished her anchor, and was now getting under weigh for the purpose of standing up within range of the next halting ground. The mules were harnessed to the gun, and the tent and baggage packed, and after three hours of needless detention, the party commenced its third hot march along the sea-beach, whence the hills gradually recede. Bedouin goat-herds occupied many wells of fresh water, which were denoted by clumps of date trees entwined by flowering convolvuli, whose matted tendrils fix the movable sands of the shore; and late in the forenoon the camp was formed at the pool of Sagállo, only three miles from the former ground, but affording the last supply of water to be obtained for thirty more.

An extensive and beautiful prospect of the western portion of the Bay of Tajúra had now opened, bound in on all sides by a zone of precipitous mountains, in which the gate leading into Goobut el Kharáb was distinctly marked by a low black point, extending from the northern shore. The schooner's services were volunteered to admit of a nearer inspection of the "basin of foulness;" but no sooner had she stood out to sea than signal guns fired from the camp announced the arrival of another packet from Shoa. The courier had been forty-four days on the journey, and the tidings he brought respecting the road, although highly satisfactory, added yet another instance to the many, of the small reliance that can be placed on information derived from the Danákil, who, even when disinterested, can rarely indeed be induced to utter a word of truth.

The strong party feeling entertained towards Mohammad Ali by the magnates of Tajúra, now vented itself in divers evil-minded and malicious hints, insinuating the defection of the absentee, who had been unavoidably detained by business, some hours after the last of the sea-port heroes had joined. "Where now is your friend Ali Mohammad?" "Where is the man who was to supply water on the road?" were the taunting interrogatories from the mouths of many; but come the son of Ali Abi did, to the confusion of his slanderers, long ere the sun had set, bringing secret intelligence that he had sent to engage an escort from his own tribe; and the whole party being now at last assembled, it was resolved in full conclave, that as not a drop of water could be procured for three stages in advance, the entire of the next day should be devoted to filling up the skins, which done, the caravan should resume its march by night—a manœuvre that savoured strongly of a design to favour the clandestine return to Tajúra of certain of the escort, who had still domestic affairs to settle.

A most unprofitable discussion, which was prolonged until eleven the following night, had for its object to persuade the transmission of baggage in advance to the Salt Lake, in consequence of the carried supply of water being, after all, considered insufficient for three days' consumption. But the proposal was negatived upon prudent grounds, the honesty of the intentions by which it had been dictated, seeming at best, extremely questionable, and no one feeling disposed to trust the faithless guides further than they could be seen, with property of value.

Scarcely were the weary eyes of the party closed in sleep, than the long 32-pounder of the "Constance," proclaiming the midnight hour, sounded to boot and saddle. The Babel-like clamour of loading was at length succeeded by a lull of voices, and the rumbling of the galloper wheels over the loose shingle, was alone heard in the still calm of the night, above the almost noiseless tread of the cushion-footed camels, which formed an interminable line. The road, lit by the full moon, shining brightly overhead, lay for the first two or three miles along the beach, and then, crossing numerous watercourses, struck over the southern shoulder of Jebel Goodah, the distance from whose lofty peak each march had reduced.

Blocks and boulders varying in size from an 18 pound shot, to that of Ossa piled upon Pelion, aided by deep chasms, gullies, and water-ways, rendering the ascent one of equal toil and peril, cost the life of a camel, which fell over a precipice and dislocated the spine; whereupon the conscientious proprietor, disdaining to take further heed of the load, abandoned it unscrupulously by the wayside. Gáleylaféo, a singular and fearful chasm which was navigated in the first twilight, did not exceed sixty feet in width; its gloomy, perpendicular walls of columnar lava, towering one hundred and fifty feet overhead, and casting a deep deceitful shadow over the broken channel, half a mile in extent.

Dawn disclosed the artillery mules in such wretched plight from their fatiguing night's labour, that it was found necessary to unlimber the gun, and place it with its carriage on the back of an Eesah camel of Herculean strength, provided for the contingency by the foresight of Mohammad Ali; and although little pleased during the imposition of its novel burthen, the animal, rising without difficulty, moved freely along at a stately gait. The same uninteresting volcanic appearance characterised the entire country to the table-land of Warelissán, a distance of twelve miles. Dreary and desolate, without a trace of vegetation saving a few leafless acacias, there was no object to relieve the gaze over the whole forbidding expanse. In this barren unsightly spot the radiation was early felt from the masses of black cindry rock, which could not be touched with impunity. The sand soil of the desert reflecting the powerful beams of the sun, lent a fearful intensity to the heat, whilst on every side the dust rose in clouds that at one moment veiled the caravan from sight, and at the next left heads of camels tossing in the inflamed atmosphere among the bright spear-blades of the escort. But on gaining the highest point, a redeeming prospect was afforded in an unexpected and most extensive bird's eye view of the estuary of Tajúra, now visible in all its shining glory, from this, its western boundary. Stretching away for miles in placid beauty, its figure was that of a gigantic hour-glass; and far below on its glassy bosom were displayed the white sails of the friendly little schooner, as, after safely navigating the dangerous and much-dreaded portals of Scylla and Charybdis, never previously braved by any craft larger than a jolly boat—bellying to the breeze, she beat gallantly up to the head of Goobut el Kharáb.

Although Warelissán proved nearly seventeen hundred feet above the level of the blue water, a suffocating south-westerly wind, which blew throughout the tedious day, rendered the heat more awfully oppressive than at any preceding station. The camp, unsheltered, occupied a naked tract of table-land, some six miles in circumference, on the shoulder of Jebel Goodah—its barren surface strewed with shining lava, and bleached animal bones; sickly acacias of most puny growth, sparingly invested with sun-burnt leaves, here and there struggling through the fissures, as if to prove the utter sterility of the soil; whilst total absence of water, and towering whirlwinds of dust, sand, and pebbles, raised by the furnace-like puffs that came stealing over the desert landscape, completed the discomfiture both of man and beast.

During the dead of night, when restless unrefreshing slumbers on the heated ground had hushed the camp in all its quarters, the elders, in great consternation, brought a report that the Bedouin war-hawks, who nestle in the lap of the adjacent wild mountains, were collecting in the neighbourhood with the design of making a sudden swoop upon the káfilah, for which reason the European escort must be prepared for battle, and muskets be discharged forthwith, to intimidate the lurking foe. They were informed, in reply, that all slept upon their arms, and were in readiness; but Mohammad Ali came shortly afterwards to announce that matters had been amicably adjusted with the aid of a few ells of blue cloth; and under the care of a double sentry, the party slept on without further disturbance until two in the morning, prior to which hour, the moon, now on her wane, had not attained sufficient altitude to render advance practicable.

The aid of her pale beams was indispensable, in consequence of the existence of the yawning pass of Rah Eesah, not one hundred yards distant from the encampment just abandoned, but till now unperceived. It derives its appellation, as "the road of the Eesahs," from the fact of this being the path usually chosen by that hostile portion of the Somauli nation, on the occasions of their frequent forays into the country of the Danákil, with whom, singularly enough, an outward understanding subsists. Its depths have proved the arena of many a sanguinary contest, and are said, after each down-pouring of the heavens, to become totally impassable, until again cleared of the huge blocks of stone, the detritus from the scarped cliffs, which so choke the bed of the chasm, as to impede all progress. The labour of removing these, secures certain immunities to the wild pioneers, who levy a toll upon every passing caravan, and who in this instance were propitiated, on application, by the division of a bale of blue cotton calico, a manufacture here esteemed beyond all price.

A deep zigzagged rent in the plateaux, produced originally by some grand convulsion of nature, and for ages the channel of escape to the sea of the gathered waters from Jebel Goodah, winds like a mythological dragon through the bowels of the earth, upwards of three miles to the southward. Masses of basalt of a dark burnt brown colour, are piled perpendicularly on either side, like the solid walls of the impregnable fortresses reared by the Cyclops of old; and rising from a very narrow channel, strewed with blocks of stone, and huge fallen fragments of rock, tower overhead to the height of five or six hundred feet. One perilous path affords barely sufficient width for a camel's tread, and with a descensus of one foot and a half in every three, leads twisting away into the gloomy depths below, dedicated to the son of Chaos and Darkness, and now plunged in total obscurity.

It was a bright and cloudless night, and the scenery, as viewed by the uncertain moonlight, cast at intervals in the windings of the road upon the glittering spear-blades of the warriors, was wild and terrific. The frowning basaltic cliffs, not three hundred yards from summit to summit, flung an impenetrable gloom over the greater portion of the frightful chasm, until, as the moon rose higher in the clear vault of heaven, she shone full upon huge shadowy masses, and gradually revealed the now dry bed, which in the rainy season must oftentimes become a brief but impetuous torrent.

No sound was heard save the voice of the camel-driver, coaxing his stumbling beasts to proceed by the most endearing expressions. In parts where the passage seemed completely choked, the stepping from stone to stone, accomplished with infinite difficulty, was followed by a drop leap, which must have shaken every bone. The gun was twice shifted to the back of a spare camel, provided for the purpose; and how the heavy laden, the fall of one of which would have obstructed the way to those that followed, kept their feet, is indeed subject of profound astonishment. All did come safely through, however, notwithstanding the appearance of sundry wild Bedouins, whose weapons and matted locks gleamed in the moonbeam, as their stealthy figures flitted in thin tracery from crag to crag. A dozen resolute spirits might have successfully opposed the united party; but these hornets of the mountains, offering no molestation, contented themselves with reconnoitring the van and rear-guards from heights inaccessible through their natural asperity, until the twilight warned them to retire to their dens and hiding places; and ere the sun shone against the summits of the broken cliffs, the straggling caravan had emerged in safety from this dark descent to Eblis.

Goobut el Kharáb, with the singular sugar-loaf islet of Good Ali, shortly opened to view for the last time, across black sheets of lava, hardened in their course to the sea, and already rotten near the water's edge.

The schooner, although riding safely at anchor near the western extremity, was altogether concealed by precipitous walls that towered above her raking masts, and kept the party in uncertainty of her arrival. Crossing the lone valley of Marmoríso, a remnant of volcanic action, rent and seamed with gaping fissures, the road turned over a large basaltic cone, which had brought fearful devastation upon the whole surrounding country, and here one solitary gazelle browsed on stubble-like vegetation scorched to a uniform brown. Skirting the base of a barren range, covered with heaps of lava blocks, and its foot ornamented with many artificial piles, marking deeds of blood, the lofty conical peak of Jebel Seeáro rose presently to sight, and not long afterwards the far-famed Lake Assál, surrounded by dancing mirage, was seen sparkling at its base.

The first glimpse of the strange phenomenon, although curious, was far from pleasing. An elliptical basin, seven miles in its transverse axis, half filled with smooth water of the deepest cærulean blue, and half with a solid sheet of glittering snow-white salt, the offspring of evaporation—girded on three sides by huge hot-looking mountains, which dip their bases into the very bowl, and on the fourth by crude half-formed rocks of lava, broken and divided by the most unintelligible chasms,—it presented the appearance of a spoiled, or at least of a very unfinished piece of work. Bereft alike of vegetation and of animal life, the appearance of the wilderness of land and stagnant water, over which a gloomy silence prevailed, and which seemed a temple for ages consecrated to drought, desolation, and sterility, is calculated to depress the spirit of every beholder. No sound broke on the ear; not a ripple played on the water; the molten surface of the lake, like burnished steel, lay unruffled by a breeze; the fierce sky was without a cloud, and the angry sun, like a ball of metal at a white heat, rode triumphant in a full blaze of noon-tide refulgence, which in sickening glare was darted back on the straining vision of the fainting wayfarer, by the hot sulphury mountains that encircled the still, hollow, basin. A white foam on the shelving shore of the dense water, did contrive for a brief moment to deceive the eye with an appearance of motion and fluidity; but the spot, on more attentive observation, ever remained unchanged—a crystallized efflorescence.

As the tedious road wound on over basalt, basaltic lava, and amygdaloid, the sun, waxing momentarily more intensely powerful, was reflected with destructive and stifling fervour from slates of snow-white sea limestone borne on their tops. Still elevated far above the level of the ocean, a number of fossil shells, of species now extinct, were discovered; a deep cleft by the wayside, presenting the unequivocal appearance of the lower crater of a volcano, situated on the high basaltic range above, whence the lava stream had been disgorged through apertures burst in the rocks, but which had re-closed after the violence of the eruption had subsided.

Dafári, a wild broken chasm at some distance from the road, usually contains abundance of rain water in its rocky pool, but having already been long drained to the dregs, it offered no temptation to halt. Another most severe and trying declivity had therefore to be overcome, ere the long and sultry march was at an end. It descended by craggy precipices many hundred feet below the level of the sea, to the small close sandy plain of Mooya, on the borders of the Lake—a positive *Jehannam*, where the gallant captain of the "Constance"* had already been some hours esconced under the leafless branches of one poor scrubby thorn, which afforded the only screen against the stifling blast of the sirocco, and the merciless rays of the refulgent orb over-head.

Adyli, a deep mysterious cavern at the further extremity of the plain, is believed by the credulous to be the shaft leading to a subterranean gallery which extends to the head of Goobut el Kharáb.

Foul-mouthed vampires and ghouls were alone wanting to complete the horrors of this accursed spot, which, from its desolate position, might have been believed the last stage in the habitable world. A close mephitic stench impeding respiration, arose from the saline exhalations of the stagnant lake. A frightful glare from the white salt and limestone hillocks threatened destruction to the vision; and a sickening heaviness in the loaded atmosphere, was enhanced rather than alleviated by the fiery breath of the parching north-westerly wind, which blew without any intermission during the entire day. The air was inflamed, the sky sparkled, and columns of burning sand, which at quick intervals towered high into the dazzling atmosphere, became so illumined as to appear like tall pillars of fire. Crowds of horses, mules, and fetid camels, tormented to madness by the dire persecutions of the poisonous gad-fly, flocked recklessly with an instinctive dread of the climate, to share the only bush; and obstinately disputing with their heels the slender shelter it afforded, compelled several of the party to seek refuge in noisome caves formed along the foot of the range by fallen masses of volcanic rock, which had become heated to a temperature seven times in excess of a potter's kiln, and fairly baked up the marrow in the bones. Verily! it was "an evil place," that lake of salt: it was "no place of seed, nor of figs, nor yet of vines; no, nor even of pomegranates; neither was there any water to drink."

In this unventilated and diabolical hollow, dreadful indeed were the sufferings in store both for man and beast. Not a drop of fresh water existed within many miles; and, notwithstanding that every human precaution had been taken to secure a supply, by means of skins carried upon camels, the very great extent of most impracticable country to be traversed, which had unavoidably led to the detention

* Lieut. Wilmot Christopher.—J. R.

of nearly all, added to the difficulty of restraining a multitude maddened by the tortures of burning thirst, rendered the provision quite insufficient; and during the whole of this appalling day, with the mercury in the thermometer standing at 126° under the shade of cloaks and umbrellas—in a suffocating Pandemonium, depressed five hundred and seventy feet below the ocean, where no zephyr fanned the fevered skin, and where the glare arising from the sea of white salt was most painful to the eyes; where the furnace-like vapour exhaled almost choking respiration, created an indomitable thirst, and not the smallest shade or shelter existed, save such as was afforded, in cruel mockery, by the stunted boughs of the solitary leafless acacia, or, worse still, by black blocks of heated lava, it was only practicable, during twelve tedious hours, to supply to each of the party two quarts of the most mephitic brick-dust-coloured fluid, which the direst necessity could alone have forced down the parched throat, and which, after all, far from alleviating thirst, served materially to augment its insupportable horrors.

It is true that since leaving the shores of India, the party had gradually been in training towards a disregard of dirty water—a circumstance of rather fortunate occurrence. On board a ship of any description the fluid is seldom very clean, or very plentiful. At Cape Aden there was little perceptible difference betwixt the sea water and the land water. At Tajúra the beverage obtainable was far from being improved in quality by the taint of the new skins in which it was transferred from the only well; and now, in the very heart of the scorching Tehâma, when a copious draught of *aqua pura* seemed absolutely indispensable every five minutes, to secure further existence upon earth, the detestable mixture that was at long intervals most parsimoniously produced, was the very acmé of abomination. Fresh hides, stripped from the rank he-goat, besmeared inside as well as out with old tallow and strong bark tan, filled from an impure well at Sagállo, tossed, tumbled, and shaken during two entire nights on a camel's back, and brewed during the same number of intervening days under a strong distilling heat—poured out an amalgamation of pottage of which the individual ingredients of goat's hair, rancid mutton fat, astringent bark, and putrid water, were not to be distinguished. It might be smelt at the distance of twenty yards, yet all, native and European, were struggling and quarrelling for a taste of the recipe. The crest-fallen mules, who had not moistened their cracked lips during two entire days, crowding around the bush, thrust their hot noses into the faces of their masters, in reproachful intimation of their desire to participate in the filthy but tantalising decoction; and deterred with difficulty from draining the last dregs, they ran franticly with open mouths to seek mitigation of their sufferings at the deceptive waters of the briny lake, which, like those of Goobut el Kharáb, were so intensely salt, as to create smarting of the lips if tasted.

Slowly flapped the leaden wings of Time on that dismal day. Each weary hour brought a grievous accession, but no alleviation, to the fearful torments endured. The stagnation of the atmosphere continued undiminished; the pangs of thirst increased, but no water arrived; and the sun's despotic dominion on the meridian, appeared to know no termination.

But the longest day must close at last, and the great luminary had at length run his fiery and tyrannical course. String after string of loaded camels, wearied with the passage of the rugged defile of Rah Eesah, were with infinite difficulty urged down the last steep declivity, and at long intervals, as the shadows lengthened, made their tardy appearance upon the desert plain; those carrying water, tents, and the greater portion of the provisions most required, being nevertheless still in the rear when the implacable orb went down, shorn of his last fierce ray. The drooping spirits of all now rose with the prospect of speedy departure from so fearful a spot. The commander of the friendly schooner, which had proved of such inestimable service, but whose protecting guns were at length to be withdrawn, shortly set out on his return to the vessel with the last despatches from the Embasy, after bidding its members a final farewell; and in order to obtain water, any further deprivation of which must have involved the dissolution of the whole party, no less than to escape from the pestilential exhalations of the desolate lake, which, as well during the night as during the day, yielded up a blast like that curling from a smith's forge—withering to the human frame—it was resolved as an unavoidable alternative, to leave the baggage to its fate, and to the tender mercies of guides and camel-drivers, pushing forward as expeditiously as possible to Goongoonteh, a cleft in the mountains that bound the opposite shore, wherein water was known to be abundant. Pursuant to this determination, the European escort, with the servants, followers, horses, and mules, were held in readiness to march so soon as the moon should rise above the gloomy lava hills, sufficiently to admit of the path being traced which leads beyond the accursed precincts of a spot, fitly likened by the Danákil to the infernal regions.

Dismal, deadly, and forbidding, but deeply interesting in a geological point of view, its overwhelming and paralysing heat precluded all possibility of minute examination, and thus researches were of necessity confined to the general character of the place. Latitude, longitude, and level were, however, accurately determined, and many were the theories ventured, to account for so unusual a phenomenon.

Of two roads which lead to Goongoonteh from the shores of the dreary Bahr Assál, one skirts the margin of the lake by a route utterly destitute of fresh water; whilst the other, although somewhat more circuitous, conducts over high lava banks stretching some distance inland to Haliksitan, and past the small well of Hanlefánta, where the drained pitcher of the fainting wayfarer may be refilled. On finally quitting the bivouac under the scraggy boughs of the dwarf acacia, where the tedious and most trying day had been endured—which each of the half-stifled party did with an inward prayer that it might never fall to his lot to seek their treacherous shelter more—a fierce dispute arose amongst the leaders of the caravan as to which path should be adopted. "What matters it," urged the intolerant Mohammadan from Shoa, who had accidentally been found starving at Ambábo, and been since daily fed by the Embassy—"what matters it if all these Christian dogs should happen to expire of thirst? Lead the Kafirs by the lower road, or, *Allahu akbar*, God is most powerful, if the waters of the well prove low, what is to become of the mules of the Faithful?"

But the breast of the son of Ali Abi fortunately warmed to a more humane and charitable feeling than the stony heart of the "red man." With his hand upon the hilt of his creese, he swore aloud upon the sacred Korán to take the upper path, and stoutly led the way, in defiance of all, after Izhák and the ruthless bigots in his train had actually entered upon a route, which the event proved must have involved the destruction of all less inured than the savage to the hardships of the waste wilderness.

'Twas midnight when the thirsty party commenced the steep ascent of the ridge of volcanic hills which frown above the south-eastern boundary of the fiery lake. The searching north-east wind had scarcely diminished in its parching fierceness, and in hot suffocating gusts swept fitfully over the broad glittering expanse of water and salt, whereon the moon shone brightly—each deadly puff succeeded by the stillness that foretells a tropical hurricane—an absolute absence even of the smallest ruffling of the close atmosphere. Around, the prospect was wild, gloomy, and unearthly, beetling basaltic cones and jagged slabs of shattered lava—the children of some mighty trouble—forming scenery the most shadowy and extravagant. A chaos of ruined churches and cathedrals, *eedgahs*, towers, monuments, and minarets, like the ruins of a demolished world, appeared to have been confusedly tossed together by the same volcanic throes, that when the earth was in labour, had produced the phenomenon below; and they shot their dilapidated spires into the molten vault of heaven, in a fantastic medley, which, under so uncertain a light, bewildered and perplexed the heated brain. The path, winding along the crest of the ridge, over sheets of broken lava, was rarely of more than sufficient width to admit of progress in single file; and the livelong hours, each seeming in itself a century, were spent in scrambling up the face of steep rugged precipices, where the moon gleamed upon the bleaching skeleton of some camel that had proved unequal to the task—thence again to descend at the imminent peril of life and limb, into yawning chasms and dark abysses, the forbidding vestiges of bygone volcanic agency.

The horrors of that dismal night set the efforts of description at defiance. An unlimited supply of water in prospect, at the distance of only sixteen miles, had for the brief moment buoyed up the drooping spirit which tenanted each way-worn frame; and when an exhausted mule was unable to totter further, his rider contrived manfully to breast the steep hill on foot But owing to the long fasting and privation endured by all, the limbs of the weaker soon refused the task, and after the first two miles, they dropped fast in the rear.

Fanned by the fiery blast of the midnight sirocco, the cry for water, uttered feebly and with difficulty by numbers of parched throats, now became incessant; and the supply of that precious element brought for the whole party falling short of one gallon and a half, it was not long to be answered. A tiny sip of diluted vinegar for a moment assuaging the burning thirst which raged in the vitals, and consumed some of the more down-hearted, again raised their drooping souls; but its effects were transient, and after struggling a few steps, overwhelmed, they sunk again, with husky voice declaring their days to be numbered, and their resolution to rise up no more. Dogs incontinently expired upon the road; horses and mules that once lay down, being unable from exhaustion to rally, were reluctantly abandoned to their fate; whilst the lion-hearted soldier, who had braved death at the cannon's mouth, subdued and unmanned by thirst, finally abandoning his resolution, lay gasping by the way side, and heedless of the exhortation of his officers, hailed approaching dissolution with delight, as bringing the termination of tortures which were not to be endured.

Whilst many of the escort and followers were thus unavoidably left stretched with open mouths along the road, in a state of utter insensibility, and apparently yielding up the ghost, others, pressing on to arrive at water, became bewildered in the intricate mazes of the wide wilderness, and recovered it with the utmost difficulty. As another day dawned, and the round red sun again rose in wrath over the Lake of Salt, towards the hateful shores of which the tortuous path was fast tending, the courage of all who had hitherto borne up against fatigue and anxiety began to flag. A dimness came before the drowsy eyes, giddiness seized the brain and the prospect ever held out by the guides, of quenching thirst immediately in advance, seeming like the tantalising delusions of a dream, had well nigh lost its magical effect; when, as the spirits of the most sanguine fainted within them, a wild Bedouin was perceived, like a delivering angel from above, hurrying forward with a large skin filled with muddy water. This most well-timed supply, obtained by Mohammad Ali from the small pool at Hanlefánta, of which, with the promised guard of his own tribe, by whom he had been met, he had taken forcible possession in defiance of the impotent threats of the ruthless "red man," was sent to the rear. It admitted of a sufficient quantity being poured over the face and down the parched throat, to revive every prostrate and perishing sufferer; and at a late hour, ghastly, haggard, and exhausted, like men who had escaped from the jaws of death, the whole had contrived to straggle into a camp, which, but for the foresight and firmness of the son of Ali Abi, few individuals indeed of the whole party would have reached alive.

A low range of limestone hillocks, interspersed with strange masses of coral, and marked by a pillar like that of Lot, encloses the well of Hanlefánta, where each mule obtained a shield full of water. From the glittering shores of the broad lake, the road crosses the saline incrustation, which extends about two miles to the opposite brink. Soiled and mossy near the margin, the dull crystallised salt appears to rest upon an earthy bottom; but it soon becomes lustrous and of a purer colour, and floating on the surface of the dense water, like a rough coarse sheet of ice, irregularly cracked, is crusted with a white yielding efflorescence, resembling snow which has been thawed and refrozen, but which still, as here, with a crisp sound, receives the impress of the foot. A well trodden path extends through the prismatic colours of the rainbow, by the longitudinal axis of the ellipse, to the north-eastern extremity of the gigantic bowl, whence the purest salt is obtainable in the vicinity of several cold springs, said to cast up large pebbles on their jet, through the ethereal blue water.

A second low belt of hills, gypsum and anhydrite, succeeded by limestone overstrewed with basaltic boulders, forms the western bank of the molten sea, and opens into a mountain ravine. Taking its source at Allooli, the highest point of the Gollo range, this torrent strives to disembogue into the extremity of the lake, although its waters seldom arrive so far, save during the rainy season. The high basaltic cliffs that hem in the pebbly channel, approximating in the upper course as they increase rapidly in altitude, form a narrow waist, where the first running stream that had greeted the eye of the pilgrims since leaving the shores of Asia, trickled onwards, leaving bright limpid pools, surrounded by brilliant sward.

Bowers, for ever green, enlivened by the melodious warbling of the feathered creation, and the serene and temperate air of the verdant meadows of Elysium, were absent from this blessed spot, but it was entered with feelings allied for the moment to escape from the horrors of purgatory to the gates of Paradise; and under the shade cast by the overhanging cliffs, which still warded off the ardent rays of the ascending sun, it was with thankful hearts that the exhausted party, after the terrors of such a night, turned their backs upon the deadly waters of the stagnant lake, to quaff at the delicious rivulet of Goongoonteh an unlimited quantity of cool though brackish fluid.

Here terminated the dreary passage of the dire Teháma—an iron-bound waste, which, at this inauspicious season of the year, opposes difficulties almost overwhelming in the path of the traveller. Setting aside the total absence of water and forage throughout a burning tract of fifty miles—its manifold intricate mountain passes, barely wide enough to admit the transit of a loaded camel, the bitter animosity of the wild bloodthirsty tribes by which they are infested, and the uniform badness of the road, if road it may be termed, everywhere beset with the huge jagged blocks of lava, and intersected by perilous acclivities and descents—it is no exaggeration to state, that the stifling sirocco which sweeps across the unwholesome salt flat during the hotter months of the year, could not fail, within eight and forty hours, to destroy the hardiest European adventurer. Some idea of the temperature of this terrible region may be derived from the fact of fifty pounds of well-packed spermaceti candles having, during the short journey from Tajúra, been so completely melted out of the box as to be reduced to a mere bundle of wicks. Even the Danákil, who from early boyhood have been accustomed to traverse the burning lava of the Tehámа, never speak of it but in conjunction with the devouring element, of whose properties it partakes so liberally, and when alluding to the Lake of Salt, invariably designate it "Fire."

Goongoonteh, a deep gloomy zig-zagged fissure, of very straitened dimensions, is hemmed in by craggy lava and basaltic walls, intersected by dykes of porphyry, augitic greenstone and pistacite, with decomposed sulphate of iron, all combining to impart a strangely variegated appearance. Scattered and inclined in various directions, although towering almost perpendicularly, they terminate abruptly in a rude pile of rocks and hills, through a narrow aperture in which the path to the next halting ground at Allooli, where the torrent takes its source, strikes off at an angle of 90°.

Huge prostrate blocks of porphyry and basalt, which have been launched from the impending scarps, and now reduce the channel to this narrow passage, are in places so heaped and jammed together by some mighty agency, as to form spacious and commodious caverns. In the rainy season especially, these doubtless prove of wonderful convenience to the wayfarer; and no tent arriving until late the following day, the re-assembled party were fain to have recourse to them for shelter against the fierce hot blast from the Salt Lake, which, unremitting in its dire persecutions, now blew directly up the ravine. But the rocks soon became too hot to be touched with impunity, and the oblique rays of the sun, after he had passed the meridian, darting through every aperture, the caves were shortly converted into positive ovens, in which the heat, if possible, was even more intolerable than ever. Unlike former stations, however, there was, in this close unventilated chasm, a luxurious supply of water to be obtained from the living rill which murmured past the entrance, and although raised to the temperature of a thermal spring by the direct influence of the solar rays, and withal somewhat brackish to the taste, it was far from being pronounced unpalatable.

Notwithstanding that the neighbourhood afforded neither the smallest particle of forage nor of fuel, it became necessary, in consequence of the non-arrival of one-half of the camels, no less than from the exhaustion of many of the party, to halt a day in the hot unhealthy gully; and this delay afforded to the treacherous creese of the lurking Bedouin an opportunity of accomplishing that which had only been threatened by drought and famine. The guides objected strongly to the occupation of the caves after nightfall, on account of the many marauding parties of Eesah and Mudaïto, by whom the wady is infested, every one, as a measure of precaution, slept in the open air among the baggage, half a mile lower down the ravine, where the caravan had halted. The dry sandy bed of the stream was here narrow, and the cliffs—broken for a short distance on either side into hillocks of large distinct boulders—again resumed their consistency after an interval of one hundred yards, and enclosed the camp in a deep gloom.

It had been intended to march at break of day to Allooli, the source of Wady Goongoonteh; but the absence of several of the camels, which had gone astray during the nocturnal confusion, caused delay in this den of iniquity until ten o'clock.

The last rains having washed away an artificial bank of stones which had formerly facilitated the ascent of the difficult and dangerous passage leading from Goongoonteh into the Wady Kélloo—as the upper course is denominated—a delay of two hours was at first starting experienced in the bed of the torrent, during which all were on the alert. Two huge pointed rocks abutting on opposite angles of the acute zig-zag, reduced it to a traversed waist, so narrow, that room for the load to pass was only afforded when the long-legged dromedary swung its unwieldy carcass alternately from side to side—the steepness of the acclivity rendering it very frequently necessary to perform this inconvenient evolution upon the knees. Many became jammed, and were unladen before they could regain an erect position; whilst others were, with infinite difficulty, by the united efforts of a dozen drivers, who manned the legs and tail, saved from being launched with their burthens over the steep side of the descent, which consisted of a treacherous pile of loose rubbish.

To the surprise of every spectator the train passed through the defile without any material accident, and thence proceeded to pick their steps among the rocks, pools, and fissures, which abound in every mountain torrent whose course is short and precipitous. Flanked by perpendicular sheets of basalt and porphyry, of unwholesome sulphury appearance, beneath which many deep pools of cool water had collected, the tortuous road was at intervals enlivened by clumps of the *doom* palm, environed by patches of refreshing green turf—sights from which the eye had long been estranged. Nine miles of gradual ascent brought the caravan safely to the encamping ground at the head of the stream—a swamp surrounded by waving palms and verdant rushes, occupying high table-land, and affording abundance of green forage to the famished cattle. Most fortunately the sky had proved cloudy, or the march, performed during the hottest hours of the day, would indeed have been terrific.

Hence to Sagállo, the dismal country is in the exclusive occupation of a wandering race of the Danákil, who, notwithstanding that the Sultán of Tajúra claims the sovereignty of the entire waste, only acknowledge his impotent authority during their occasional temporary sojourn among the huts of that sea-port. The guides asserted, with many imprecations, that from time immemorial few káfilahs had ever halted at Allooli without losing one or more of its members by the Adrúsi creeses, or by those of the Eesah; and on the bank opposite to the shady clump of *doom* palms, under whose canopy the residue of the day was passed, numerous cairns, consisting of circular piles of stone, similar to those left at Goongoonteh to commemorate the outrage of the preceding night, stood memorials of the dark deeds that had been perpetrated.

During about three years the road from Abyssinia to the sea coast was completely closed by hordes of these ruffian outcasts, who continued their murderous depredations on every passer-by, until Loheïta, the present Akil of the Débeni, a young, daring, and warlike chieftain, succeeding to the rule on the demise of his father, routed the banditti after a severe struggle, and re-opened the route. The Wady Kélloo is, however, permanently infested by parties of wild Bedouins, who skulk about the rocky passes, lie in wait for stragglers from the caravan, assassinate all who fall into their ruthless clutches, and, when time permits, further gratify their savage propensities by mangling and mutilating the corse.

Although Allooli was represented to be even more perilous than Goongoonteh, it possessed, in point of locality, immense superiority; and every advantage that could be devised was taken of its capabilities for defence. The baggage, formed in a compact circle on an open naked plain, was surrounded by a line of camels, and the mules and horses were placed in the centre next to the beds of the party. Guards and sentinels patrolled under an officer of the watch; and at the solicitation of the Ras el Káfilah, who was exceedingly anxious to avoid the inconvenient consequences of a blood feud, a musket was discharged every hour at the relief of sentries, in order to intimate to the evil-minded that all within the breast-work were not asleep.

Notwithstanding the presence, in the immediate neighbourhood, for several days previously, of a large band of Eesah, the hot night passed without any alarm. The non-arrival, until long after day-break, of the camels lost at Goongoonteh, added to the length of the next march, obliging the abandonment of the intention entertained, to speed beyond the pale of this site of assassination, the party halted on the 10th. Allooli stands two hundred and twenty-eight feet above the sea, and, although intensely hot, and its waters saline, it proved a paradise when compared with every preceding station. Here animal life was once more abundant. A horde of pastoral savages, who from time to time appeared on the adjacent heights, were made acquainted with the effect of rifle bullets, by the slaughter from the tent door of sundry gazels that visited the swamp; and the venison afforded a most seasonable accession to the empty larder, which was further replenished from the trees overhead, whose fan-like leaves gave shelter to a beautiful variety of the wood pigeon.

Shortly after midnight the march was resumed by the moon's light over a succession of small barren terraces, confined by conical and rounded hills. In the lone valley of Henráddee Dowár, which opens into the wide level plains of Gurgúddee, there stood by the way-side a vast pile of loose stones, half concealed among the tall jaundice-looking flowers of the senna plant.

Gurgúddee, eight miles in length, and stretching on either hand to the far horizon, is bounded by steep mountain ranges, whence an alluvial deposit, washed down by the rains, presented over the whole of the level plain a surface of cracked and hardened mud, like that of a recently-dried morass. From the southern side, where the clayey tract is thickly clothed with stunted tamarisk and *spartium*, a road strikes up the valley in a north-westerly direction to the Mudaïto town of Aussa, distant some three days' journey for a caravan. As the day dawned, the steeple necks of a troop of ostriches were perceived nodding in the landscape, as the gigantic birds kicked the dust behind their heavy heels; and a herd of graceful gazels were seen scouring towards a belt of stony hillocks which skirted the dry pebbly bed of a river, that expends its waters on the sun-dried plain.

The caravan halted early at Bedi Kurroof, after a march of sixteen miles, and the camp was formed on a stony eminence of basalt and lava, affording neither tree nor shade. A day of fierce heat succeeded. There was no forage for the cattle; the water was of the most brackish description; and the spot being of old infested by Bedouins, the party passed a restless and watchful night.

A legend of blood too was attached to this wild bivouac, as to most others on the road.

Some hours before dawn on the 12th, the káfilah was again loaded and in motion across a low belt of stony eminences which gradually descend to the Kori Wady, a long water-course, varying in width from two to four hundred yards. Threading the moist channel of this stream, where the foot often absolutely left an impression on the sand, and passing the watering pool of Leilé, the road ascended a deep valley to the halting ground at Suggadéra, in the country of the Danákil Débeni. The entire borders were flanked by dwarf palms and drooping tamarisk, bounded by low hills with cliffs of conglomerate and sandstone, which disclosed dykes of porphyry at an acute angle. Flocks of goats, diligently browsing on the fat pods which fall at this season from the acacia, were tended by ancient Bedouin crones in greasy leathern petticoats, who plaited mats of the split date-leaf; whilst groups of men, women, and children, lining the eminences at every turn, watched the progress of the stranger party.

A pastoral race, and subsisting chiefly upon the fermented juice of the palm, and upon the milk derived from numerous flocks of sheep and goats, or from a few breeding camels, the Débeni, a division of the Danákil, are during certain months of the year engaged in the transportation of salt from the deadly Bahr Assál to the Mudaito town of Aussa, where it is bartered for grain. Architecture affords no term applicable to a structure of any kind inferior to a hut or hovel, or it might with propriety be applied to the base jumble of rough stone and shavings of the date stalk, tenanted by these nomade savages, who are divided into clans, and have no fixed habitations. Nevertheless, there was something cheering in the aspect even of these frail edifices, the first human tenement which had greeted the eye since leaving the sea-coast, now ninety miles distant. Bare, desolate, and fiery, the entire intervening tract, although infested by the lurking robber and the midnight assassin, may be pronounced, in all its sultry parts, utterly unfitted for the location of man.

Water of rather an improved description was obtained at Suggadéra, under basaltic rock, stained green by carbonate of copper. But not a particle of forage was to be had; and the heat, reflected from a pebbly hill beneath which the tent was erected, brought the mercury in the thermometer to 118°, during the greater portion of the day; and the evil appearance of the place, surrounded by gloomy hills cast into the deepest shadow, led to the maintenance of a vigilant watch during the dark night.

Although disturbed at the early hour of 2 A.M., and denied further repose save on the bare ground, the loss of a camel, which was not recovered until late, so far retarded advance, that only four miles were achieved on the 13th. The road continued to wind with a gentle ascent along the bed of the Wady Kóri, the hills gradually diminishing in apparent height until they merged into the elevated plain of Murrah, which exhibited pebbles of pink quartz, with a few scanty tufts of sweet-scented grass, yellow and withered. Here, at the distance of two miles from a puddle of dirty rain-water, in defiance of the impotent Ras el Káfilah, the camel-drivers, who studiously avoid trees and the vicinity of a pool, resolved to halt, as being a place after their own hearts.

In the dry water-course just left, the chirruping of some solitary hermit bird, and the bursting bud of a certain dwarf shrub which clothed the borders, agreeably reminded the traveller of more favoured climes. But most completely was the illusion dispelled by the forbidding aspect of the sultry plain of Murrah. Monotonous fields strewn with black boulders, glaring in the sun, distressed the gaze wheresoever it was turned—each cindery mass seeming as though it had been showered down during a violent eruption of some neighbouring volcano; although, on nearer inspection, it proved to be the time-worn fragment of an extensive lava sheet. The bare stony plain was decorated with numerous cairns, marking deeds of treachery and blood; and at the distance of twenty miles rose a lofty range of hot table-land, behind which the Abyssinian river Háwash is lost in the great lake of Aussa.

The presence of the watering-place of a host of wild Bedouins, whose appearance was far from prepossessing, again induced the cautious elders to anticipate an attack; and the camp occupying a very unfavourable position for warlike operations, no little difficulty was experienced in making defensive dispositions. A gloomy black hill threw its impenetrable shadow immediately in front; and on the flank a pile of half-ruined sheepfolds, constructed of blocks of lava, afforded extensive concealment. The night, however, passed away without any alarm, and the intense heat of the day giving place to a somewhat cooler atmosphere, admitted of sleep by turns in some comfort—parties of the Danákil escort contriving, by chanting their wild war chorus, to keep their heavy eyes longer open than usual.

An hour after midnight the loading commenced, and the steep rocky hill having been surmounted by a path strewed with loose stones, a terrace of slow ascent, presenting the same dreary appearance of rocks and lava boulders, continued during the residue of the moon's reign. At break of day, however, the aspect of the country began rapidly to mprove. Gaining the higher and more salubrious level of Gulámo, the bare sterile land, strewed with black blocks of lava which tore the feet with their jagged edges, was fast giving place to sandy plains covered with dry yellow grass—a most welcome prospect for the exhausted cattle. Heretofore, saving in the wadys, no tree had been seen except small stunted leafless acacias, few and far between, and scarcely deserving of the name. Several small ravines were now choked with continuous groves, and a mountain stream termed Chekaïto, which rises in the country of the Eesah, and in the rainy season disembogues into the lake of Aussa, was thickly clothed on both sides with green belts of tamarisk, wild caper, and other wood, overhung with creeping parasites, and affording food and shelter to birds. The pensile nests of the long-tailed loxia depended from the boughs; and whilst the stems, covered with drift to the height of fifteen feet, gave evidence of a headlong course during the rains, water, even at this season, was here and there to be obtained.

After five times crossing the serpentine bed to the point of junction with the Sagulli, where ostriches cropped the grass around numerous deserted sheep-pens, the caravan finally halted at Duddee, no great distance from Ramudéle. For days together the pilgrimage had led across dreary and desolate wastes, and through sterile ravines where no verdure relieved the eye, no melody broke upon the ear, and so few living creatures were to be seen, that the unwonted appearance of a solitary butterfly which had become bewildered in the desert, was duly hailed as an event. The general character is that of a stern wilderness. parched by the intolerable heat of a vertical sun blazing in fierce refulgence over the naked landscape, of which the chief varieties consist in immense plains of dry cracked mud, or in barren rocks towering towards an unclouded and burning sky. The utter sterility of the soil is rather marked than alleviated by occasional sickly plants of most puny growth, and by the scanty verdure of the few valleys wherein water is to be found, generally in a state of stagnation. But at Duddee forage and fuel were abundant. The water obtained by digging in the channel of the stream was no longer brackish. The heat, although the thermometer rose to 110°, was infinitely more endurable than it had hitherto proved; and the insatiable thirst by which all had been incessantly tormented on the lower ground, had well nigh disappeared.

A march of twelve miles over a succession of grassy plains, untenanted by man or beast, but presenting the first cone of the termites that had been seen, brought the party on the 15th to the enclosed valley of Gobaad, a thousand and fifty-seven feet above the ocean. Volcanic ashes, jasper chalcedony, and quartz, strewed the sandy route, low volcanic ridges, of comparatively recent formation, intersecting the landscape from west to east. The encamping ground, among heaps of hard gravel, near which water was good and plentiful, had only two years previously formed the scene of the discomfiture of the Ras el Káfilah and his party, who had been plundered of all they possessed by two hundred and fifty mounted Eesah.

To Gobaad, from the shores of Lake Abhibbab, which is formed by the waters drained from Abyssinia, it is said to be one easy day's journey for the pedestrian.

Ascending by an extremely bad road the broken range to the southward, which commands a fine prospect over the valley of Gobaad, the káfilah reached Sankul on the 17th. It forms the focus of several small dales converging from the table-land, and shut in from all breeze by the surrounding steep black cliffs of basalt, passing into hornblende. A small cave near the encamping ground was occupied by a colony of industrious bees, and at the only well flocks of the diminutive black-faced Berbera sheep were drinking from a trough formed of an ox-hide stretched between four stakes, to which the water was transferred in gourds by greasy Bedouin shepherdesses. The evening was passed in perpetual wrangling with these matrons during the operation of filling up the water-skins; and sad presage was afforded of a coming day of drought, which the exhausted and sinking cattle of the caravan were hourly waxing less able to endure.

The next march led over the high table-land of Hood Ali, a stony level thickly studded with dry grass, and extending in one monotonous plateau far as the eye could reach. The fetid carrion-flower here presented its globular purple blossoms among the crevices, and a singular medicinal plant, termed Lab-lubba, was detected by the keen eye of a savage who had before evinced a latent taste for botanical studies. The usual encamping ground at Arabdéra was found to be pre-occupied by a nomade tribe of Bedouin goat-herds, who monopolised the scanty water. Descending the range, therefore, the bluff brow of which commanded an extensive prospect over the wide level valley of Dullool, the káfilah halted at Suggagédan. This arid spot in the strand-like waste was covered with masses of lava and with blocks of basalt from the adjacent hills. It was parched by a burning atmosphere, and afforded no water whatever—calamities which resulted in the abandonment of a horse and two of the mules that were no longer able to bear up against thirst and fatigue; whilst many others now dragged their weary limbs with difficulty, and seemed but too well disposed to follow the example.

Dullool is one thousand two hundred and twenty-eight feet above the level of the sea—a perfect flat, covered with alluvial deposit, and studded with extensive tracts of coarse dry grass in tufts, among which, as the almost interminable string of camels crossed the following morning, both ostriches and gazels were described. It is bounded by a bold mountain range, and the further extremity of the plain, towards the foot of Jebel Márie, is perfectly bare, stretching away to the westward, in one uninterrupted sheet of hard compact mud, which imparts the aspect of the Runn of Cutch. A herd of wild asses, precisely similar to those found on the Indian salt desert, materially enhanced the resemblance; and the sun, which had now attained considerable altitude, casting his rays in a full blaze over the naked plain, called up the dancing mirage that was alone wanting to complete the picture.

On this level expanse, which terminated in a *cul-de-sac*, shut in by high basaltic walls, inaccessible either to man or beast, the Adaïel affect to ride down the wild ass, upon lean mules forsooth, and to rip up the quarry with their creeses. There had been much vainglorious talking upon this head, but it ended, like every Danákil boast, in nothing. The hawk-eye of the Ogre detected an out-lying mare among the ravines at the foot of the range, and he dashed off the road with such lightning speed, that the animal narrowly escaped being hemmed into a corner; but once on the broad desert, and she tossed her arched neck, kicked up her wanton heels, and laughed at the absurd efforts of her impotent pursuers.

Loose stones again strewed the approach to the Márie range, which is of trap formation, of a slaty texture, stained with red iron oxide, and intersected by veins of iron clay. A breach in the hills, here about a thousand or twelve hundred feet high, formed a steep sloping ridge of lava rocks, containing quantities of carbonate of lime, disposed in rhomboids and hexagonal sheets. In this nook, surrounded by a thick jungle of acacia, were sundry basins filled with clear water, to which the solar rays had not penetrated. They afforded most refreshing draughts; and the skins having been replenished, the encampment was formed at Dawáylaka, a full mile beyond. Márie is not a word either of Arabic or Dankáli derivation; whence it seems not improbable that this bold range of hills may in days of yore have been named by some wanderer from the West.

A fine fresh morning succeeded to a very sultry night, passed upon the hard hot stones; and at break of day, the cattle having been taken to the pools, where, at so early an hour, they would drink but little, the skins were again replenished, and the caravan pursued its march to Oomergooloof, which can boast of no water at any season. Of two roads, the lower, but more level, was adopted, in consequence of the exhaustion of the beasts of burthen. It led across a dry desert plain of six miles, over which the delusion of mirage was complete.

Immediately opposite to Oomergooloof is a projecting spur from Jebel Oobnoo, a lofty range visible to the westward; and thus divides the plain into two valleys, whereof the southernmost is denominated Wady Arfa. The Márie range here towers overhead, steep and precipitous, to the height of about nine hundred feet; stupendous masses of rock which have been detached from the summit, and strew the entire base, corroborating the assertion of the Danákil that earthquakes are frequently felt in the vicinity. Nomade tribes with their families and flocks, having settled at the wells which exist at a distance, had compelled the gentler portion of the Libyan creation to resort to regions more blest with water; and not even a desert-loving gazel was espied during the march of twelve miles.

Several herds of cattle pertaining to the Issehirába Mudaïto, grazed in the neighbourhood; and these were said to derive their supply of water from pools formed by a cluster of hot sulphureous springs at the further extremity of the plain, which, with a loud noise, rise bubbling from the earth at a boiling temperature. Possessing marvellous medicinal properties, they are believed to be a panacea for every malady: but the tribe not being on terms with the Danákil, these thermal wells could not be visited, neither could water be obtained either for man or beast. A few Mudaïto females, with their children, strolled into the camp to sell sheep, and stare at the Feringees; but the Ras el Káfilah would scarcely permit them to be spoken to, and was in a nervous fidget until they departed. Avowing that these greasy dames had come for no other purpose than to spy out the nakedness of the land, and that the creeses of their liege lords would prove troublesome during the night, he strictly interdicted all wandering beyond camp limits, and insisted upon the discharge of several volleys of musquetry in addition to the cartridge expended at guard-mounting, and at every relief of sentries.

The sky having become gradually overcast towards evening, a deluge, equally to be desired and dreaded, was deemed close at hand, but the threatening aspect passed off with the hot blast of the Shimál, accompanied by a cloud of dust, and followed by a close oppressive night. Skirting the Márie range to a tract thickly strewed with rounded masses of lava and basalt, the detritus from the adjacent hills, the road now wound over a volcanic ridge which divides the valley of Dulool from that of Amádoo, running exactly parallel to it. In this latter the caravan halted on the 21st, about a mile from a large pool of rain-water, occupying a rocky nook formed by huge blocks of basalt. The stagnant green fluid was far more palatable than it looked, although troubled by a legion of horned cattle, asses, goats, and sheep, the property of the Galeyla Mudaïto, who were encamped in great force in the neighbourhood, and looked what they are said to be—most desperate villains.

From Amádoo, Aussa was represented to be only one day's journey for a swift mule, and two for a caravan of laden camels, the road branching off across Wady Arfa, and over the Jebel Oobnoo range, by which the extensive valley is bounded. At this point, moreover, had ceased the pretended influence of Mohammad ibn Mohammad, Sultán of Tajúra, the utter futility of propitiating whom had long been sufficiently apparent. Although in the eyes of the unitiated it was no difficult matter to invest this avaricious imbecile with supreme authority over a fiery desolate tract, in most parts obviously unfitted for human location, his own immediate retainers did not now conceal that Mirsa Dukhán, and the Gollo mountains near the Salt Lake, bound even his nominal jurisdiction. He is in fact Sultán of the sultry strand whereon his frail tenement is erected; for the few lawless warriors beyond, over whom he would assert supremacy, are universally thieves and murderers, who disdain all fixed abode, disclaim all mortal control, and acknowledge their own unbridled inclinations as their only master.

Aussa, formerly an important town, was, less than a century ago, the capital and principal seat of the united tribes of the Mudaïto, who extend thence to Ras Billool, and are represented to be countless as the hairs of a Danákil head. Regarded as the seat of wisdom and learning, and governed in the latter days of its strength by Yoosuf Ali ibn Ajdáhis, a brave and martial sultán, whose armoury boasted of many matchlocks, and of several small pieces of cannon, it long flourished in powerful independence—a bright spot of beauty in a waste of barrenness. But the sun of its prosperity at length set; and the predatory hostilities long exercised towards the various united tribes of Adaïel, leading to a general invasion on the part of the Ado-himéra, the prince was slain, the stronghold of the "red house" sacked, and its garrison put to the sword: nor in these degenerate days is this once important place more than an extensive encampment, whereat is held a perpetual fair, frequented by all the tribes of Danákil, Ecsah, Somauli, and Mudaïto.

The site of Aussa, a wide-stretching valley, described to be from eight to ten days' journey across for a caravan, is hemmed in by lofty mountains, and fertilised in all its extended quarters by the Lakes Guraaïd, Abhibbab, Hilloo, and Dugód—the first situated a little to the eastward of the town, and the last by far the largest of the four. These vast stagnant basins in the plain receive the Abyssinian waters of the Háwash and its tributaries, in addition to the contributions of all the streams from Jebel Oobnoo and other collateral ranges—the abundance of fluid thus lost upon volcanic formation, so enriching the soil as to enable this district to produce wheat, juwarree, barley, Indian corn, pepper, and tobacco, in quantities sufficient for the supply of the entire coast.

The Háwash may be conjectured to have experienced interruption in its course to the Bahr Assál and Bay of Tajúra, at the same period that volcanic agency divided the waters of the great Salt Lake from Goobut el Kharáb. Miles around the wonted boundaries of each lagoon now become annually inundated during the spreading of the great freshes; and, as the floods, carried off by absorption and evaporation, again recede, the soil is covered with a fertilizing sediment—a fat alluvial deposit, which with little labour yields an ample return. Even the lazy and listless Danákil, who neither sows nor reaps elsewhere, is here induced to turn agriculturist; but not a single acre of ground in any direction is to be found under the plough from the sea even to the mountains of Abyssinia—a distance exceeding three hundred and fifty miles.

Pastoral as well as agricultural pursuits engage the population of Assua; but whilst the cultivating, portion of the inhabitants are permanent residents on the soil, the shepherds are annually driven away by the gad-fly, which attacks the flocks from the setting in of the rainy season until the termination of the fast of Ramzan, when the waters have again subsided, and the herdsman, descending from the mountains, returns to his occupation in the valley. An extensive commercial intercourse is moreover carried on with Tajúra. Salt from the Bahr Assál, blue calico, which is in high demand for the caps universally worn by the married Bedouin females, zinc, pewter, and brass or copper wire, used both for personal ornament, and for the decoration of weapons, are bartered for the produce of the luxuriant soil—some few caravans crossing the Háwash, and pursuing their journey along the western bank to Dowwé, on the frontier of the Wollo Galla, in order to purchase slaves; or striking into the main road at Amádoo, and so prosecuting their way to Shoa for a similar purpose.

Mudaïto tribes occupy the entire plain of Aussa, but they are now divided into five distinct nations. The Assa-himéra are under the rule of Humferi, a descendant of the ancient house of Ajdháhis, who preserves the empty title of Sultán, and resides at the decayed capital. Eastward are the Issé-hirába, governed by Das Ali, an independent chief, and the

Galeyla, under Daamer Ibrahim; south are the Dár, who own allegiance to Akil Digger Myárgi; and west are the ferocious Koorhá, under the sway of Yoosuf Aboo Bekr, who, also with the title of Akil, resides at Alta, and wages war indiscriminately on both Assa-himéra and Débenik-Wóema.

South-westward of the valley of Aussa are the independent Adaïel tribes Hurruk Bodaïto, over whom presides Gobuz Elincha, a powerful chieftain, who has espoused Léni, daughter of Birru Lubo, the Prince of Argobba, and through whose territories lies the high caravan route to Dowwé, with which the traffic is considerable. On the north, the Mudaïto are bounded by the distinct nation styled Hírto, under the rule of Yingool Ali—Mohammadans, deriving their origin from the Arab invaders of the seventh century, and speaking a language not very dissimilar from the Adaïel, who claim the same descent.

After a march of three miles on the 22nd, over a stony table-land thickly strewed with the never-ending basaltic boulders, the caravan entered the territory of the Danákil tribe Wóema, under the uncle and father of Mohammad Ali. A desolate hollow passed on the way, which appeared in the rainy season to form an extensive pond, was enlivened by four bee-hive-shaped wigwams, placed as usual on the site where large hot stones were most abundant, and tenanted by goatherds, whose numerous flocks were being driven forth to graze by the Bedouin females. Their supply of water is derived from a sequestered pool, occupying a deep narrow precipitous ravine, which abounds in the Hyrax, and boasts of a few trees not dissimilar from the Casuarina. Bearing the euphonous title of Korandúdda, this gully wound at the foot of the high terrace selected for the encampment—another right dreary plain, covered with volcanic pebbles, among which the dry yellow grass peeped out in scanty tufts.

No traveller through the bleak barren country of the Adaïel can fail to appreciate the simile of "the shadow of a rock in a weary land;" for a tree is indeed a rare phenomenon—and when a few leafless branches do greet the eye, they are studiously shunned, upon the same principle that induces the savage to eschew the immediate vicinity of water. A few straggling acacias occupied the valley of Fiáloo, half a mile to the southward, which is the usual encamping ground, and here were large herds of cattle, eccentrically marked and brindled, and glorying in superb horns raking gracefully from the brow. A fat ox was purchased without difficulty; together with a supply of fresh milk, which, if not improved by confinement in a greasy skin bag, proved, nevertheless, an extraordinary luxury.

One of the retainers of Mohammad Ali was now despatched to acquaint Ali Abi of the arrival of the káfilah. It had all along been promised that after entering the territories of the old Sheïkh, every danger was to cease, but the goal now gained, the country proved to be a perfect nest of hornets. The thieving propensities of the Galeyla Mudaïto having been lately exercised upon the Wóema, it had been resolved to inflict summary chastisement, and rag-a-muffins were collecting from all quarters, preparatory to a "*goom.*" From morning till night the camp and tent were unceasingly thronged with scowling knaves, amongst whom were several of the Eesah—their heads decorated with white ostrich plumes in token of having recently slain an antagonist in single combat, or more probably murdered some sleeping victim.

A dense cloud of dust rolling along from the north-eastward, closed the day. Revolving within its own circumference, and advancing on a spiral axis, it burst in full force in the very centre of the camp. The tent fell on the first outpouring of its wrath, and the consistency being so dense as to render it impossible to keep the eyes open, the party were fain to take refuge beneath tarpaulins, and stretched upon the ground, to listen with quick and difficult respiration, until the whirlwind had expended its violence among chairs, table, and bottles.

Many significant glances had been exchanged over portions of the baggage that had unavoidably been exposed; but a night of redoubled vigilance was cut short by a summons to relinquish sleep and bedding at two in the morning, and a march of sixteen miles over a vas, alluvial flat conducted past the Bedouin station of Ulwúlli to Barurúddat on the plain of Kelláli. The road led along the base of the low range of Jebel Eesah, through abundance of coarse grass concealing lava pieces and volcanic detritus, the prospect being bounded by distant blue mountains towering to the peak of Kúffal Ali. A *korhaan* rose at intervals, wild and noisy as his chattering kindred in the south, but few other signs of animated nature enlivened the long sultry march. In the grey of the morning, a solitary Bedouin horseman ambled past with some message to the savages at Amádoo, and from him was obtained the disagreeable intelligence, which subsequently proved too true, that not a drop of water existed over the whole of the wide plain within a day's journey, and that the station beyond was thronged with tribes, collected with their flocks and herds from all the country round, at this the only oasis.

After a hot dusty day the sky was again overcast, and sufficient rain fell to render every one wet and uncomfortable, without filling the pools, or checking the dire persecutions of a host of cattle ticks, which covered every part of the ground. Absence of water led to another midnight march, and the moon affording little light, the road was for some time lost, though eventually recovered by the sagacity of a female slave of Mohammad Ali's, when all the lords of the creation were at fault.

The road continued to skirt the low Eesah range for several miles* to the termination of the plain, which becomes gradually shut in by rounded hills enclosing a dell choked with low thorns, and tenanted by by the *galla-fiela*†, a strange species of antelope, having a long raking neck, which imparted the appearance of a lama in miniature. As the day broke, flocks and herds were observed advancing from every quarter towards a common focus, and on gaining the brow of the last hill overhanging the halting-ground, a confused lowing of beeves and bleating of sheep arose from the deep ravine below, whilst the mountain sides were streaked with numberless white lines of cattle and goats descending towards the water.

Arriving at the Wady Killulo, a most busy scene presented itself. Owing to the general want of water elsewhere throughout the country, vast numbers of flocks and herds had assembled from far and wide, and they were tended by picturesque members of all the principal tribes of Danákil composing the Debenik-Wóema, as well as from the Eesah, the Mudaïto, and their subordinate subdivisions. Dogs lay basking on the grassy bank beside their lounging masters; women, screaming to the utmost of their shrill voices, filled up their water skins with an ink-

* The reader who may not feel thoroughly satiated with miles and furlongs, as embodied in this narrative, is referred to the Appendix, where they will be found detailed in a tabular form.

† *i. e.* camel-goat.

black fluid stirred to the consistency of mire, and redolent of pollution thousands of sheep, oxen, and goats, assembled in dense masses in and around the dark, deep pools, were undergoing separation by their respective owners, before being driven to pasture; and with the long files that ascended and descended the mountain side in every direction, imparted the bustling appearance of a great cattle fair.

The temporary mat huts of all these nomade visitors who boasted of habitations were erected at a distance on the table-land to the south-westward of this important wady, which occupies a rugged rocky chasm opening upon the Kelláli plain, and, receiving the drainage of all the southern portion of the Oobnoo range, disembogues during the rainy season into the lake at Aussa. Even during this the hottest portion of the year, when the entire country elsewhere is dry, its rocky pools embedded in soft limestone, tainted with sulphuretted hydrogen, and abounding in rushes and crocodiles, afford an inexhaustible supply, without which the flocks and herds of the entire arid districts by which it is surrounded, could not exist.

To it the horses and mules of the Embassy were indebted for a new lease of life, short though it proved to many. Two of the former and eleven of the latter had already been left to the hyænas, in addition to the animal feloniously abstracted by the Hy Somauli, of the recovery of which Mohammad Ali affected to be sanguine.

The second knot in the string of the tedious journey had been unloosed by arrival at Killulloo, which is considered exactly half way from the sea-coast to the frontier of Abyssinia. But although the worst portion of the road was now behind, the Embassy was destined to waste many days of existence in this vile spot, amidst annoying debates and discussions, most trying to the patience, which threatened to terminate so unpleasantly as well nigh to result in the abandonment of the baggage, as affording the only prospect left of ever reaching the destination.

Throughout this period of irksome detention, the thermometer stood daily at 112°, and the temperature of the small tent, already sufficiently oppressive, was rendered doubly unbearable by the unceasing obtrusions of the wild, dirty, unmannerly rabble who filled the ravine.

On the last day of the month, after nearly a week's tedious detention in an insalubrious and soul-depressing spot, surrounded by black basaltic rocks, where little forage could be obtained, where water, although abundant, was extremely bad, and where the persecutions of prying savages, from whom there was no escape, were unceasing, the Embassy was again permitted to resume its march.

The road wound up the Killulloo Wady, and thence over a barren rise strewed with obsidian, and with stones, the common pest of the country, to Waramilli. An interesting sight was presented in the line of march of a tribe proceeding in quest of water to the northward—a long line of dromedaries, horned cattle, oxen, sheep, and goats, interspersed by women and children, scantily clad in leathern petticoats, and laden with the rude date matting of portable wigwams, or the still ruder implements of household gear. Whilst the females thus bore heavy burdens slung across their breasts or led the files of camels, upon which rocked the long, raking, ship-like ribs of the dismantled cabin, the lazy lords sauntered ungallantly along, encumbered with naught save the equipment of spear and buckler, the ferocious aspect of all giving ample presage of the intentions entertained towards any party less formidable than themselves.

Total absence of water on the route usually pursued had determined the Ras el Káfilah, after much discussion and deliberation, to adopt the lower and shorter road, which, in consequence of the frequent forays of the Galla, had been for some years closed to caravans. But notwithstanding that so much invaluable time had been lost at Killulloo under such provoking circumstances, and that the march finally made thence fell short of seven miles, he again persisted in halting, thus affording to Hajji Ali Mohammad and Wayess ibn Hagaïo an opportunity of rejoining with a party of troublesome Bedouins.

Waramilli is the usual encamping ground of a section of the Gibdosa Adaïel, but their place was fortunately empty. Completely environed by low hills, it proved insufferably hot; and no water was obtainable nearer than Wady Killulloo, now distant more than two miles from the bivouac.

Two windy nights, during which it blew a perfect hurricane, were passed in unabated vigilance, owing to the number of ruffians lurking about the broken ground, the waters whereof tumble in the rainy season into the rugged chasm of Killulloo. At an early hour on the 2nd of July, a voice went through the camp, summoning the slothful camel-drivers to bestir themselves.

A march of fifteen miles over a country more level than usual, though sufficiently rough and stony withal, led through the Doomi valley to Noga Koomi. An abutting prong of land, under which the road wound, was adorned with a cluster of bee-hive-shaped huts styled Koriddra, and at its base the *Balsamodendron Myrrha* grew abundantly, the aromatic branches furnishing every savage in the caravan with a new tooth-brush, to be carried in the scabbard of the creese. The encampment occupied a wide, dreary plain, bounded by the high mountain range of Jebel Feeóh; and although water was said to exist in the neighbourhood, it proved too distant to be accessible.

It rained heavily during the greater part of the night, and an early summons to raise found the party again drenched to the skin.

Another march of fifteen miles brought the caravan to Meinha-tolli, where some hollows had been filled by the recent heavy fall of rain; but large droves of horned cattle having soiled in them, the muddy water was so strongly tainted as to be barely drinkable under any disguise. The country throughout bears signs of violent volcanic eruption of later times, which has covered one portion with lava, and another with ashes and cinders. At the outset the road led over the usual basaltic ground, strewed with fragments of obsidian, but after crossing Arnoot, a deep ravine choked with refreshing green bushes, in which the exhausted beasts obtained a most welcome supply of muddy water, the stony valleys gave place to sandy plains, clothed with short yellow grass, and intersected by low ranges of hills.

Thunder and lightening, with severe squalls and heavy rain, again closed the day—and great confusion and discomfort was occasioned by a sudden whirl of wind, followed by the fall upon the party, of the saturated tent, from the wet folds of which escape was not easily affected. A dreary night succeeded. The watery moon shed but a dull and flitting light over the drenched camp; and the pacing officer of the watch, after an hour's exposure to the pitiless hurricane, calling up his relief, threw himself with aching bones upon the inundated bed.

The rainy season having now fairly set in, it was believed that the pools on the upper road would furnish a sufficient supply of water, and the course was accordingly shaped towards it. Emerging upon the extensive plain of Merihán, bounded to the westward by the lofty peaked range of Feéoh, the route skirted the Bundoor hills, thickly clothed with grass, and varying in height from six hundred to a thousand feet.

Wayess, the chief of the Wóema, formerly held his head quarters in this neighbourhood, at Hagaïo-dera-dubba; but the Eesah Somauli making frequent inroads, and at last sweeping off all the cattle of the tribe, it was abandoned. The hill ranges on both sides have sent lava streams almost to the middle of the plain, but generally it is covered with a fine light coloured soil, strewn with volcanic ashes and small fragments of obsidian—the grass, improved by the recent showers, having partially acquired a greenish tint. A singular detached hill composed of fresh-water limestone, contained a few impressions of small spiral shells, whilst the surrounding rocks exhibit the usual cellular basalt.

Wady Bundoora, clothed in the thicket of verdant bushes, had been selected as the halting ground, and its appearance promised a copious supply of water; but every pool proved dry, and the march was therefore continued to Madéra dubba—a second and similar ravine, which was confidently expected to afford the desired element. Disappointment was however again in store, and the rain not having extended thus far, the usual reservoirs were referred to in vain. Worse than all, information was here received that not a drop of water would be found at the next station; whilst, owing to the wear and tear of skins, added to the too confident anticipations indulged, barely a sufficient supply for even one day accompanied the káfilah.

It had been determined under these untoward circumstances, to move on at midnight; but after an insufferably hot day, rain again interfered. Unfortunately it did not fall in sufficient quantities to be of much utility; a few pints caught in tarpaulins, which, with all available utensils, were placed for the reception of the precious fluid, proving very inadequate to the wants of the thirsty party. At 3 A.M. the caravan advanced down the valley, with cool refreshing weather, and a fine moon shining brightly overhead. From the summit of a tumulus of black lava, marking the point where the undulations of the Bundoora hills trend towards the mountains of the Ittoo Galla, an extensive view was obtained with the dawning day, over a country bearing the most extraordinary volcanic character—huge craters on the one hand towering to the clouds, whilst on the other sank the wide valley of Kordité, through which lay the high road to the desolate plains of Errur.

A few pools of muddy rain water by the way side were eagerly drained by the sinking cattle, but a deep ravine, bordered with green trees and bushes, was explored to no purpose; and after crossing the fine open plain of Eyrolúf, abounding in gazels and swine, the road led round the base of a remarkable cone, styled Jebel Hélmund, which had long been in view. Isolated, and four hundred feet in height, with a crater opening to the north-eastward, which would seem at no very remote period to have discontinued its eruption, it is surrounded by a broad belt of lava, some three miles in diameter. This has formed towards the plain a black scarped wall, rising from fifteen to twenty feet, of which the wooded crevices teemed with quail, partridges, and guinea-fowl, and were said to be so many great dens of lions.

The sultry afternoon was already far advanced, when the weary eye was refreshed by a glimpse of the verdant plain of Sultélli, a perfectly level expanse, so ingeniously overgrown with pale green vegetation as to furnish an exact representation of a wide lake covered with floating duckweed, around which numerous camels were busily browsing on the rank herbage. During the greater part of the year, this plain presents one vast and delightful sheet of water; but the fairy form of the light-footed gazel was presently seen bounding over the delusive surface, and although clothed throughout with the most tantalising verdure, it yet proved perfectly dry. The camels were milch females, capable of subsisting for days, and even for weeks together, without drinking, whilst their milk serves to quench the thirst of their unwashed Bedouin attendants. Beedur, the chief of a section of the Débeni, who resides in this spot during the rainy season, had long since decamped with his clan to more distant pastures.

Every hollow in the rich black soil abounding with shells, was vainly explored; and after a seventeen mile march, the party, weary and thirsty, were fain to encamp on the opposite side, and giving up the search as fruitless, to rest satisfied with the nauseous contents of water-skins filled at the putrid pools of Meinha-tólli—a second, and if possible, a worse edition of the impurities brewed at the Salt Lake. Both amongst men and cattle the utmost distress prevailed. A suffocating blast blew incessantly; heat the most intense was reflected from the adjacent black rocks; and nearly all of the horses and mules were so completely exhausted that there appeared no prospect of dragging them other sixteen miles to the nearest reservoir.

But towards midnight the beneficent flood-gates of Heaven were providentially opened, and a violent storm bursting over the camp, in less than half an hour filled every ravine and hollow to overflowing, and afforded a plentiful and truly seasonable, although transient, supply. Tearing up their pickets from the saturated soil, the dying animals thrust into the turbid stream that rolled through the encampment their hot noses, which for two entire days and nights had been strangers to moisture, and filled their sunken flanks almost to bursting. Cackling troops of guinea-fowl flocked to the pools from the adjacent heights. Embankments were thrown up, and wells excavated; and European, Danákil, and camp-follower—Christian, Moslem, and Hindoo—all drenched to the skin, falling together upon their knees in the posture of thanksgiving, sucked down the first copious draught of palatable water that had been enjoyed since leaving Fiáloo.

Singular and interesting indeed is the wild scenery in the vicinity of the treacherous oasis of Sultélli. A field of extinct volcanic cones, vomited forth out of the entrails of the earth, and encircled each by a black belt of vitrified lava, environs it on three sides; and of these, Mount Abida, three thousand feet in height, whose yawning cup, enveloped in clouds, stretches some two and a half miles in diameter, would seem to be the parent. Beyond, the still loftier crater of Aiúlloo, the ancient landmark of the now decayed empire of Æthiopia, is visible in dim perspective; and looming hazily in the extreme distance, the great blue Abyssinian range towards which the steps of the toil-worn wayfarers were directed—now for the first time visible—arose in towering grandeur to the skies.

The well-timed deluge of rain which fell during the night had been so eagerly drunk by the thirsty desert, that when the morning dawned the only traces of the storm were presented in numberless channels left by the torrent, with here and there a muddy pool, around which the guinea-fowl were still rallying in clamorous troops. Every portion of the road having been saturated, and rendered far too heavy for the jaded camels, advance was of necessity delayed until noon; by which time they had become sufficiently dry to admit of the resumption of the journey. A bare alluvial plain, skirting the base of the Kóomi range, led to a few acacia trees of larger growth than had heretofore been seen. They occupied a hollow styled Ras Mittur, which is the point of union of the two roads from Killulloo, and hence the caravan struck off across a grassy plain, abounding in herds of graceful *mhorr*. The course followed the eastern border of the field of truncated cones; and in the fresh green hue of the bushes with which the cindery tract is studded, was afforded abundant proof of the fertility of its decomposed lava.

Passing a cluster of Bedouin huts, whose inmates were watching their grazing herds of milch camels, the road next threaded a narrow belt of verdant jungle. This suddenly opened into the wide plain of Moolu, studded in every direction with flocks of sheep and goats, assembled from all parts of the country round; and in a deep hollow in the very centre lay the attraction—the oasis of Yoor Eraïn Mároo, a noble sheet of water surrounded by a belt of hillocks, and measuring during even this season of drought, a full mile in diameter. It was indeed a sight most refreshing to the eye. Troops of water-fowl of various plumage sailed over its glassy surface. Birds chattered amongst the autumnal branches of the numerous trees, whose tall stems, half immersed, rose thickly in the centre, and the cool waters of the basin afforded to the weary and travel-stained pilgrims the first unlimited supply of pure liquid that had been revelled in since bidding adieu to the shores of India.

In so sultry a land, where, throughout the desert and forbidding plain, Nature has dispensed the first necessary of life with a most niggard hand, those pools which have not a constant supply of running water soon become adulterated by various decompositions of organic and inorganic matter. Wacke cannot resist any long exposure, and hence fluid in contact with it imbibes oxide of iron and muriate of soda; whilst flocks and herds, corrupting the element in a still more offensive manner, not only impart a fetid taste and smell, but stir up the deposited mud, which emits a volume of sulphuretted hydrogen. Yoor Eraïn Mároo was free from all these impurities, and its reservoir affords a never-failing supply; but the surrounding country is said during the wet season to be extremely unhealthy, violent storms and incessant rain compelling the shepherds to abandon the plains and wadies, and retreat with their flocks to the neighbouring mountains.

From Waramilli to Moolu, the country is chiefly occupied by the sub-tribes of Débeni, under the chiefs Beedur and Boo Bekr Sumbhool, the latter of whom usually resides at Hamoosa, and the former at Doomi or Sultélli. Although not always to be trusted, these wanderers appear under a friendly garb; but the lion-hearted guides and escort, with Izhák at their head, had, from the first moment of leaving Killulloo, been doing their utmost to impress upon the minds of their audience the extreme danger to be apprehended on this portion of the road, from the various wild hordes now adjacent, whom they painted as perfect dare-devils.

Neither fuel nor water could be discovered at the ground selected for the bivouac, but a small supply of the latter requisite was obtained on the way, from a muddy brook trickling over the charred surface of the soil, and filling the gaping cracks and crevices on its progress towards the lower ground.

Betimes in the morning the march was resumed across an alluvial plain, which a few days later in the season would probably have presented a swamp impassable to camels; but no difficulties were now experienced, and the caravan passed merrily on towards a conspicuous barn-shaped hill, which had been visible for many miles. At its base, among sundry other cairns, stood a mound of loose stones encircled by a thorn fence, and almost concealed under the forest of withered boughs that decked every part. Beneath this grotto reposed the sainted bone Othmán, a celebrated Tukhaïel sheikh of days long gone.

Picturesque clumps of magnificent camel-thorns of ancient growth here studded the face of the landscape, and, covered with golden blossoms, perfumed the entire atmosphere. The myrrh tree flourished on the hill-side, and the "*garsee*" was first found under a load of fruit resembling the "*leechee.*" The bright crimson pulp possesses an agreeable acidity, and the kernel that it envelopes pleases the Danákil in a mess of sheep's-tail fat. No wood had hitherto been seen sufficiently dense to invite the elephant; but in this covert the giant evidently existed; and the oryx, appropriately styled "*Aboo el kuroon*," "the father of horns," ranged in considerable numbers; the half-devoured carcase of one which had been slain the preceding night, attesting the presence also of the "king of beasts."

Meanwhile the caravan had reached Bardúdda, where a large pool of dirty rain-water extended strong inducements to encamp, and again left to a violent altercation between the authorities.

The outlines of the highlands of Abyssinia, which had been first indistinctly visible from Sultélli, now stood out in bold relief; and to the southward the view was bounded by the lofty hills of the Afrubba, Farsa, and Azbóti Galla, where coffee grows wild in abundance. An intermediate extensive prospect is obtained over the thickly-wooded Moolu plain, stretching some thirty miles in the direction of Errur. This latter is the residence of the old sheikh Hajji Ali Mohammad, and the head quarters of the Débeni, who take hereditary share in the waters of the valley with their brethren the Wóema. It forms, moreover, a place of resort for every wandering vagabond in the surrounding country who possess a sheep, a goat, or an ox, or has the ability and the inclination to assert his privilege of erecting a temporary cabin; and thus the recurrence of each season of drought, compelling the abandonment of less favoured pastures, pours in its migratory swarm to swell the more permanent muster upon the sultry plains of Errur, and to create the strife inseparable from a gathering of these lawless hordes.

The route on the 9th led across the flat of Halikdiggi Kabír, a continuation of the Moolu plain, extending from the Azbóti and Ittoo ranges to the mountains Aiúlloo and Abida. Twelve miles in breadth, it presents one monotonous alluvial level, treeless, but thickly covered with grass, interspersed with dwarf shrubs, and enlivened by herds of the elegant *mhorr*, amongst which the secretary bird occasionally strutted in native dignity. Baézas and zebras, too, were descried on the hills which bound the flat; and a luckless leopard being detected in the act of stealing across the expanse, the savage group pursued like demons from every quarter, and having presently hooted and hunted the terrified animal into a bush, transfixed his carcass incontinently with thirty spears.

The whole landscape was alive during this animated scene, which scarcely occupied a minute; and in due process of time the panting warriors rejoined the caravan, their necks, spears, and shields adorned with strips of the victim's tail, whilst he who by dint of superior wind and fleetness had drawn the first blood, was by his comrades publicly invested with the spotted spoils that he had won. The appearance of the party on their return, accompanied by a stray horseman who had fortuitously joined in the chase, gave birth in the bosom of the Ras to the apprehension that the Ittoo Galla were descending upon the caravan. The ranks were accordingly closed, and the Europeans again took post on the flank to be assailed, until a nearer approach revealed in the savage band the features of friends.

A descent of thirty feet over a narrow tongue of land, led into the valley of Halik-diggi Zughir, styled by the Adaïel the Great Háwash—its breadth being about two and a half miles, and the bed a perfect level, covered with fine grass, on which grazed a troop of wild asses. Bounded throughout the serpentine progress by parallel banks of corresponding height and appearance, the hollow would seem to extend from the mountains of the Ittoo Galla north to the Aiúlloo volcano. It wears the aspect of having been once the channel of a considerable stream—that of the Háwash perhaps, which river may not improbably have been diverted into its present course at the period when the extensive volcanic tract around Mount Abida was in a state of activity, and when subterranean influence must have caused extraordinary revolutions in the entire aspect of the country.

Mules, horses, and camels, in considerable numbers, were abandoned before the termination of this tedious and sultry march—fatigue, want of water, and lack of forage, having reduced all to such positive skeletons that they walked with difficulty. Ascending three successive terraces, each of fifty feet elevation, the road finally wound into the confined and waterless valley of Háo, famous for the number of parties that have at various times been surprised and cut up by the neighbouring Galla.

From the summit of the height was obtained an exhilarating prospect over the dark lone valley of the long-looked-for Háwash. The course of the shining river was marked by a dense belt of trees and verdure, which stretches towards the base of the great mountain range, whereof the cloud-capped cone that frowns over the capital of Shoa forms the most conspicuous feature. Although still far distant, the ultimate destination of the Embassy seemed almost to have been gained; and none had an idea of the length of time that must elapse ere his foot should press the soil of Ankóber. A day of intense heat was as usual followed by a heavy fall of rain, which, owing to the unaccommodating arrangement again peremptorily exacted, of striking the tent at sunset, thoroughly drenched the whole party; but before finally drawing the mat over his sleepy head, the Ras el Káfilah mounted a cone which stood in the centre of the compact circle, and proclaimed in a loud voice to all, a night of light sleep and watchfulness.

Numerous were the apprehensions now in agitation relative to the state of the formidable river in advance, whose shallow stream so easily forded during the season of drought, was not unreasonably conjectured to be swollen by the recent rains. Second of the rivers of Abyssinia, and rising in the very heart of Æthiopia, at an elevation of eight thousand feet above the sea, which it never reaches, the Háwash is fed at long intervals by niggard tributaries from the high bulwarks of Shoa and Efát, and flows like a great artery through the arid and inhospitable plains of the Adaïel, green and wooded throughout its long course, until finally absorbed in the lagoons at Aussa; and the canopy of fleecy clouds, which, as the day dawned, hung thick and heavy over the lofty blue peaks beyond, gave sad presage of the deluge that was pouring between the verdant banks from the higher regions of its source.

Passing along the face of the murderous hill, which is of wacke formation, the road descended by several sloping terraces, to the level valley through which the river winds. At first thinly wooded, the soil was covered with tall rank grass, which, in consequence of the perpetual incursions of the Galla, grew in all its native luxuriance, uncropped, whether by flock or herd. But as the path wound on, gum bearing acacias and other forest trees increased both in size and number —the jungle and undergrowth, teeming with guinea-fowl, which rose clamorously at every step, waxed thicker and thicker—groves of waving tamarisk, ringing to the voice of the bell-bird, flanked every open glade, whereon lay traces of recent inundation; and the noble trees which towered above them from the banks of the Háwash, gave evidence in their shattered branches of the presence of the most ponderous of terrestrial mammalia.

Vegetation having here assumed a luxuriance known to none of the joyless and unproductive regions hitherto traversed, it is with some difficulty that the pilgrim, anxious to behold the rare phenomenon of a running stream, forces his way through the dense thickets, which, until the foot touches the very brink of the precipitous bank, so completely screen the silent river from view, that its very existence might almost be questioned. But after a persevering struggle, further progress was at length arrested by a deep volume of turbid water, covered with drift-wood, which rolled at the rate of some three miles an hour, between steep clayey walls twenty-five feet in height, bounding a mere break through the mud and sand. The breadth of the channel fell short of sixty yards, and the flood was not yet at its maximum; but its depth and violence, added to the broad belt of tamarisk and acacia, interlaced by large creepers and parasites, which hems in both sides, promised to offer much difficulty and delay in the coming passage. Pensive willows that drooped mournfully over the troubled current were festooned with recent drift, hanging many feet above the level of the abrupt banks; and this appearance, no less than the rubbish scattered over the borders, fully proved the assertion of the natives that the water had recently been out, to the overflowing of the adjacent flat country for many miles.

The Háwash, here upwards of two thousand two hundred feet above the ocean, forms in this direction the nominal boundary of the dominions of the King of Shoa.

The rest of the day was devoted by the camel-drivers for the preparation of rafts for the transport of the baggage.

At sun-down the caravan was closely packed within a stout thorn fence, serving as a partial protection against the wild beasts and plunderers with which the dense thicket is infested—its endless depths being so entangled and interwoven that no eye could penetrate the gloom. The moonless night was passed in extreme discomfort, owing to a deluge of rain which commenced early, and fell incessantly for many hours. Deafening thunder pealed in startling claps overhead, and broad sheets of fire lighting up the entire face of the landscape at short intervals, for a moment only disclosed the savage loneliness of the wild spot, which was the next instant shrouded in pitchy darkness.

With the dawning day, preparations were commenced for crossing the river on ten frail rafts which had already been launched—transverse layers of drift-wood rudely lashed together, being rendered sufficiently buoyant by the addition of numerous inflated hides and water-skins, to support two camel loads. The sharp creeses of the Danákil had removed many of the overlopping boughs, interlaced with creepers, which impeded transit towards the point selected for the passage, and in the course of a few hours every portion of the baggage had been deposited at the water's edge.

Casting off his garments, Mohammad Ali, always the foremost in cases of difficulty and danger, now seized the end of a rope betwixt his teeth, and, plunging into the river, swam with it to the opposite bank, where it was belayed, upon the principle of the flying bridge, to an overhanging willow—a guy which connected it with each raft serving to counteract the violence of the stream, which, in spite of the heavy rain, had fallen upwards of a foot during the night. Raiment was now discarded by every Danákil, and the work commenced in right earnest, but difficult and laborious indeed was the task before them.

Bidding adieu with light hearts to the muddy Háwash, the party resumed its march so soon as the camel furniture had become sufficiently dry; and skirting the Dubbélli lake, from the waters of which Behemoth blew a parting salute, passed the Bedouin hamlet of Mulku-kúyu in the Dófah district, to a fourth pond bearing the euphonous title of Ailabello. Prettily situated in a secluded green hollow, and presenting about the same circumference as its neighbour, below which it is considerably depressed, this pool resembles a circular walled cistern, and is obviously the basin of an old crater. Its waters, alkaline, bitter, and strongly sulphureous even to the smell, receive constant accessions from a hot mineral well at the brink, and possessing the singular detergent property of bleaching the filthiest cloth, many of the Danákil were for once to be seen in flaunting white togas. Thence the road lay over a grassy plain, covered with volcanic sand and ashes, and shut in by cones of trivial altitude, forming another field of extinct craters, many very perfect, and each environed by its individual zone—whilst the circumjacent country, embracing a diameter of eight miles, exhibited through the superincumbent soil, tracts of jet black lava.

Apprehensions being entertained of the non-existence of pools at the station suited for the encampment, still some miles in advance, a detour was made from the main road to Lé Ado, "the White Water," a very extensive lake, at which the skins were replenished and the thirst of the animals slaked. A belt of high acacia jungle embosomed this noble expanse, which exceeded two miles in diameter, the glassy surface in parts verdant with sheets of the lotus in full flower, and literally covered with aquatic fowls. Geese, mallard, whistling teal, herons, and flamingoes, with a new species of the *parra*, were screaming in all directions as they winged their flight from the point invaded, where a party of Bedouin shepherdesses deposed to having seen a troop of elephants bathing not half an hour before—the numerous prints of their colossal feet remaining in testimony, sunk fresh and heavy on the moist sands.

Prior to crossing the Háwash, the only sheep observed were the *Ovis aries laticaudata*, or Hejáz lamb, with sable head and neck, thick fatty tail, and fleece composed of hair instead of wool. This species had now, to the westward of the river, given place to the larger Abyssinian breed, with huge pendulous appendages of truly preposterous size, encumbered with fat, and vibrating to the animal's progress. Parti-coloured goats, armed with long wrinkled horns, still diversified the flocks, which were uniformly attended by small dogs with fox-like heads, spotted yellow and white, and evidently high in favour with their dark mistresses.

Numerous wigwams peeped through the extensive forest of aged camel-thorns, which borders on Lé Ado, and eventually debouches upon a succession of barren plains covered with herds of antelope. Two rough stone enclosures by the wayside were surmounted by poles, from which dangled the heads of many lions, dilapidated by time, although still enveloped in the skin, and said to have been speared on this spot many years before by the Bedouins, who exalted these trophies in commemoration of the deed.

Immediately beyond this point lay the encamping ground at Wady Azbóti, where numerous shallow pools had been filled by the recent rain, but where the ill-starred cattle were compelled to content themselves with water only, not a vestige of grass or green herb remaining in the vicinity. It had, in fact, now become a proverb, that these two essential materials to existence could never be found in the Adel wilderness in one and the same place. Vast flights of locusts, which had assisted to lay the country bare, still carrying desolation in their progress, were shaping their destructive course towards Abyssinia. They quite darkened the air at the moment that the caravan halted; and a host of voracious adjutants wheeled high above the dense cloud, at one moment bursting with meteor-like velocity through the serried phalanx, and at the next stalking over the field to fill their capacious maws with the victims which their long scythe-fashioned pinions had swept incontinently from the sky.

The groves around Azbóti afforded a welcome supply of bustard, partridges, and guinea-fowl, together with the mhorr, and pigmy antelope.

Rising tier above tier to the supremely soaring peak of Mamrat, "the Mother of Grace," with her doomed head ever canopied in clouds, the lofty mountains which fortify the royal dominions now shot like giant castles from the sandy plain, the most conspicuous features in the landscape. Volcanic impediments, such as had beset the heretofore weary path, had at length finally ceased, but the glowing sulphur hills of Sullála reared their fantastic spires on the verge of the monotonuso expanse; and high among the moie venerable witnesses to the history of the troubled lowlands, the position of Ankóber was discernible to the naked eye, with the steep Cháka range stretching beyond at still greater elevation. The luxuriant verdure which clothed the rugged sides of the nearer slopes, whilst it contrasted strongly with the aridity of the barren tracts at their base, indicated the presence of the autumnal rains; and hereof further evidence was afforded in the low grumbling of frequent thunder, echoing like distant artillery among the serrated summits, as the heavy black clouds at intervals drew their smoky mantle across the scene, and veiled the monsters from sight.

The departure of the silent spy was followed by the arrival of a most boisterous visiter from the highlands. The hazy sun, shorn of his bright beams, and looming a dull fiery globe in the dense mist, had no sooner disappeared in wrath, than a furious whirlwind tore along the desert plain, and during the gloomy twilight, the storm, which had been cradled amid the mighty mountains, descended in desolation, like an angry giant from his keep.

Black masses of cloud, rolling impetuously along the steep acclivities, settled at length over the face of the waste, for a time shrouding the very earth in its dark dank embrace, only to render more striking the contrast to the dazzling light which in another moment had succeeded. Brilliant corruscations blazed and scintillated in every quarter of the fervid heavens, hissing and spluttering through the heavy fog, or darting like fiery serpents along the surface of the ground—at one instant awfully revealing the towering peaks that frowned far in the distance—at the next flashing in a hot sulphury flame through the centre of the encampment.

Meanwhile the deep roll of thunder continued without a moment's intermission, the prolonged growl of each startling clap varying ever as it receeded in a fitful change of intonation; whilst the wailing of the blast, accompanied by the sharp rattle of hail, and the impetuous descent of torrents of rain, completed the horrors of a tempest which, now at its height, careered madly over the unbroken plain. The soil had soon swallowed the deluge to overflowing. Muddy rivulets poured through every qurrter of the flooded bivouac; and the heavy tarpaulins, which had afforded some temporary shelter, proving of little further avail, the shivering but still watchful party were exposed during many dismal hours that ensued to all the merciless fury of this unappeasable hurricane.

A cool cloudy morning succeeding to this dreary, boisterous and uncomfortable night, the caravan was in motion before sunrise across the uninteresting plain of Azbóti, in parts completely swamped, and covered towards its borders with one interminable sheet of the aloë and *lilium*, growing beneath spreading acacias upon a gravelly soil. Then commenced a belt of hammocks, formed by prominences abutting from the high land of Abyssinia—a succession of hill and dale, thickly wooded with a variety of timber, and still clothed with an undergrowth of the wild aloë, through which wary herds of Baéza threaded their way. The road soon entered the pebbly bed of a mountain stream, running easterly between precipitous basaltic cliffs towards the Háwash; but although such torrents of rain had fallen the preceding night, no water was discovered in the wooded wady of Kokaï, until reaching Dathára, nearly thirteen miles from the last encampment, where the party partook of the first crystal brook that had occurred during the entire weary journey from the sea-coast.

Three thousand feet above the ocean, with an invigorating breeze and a cloudy sky, the climate of this principal pass into Southern Abyssinia, was that of a fine summer's day in England, rather than of the middle of July between the tropics. Here for the first time during the pilgrimage, the tent was erected under the shade of a wide-spreading tamarind, which, among many other trees of noble growth, graced the sequestered spot. Above the surrounding foliage the long white roofs of many of the royal magazines were visible, perched high on the blue mountain side.

But from the summit of an adjacent basaltic knoll, which was ascended towards the close of day, there burst upon the delighted gaze a prospect more than ever alluring of the Abyssinian Alps. Hill rose above hill, clothed in the most luxuriant and vigorous vegetation. Mountain towered over mountain in a smiling chaos of disorder; and the soaring peaks of the most remote range threw their hoary heads, sparkling with a white mantle of hail, far into the cold azure sky. Villages and hamlets embosomed in dark groves of evergreens were grouped in Arcadian repose. Rich fields of every hue chequered the deep lone valley; and the sun, bidding a diurnal farewell to his much-loved plains of the east, shot a last stream of golden light, varied as the hues of the Iris, over the mingled beauties of wild woodland scenery, and the labours of the Christian husbandmen.

Gradually ascending through a hilly and well-wooded country, still a positive garden of the wild aloë, the road now led through a succession of deep glades, which opened in turn upon verdant mountain scenery; and at an early hour, after the first signs of cultivation had been afforded in the truly grateful sight of ploughs turning up the soil, the tents were erected on the open plain of Dinómali.

At this, the frontier station of Argóbba, are levied the royal import duties of ten per cent.; and a scene of noise, bustle, and confusion did not fail to ensue, such as is wont to attend the arrival of every caravan.

Loaded for the thirty-fifth and last time with the baggage of the British Embassy, the caravan, escorted by the detachment of Ayto Kátama, with flutes playing and muskets echoing, and the heads of the warriors decorated with white plumes, in earnest of their bold exploits during the late expedition, advanced on the afternoon of the 16th of July, to Fárri, the frontier town of the kingdom of Efát. Clusters of conical-roofed houses, covering the sloping sides of twin hills which form a gorge wherein the royal dues are deposited, here presented the first permanent habitations that had greeted the eye since leaving the sea-coast; rude and ungainly, but right welcome signs of transition from depopulated wastes to the abodes of man.

As well from the steepness of the rugged mountains of Abyssinia, which towered overhead, as from the pinching climate of their wintry summits, the camel becomes useless as a beast of burden; and none being ever taken beyond the frontier, many of the Wulásma's retinue now gazed at the ungainly quadrupeds for the first time.

Having thus happily shaken the Adel dust from off the feet, and taken affectionate leave of the greasy Danakil, it is not a little pleasant to bid adieu also to their scorching plains of unblessed sterility. Every change in the soil and climate of Africa is in extremes, and barrenness and unbounded fertility border on each other with a suddenness whereof the denizens of temperate climes can form no conception. As if by the touch of the magician's wand, the scene now passes in an instant from parched and arid waste to the green and lovely highlands of Abyssinia, presenting one sheet of rich and thriving cultivation. Each fertile knoll is crowned with its peaceful hamlet—each rural vale traversed by its crystal brook, and teeming with herds and flocks. The cool mountain zephyr is redolent of eglantine and jasmine, and the soft green turf, spangled with clover, daisies, and buttercups, yields at every step the aromatic fragrance of the mint and thyme.

The baggage having at length been consigned to the shoulders of six hundred grumbling Moslem porters, assembled by the royal fiat from the adjacent villages, and who, now on the road, formed a line which extended upwards of a mile, the Embassy, on the morning of the 17th, commenced the ascent of the Abyssinian Alps.

It was a cool and lovely morning, and a fresh invigorating breeze played over the mountain side, on which, though less than ten degrees removed from the equator, flourished the vegetation of northern climes. The rough and stony road wound on by a steep ascent over hill and dale—now skirting the extreme verge of a precipitous cliff—now dipping into the basin of some verdant hollow, whence, after traversing the pebbly course of a murmuring brook, it suddenly emerged into a succession of shady lanes, bounded by flowering hedge-rows.

At various turns of the road the prospect was rugged, wild and beautiful. Aigibbi, the first Christian village of Efát, was soon revealed on the summit of a height, where, within an enclosure of thorns, rest the remains of a traveller, who not long before had closed his eyes on the threshold of the kingdom, a victim to the pestilential sky of the lowlands. Three principal ranges were next crossed in succession, severally intersected by rivulets which are all tributary to the Hawash, although the waters are for the most part absorbed before they reach that stream. Lastly, the view opened upon the wooded site of Ankóber, occupying a central position in a horse-shoe crescent of mountains, still high above, which enclosed a magnificent amphitheatre of ten miles in diameter. This is clothed throughout with a splendidly varied and vigorous vegetation, and choked by minor abutments, converging towards its gorge on the confines of the Adel plains.

D'HERICOURT'S SECOND VOYAGE INTO ABYSSINIA FROM TAJURREH TO SHOA, SEPTEMBER AND OCTOBER, 1842.

(marked X on Map, page 39.)

The party started on 15th September, from Ambabo, about 10 miles to the west of Tajurreh. They passed the first night at Douloulle, 3 leagues to the south of Ambabo, where there are three wells of brackish water. The party consisted of 60 men and 150 camels. The traveller says,—"I indicated in the relation of my first voyage the different stages of the road traversed by the caravans, from the Indian Ocean to the kingdom of Shoa. I will not give again the detailed nomenclature which the interests of geography obliged me to give in my first journey, and which would only be a tedious repetition. Nobody will be astonished besides at my experiencing in retracing the circumstances of my first journey into the country of Adel, a sort of aftertaste of the discomfort which I experienced on the road. I think I may say, without being accused of exaggerating the difficulties of my works, in order to enhance their merits, that there are few journeys more fatiguing to mind and body, more perilous, and at the same time more monotonous, than that across the deserts of the Adels. Major Harris, one of the most experienced African travellers, has preserved a similar impression, and has described it in vivid but accurate colours. When I met him afterwards in Shoa, he could not disguise his astonishment when I told him that I had attempted, alone and for the second time, an expedition so little attractive. In other countries, at least, the magnificence of nature compensates for perils and fatigues; nature itself makes the desert of the Adels the most frightful residence. The country of the Adels, which is traversed in going from the Indian Ocean to Shoa, and which, descending from north-east to south-west, has a length of 130 leagues, which take a month to traverse,—this country is hilly, tossed about by volcanic action in an incredible manner. In no part of the world are there so many extinct craters, so great a flow of lava. No fertilising water traverses the burning gorges of this country, ravaged by subterraneous fires, and burned up by a tropical sun. There is little majesty or beauty in these mountains, but only a uniform mediocrity: almost always gently sloping hills with extinct craters, from which thick beds of lava have flown."

On quitting Douloulle the road leads for some time along the shore of the gulf; then, leaving the sea, it enters a gorge between the almost vertical sides of the mountains. The pathway then narrows, and is very tortuous and rocky. After traversing a little plateau, a still more arid and difficult gorge is entered. Blocks of trachyte and basalt bruise at every instant the camel's feet. Soon the animals were obliged to pass in single file, and the ascents became so steep, that it was with difficulty they could surmount it. One broke its leg.

They arrived at the banks of the Salt Lake, Mel el Assal, which is a source of great riches to the Danahil, from its salt. The spectacle of this dead sea, slowly thickening and solidifying, is a frightful spectacle under a burning sky. The lake is 217 metres below the level of the sea.

They remained there two days, whilst the Bedouins loaded their camels with salt. The first station on leaving the valley was near the source of the brackish waters of Gongonta, at the entrance of a narrow gorge of distorted rocks. Here they found the tombs of the soldiers murdered in Harris's journey.

From Gongonta they went to Allouli. From this place to Gaubade they traversed a frightful country, where they were often in want of water. At Gaubade this necessary never fails. They remained there two days to rest the camels.

On the 9th October they arrived at Omar Goulouf, 11 leagues from Gaubade. They had had no rain since quitting Ambabo.

On the 12th October they resumed their journey, and arrived without accident at Kilalou, about 15 leagues from Omar Goulouf. Here they were met by Bedouins, who told them there was no more water on the road, as far as Marc-le-Petit, 24 leagues distant. The heat continued intense. They remained a day at Kilalou to rest the camels, and let them drink their fill at the little stream, the first they had met since leaving the sea, and the waters of which were at their height, caused, the author thinks, not by local rains, but by the overflow of Lake Aoussa. They replenished their water at Kilalou, and left at six o'clock in the evening, marching at night only, on account of the heat.

On the morning of the 4th day they reached Maro-le Petit. Here a sight, animated but saddening, presented itself. The reservoir of Maro-le-Petit is an immense basin, a half league in circumference, formed by the junction of all the rain fall of the neighbouring valleys. Many hundred Adels, who had come from all parts for 8 leagues round, were watering their flocks. There were more than 50,000 head of cattle; camels, oxen, sheep, and goats. The thirsty animals rushed into the lake, and disputed with the women the muddy, green, stagnant water.

The next day they reached Maro-le-Grand. Here there is a natural basin, which is filled in the rainy season with water, which lasts until the next rainy season.

They encamped next day at Bordouda. Here they were only three days distant from the great River Aouache (Habesh) which forms the boundary of the territory of the Adels. They soon entered on a much more fertile country, consisting of a rich plain, carpeted with a virgin vegetation. The mountains of Choa here became visible, and traces of elephants were found.

The Aouache is a rapid river, its bed being 50 to 60 metres broad. It often overflows its banks, which are covered with a rich vegetation and magnificent trees, and abound with lions, leopards, panthers, elephants, hippopotami, zebras, antelopes, deer, gazelles, &c. They took several days to cross it.

Soon after crossing the stream they met with little lakes, named Léado, abounding in crocodiles and hippopotami. Four days after leaving the Aouache they reached the frontier of Shoa, at Dénemali, and proceeded to Angolola, where the King was residing.

The following was the temperature near the River Aouache (Hawash), on the 25th October, 1842 :—

8 a.m.	71 degrees.	Noon..	80 degrees.
9 „	73 „	1 p.m.	82 „
10 „	79 „	2 „	82 „
11 „	79 „	3 „	84 „

And on the Salt Lake, 23 miles south-west of Tajurra, on the 13th September, 1842 :—

9 a.m.	95 degrees.	Noon	101 degrees.
10 a.m.	98 „	1 p.m.	104 „
11 a.m.	99 „		

MESSRS. ISENBERG AND KRAPF'S ROUTE FROM TAJURREH TO ANKOBAR, IN APRIL AND MAY, 1839. (Marked X on Map, page 39).

On the 26th of April they left Tajurreh, where they had been detained 20 days, partly by their guide's illness, and partly by exorbitant demands for the hire of their camels. They at last agreed to pay 17 dollars for each camel, 25 dollars, the price of a female slave, having been demanded; and fifty dollars to Mohammed 'Alí, their guide, who had modestly asked 300. To his friend and assistant, the Arab 'Alí, they gave 15 dollars. The distance was calculated at 140 hours (about 330 miles). They had four camels for their baggage, and a mule for their own use. The whole journey, which they hoped to perform in 14 or 15 days, took up 36, and was very fatiguing. As the Danákil are migratory, there are few villages between Tajurreh and Shwá, or rather I'fát; for by that name only is the King of Shwá's territory known at Tajurreh; but there are above 50 resting-places, where the caravans usually halt Their first journey carried them to Anbábo, on the shore of the bay (Ghubbat-el Kharáb), about 4 miles W.S.W. from Tajurreh. The sea here runs up a good way inland, and forms a natural boundary between the Danákil and the Somáyil or Somális.

On the 27th they passed through Dullul and Suktá to Sagallo, about 9 miles from Anbábo. These are merely watering-places in a stony, sandy, uninhabited tract, overgrown with dwarf acacias, abounding in birds, but little infested by beasts of prey.

Sunday, 28*th April*.—Having been detained by the want of a strayed camel, they could not proceed till the afternoon: their road lay through a very sandy and stony tract, overgrown with dwarf acacias and frequented by a diminutive sort of gazelle, called in Arabic Bení Isráyil, and by hares, the only kind of game found there. This country is by no means deficient in water: at Tajurreh there is a walled cistern: on their road to this place there are spots where the traveller has but to dig a hole in the ground and he finds water. Its quality of course depends on the nature of the soil: and it has a reddish colour and unpleasantly bitter taste, which is made still worse by a certain herb which is put into their ill-tanned skins. The Danákil of this tract have many peculiarities: they are of the same race as the Shohos, and differ from them but little either in their language or features: but they are more civilised in their demeanour, and perhaps more intelligent. When they salute each other or converse together, the person addressed usually repects every sentence spoken to him, or at least the last word, which is generally abbreviated, and sometimes the last syllable only is repeated, or attention is shown by a hem in answer to every sentence. They are besotted Mahommedans, and in general very ignorant. Even the women while grinding, usually chant the words of their creed, "Lá Iláh, illá-lláh," &c., or verses of a religious cast. Their mills are much used like those used on board of Arab vessels. The women do not live much more apart from the men than in Abyssinia, nor, as it is said, is their conduct much more correct.

Monday, 29th.—They left Sagallo at midnight, and for half an hour travelled along the sea coast in a direction due W., then turning to the N.W. began to ascend an eminence, and passing the defile called Ankyeféro, after a further ascent, reached the station of Der Kelle, and at length a table-land called Wardelihán, which afforded an extensive view over the bay and the Dankalí country inland. They then travelled westward till they arrived at a spot where a few low acacias were growing at 7h. 45m. P.M., and alighted there to pass the night. On estimating their rate of travelling during the 7¾ hours, including their halts, they calculated the whole distance at a 4 hours' journey. During their ascent, they breathed a cool air, but on reaching the table-land, although the sun had only been up a short time, the heat grew very oppressive, having been increased by a S.E. wind. The plain was covered with volcanic stones.

Tuesday, 30th.—This morning they started at 3 o'clock, and descended in a south-western and southern direction through a narrow ravine, called Raïzán, which it was very laborious for the camels to pass. This brought them to the western end of the bay of Tajurreh, which, after forcing its way in a narrow channel through the mountains, here terminates in a second bay. From Wárdelihán to this point, they had travelled for the distance of about 1½ hour; which makes the distance from the point where they left the sea-shore to Wárdelihán 5 hours; so that making allowance for the windings in the road, the distance in a direct line from the spot where they left the shore to the end of the bay, is probably about 3 hours. From the end of the pass they ascended again and came to another table-land, where ashes, lava and calcined stones still more evidently show the volcanic character of the country. After proceeding for another hour in a western direction, they came in sight of the salt lake Asal, in a valley in front of them; and at 8 o'clock encamped at the caravanseraï of Daferri, on the declivity of a hill. In consequence of the excessive heat and want of water, their caravan started at 3 o'clock in the afternoon, and began to move in a south-westerly direction round the lake; but owing to the ruggedness of the ground, they were obliged to follow a very zigzag and irregular course. After crossing the valley of Marmoriso, where the caravans sometimes halt, they came to an eminence called Muyà, whence they descended a deep declivity, and reached the valley bearing the same name, at 7 o'clock.

Wednesday, 1st May.—From Muyà they set off at half-past 1 A.M., and first reached a rather elevated plain, named Halaksitán (Halaksheítáan?). On account of the ruggedness of the ground, full of chasms and gulfs, the vestiges of volcanic agency, they sought to get round the lake Asal towards the S., in a semi-circle. To effect this they had to round some mountains S. of the lake, and arrived at a resting-place at its southern extremity; but as there was no water, the caravan thought it better not to stop. They next descended to the lake, the shores of which are covered with a thick salt crust, having the appearance of ice. To this place caravans resort for salt to carry it to Habesh, of which trade the Danákil make a monopoly, claiming the right to take salt from hence as their exclusive privilege. The lake is nearly oval, its length from N. to S. about 2 hours, and its greatest breadth from E. to W. perhaps 1 hour. It is about 2 hours distant in a direct line from the western end of the bay at Tajurrah. After leaving the lake the caravan entered a valley towards the W., which ran between moderately high mountains, first westward, then S.W., and at 10 o'clock they alighted at a halting-place called Gwagate, where there is water.

Thursday, 2*nd*.—On this morning they did not set off till sunrise, at ½ past 5. Their road ran first W., then S. and S.W., through the valley of Kallu, which by its abundance of water and verdure, strongly brought to their recollection the valley of Samhar, but its mountains are higher and more thickly wooded. Towards ½ past 8 they arrived at their halting-place Alluli, after having travelled over a distance of 2 hours. This evening 'Alí, the Arab, informed them that the principal Danákil tribes between Tajurreh and Shwá are these: Debenik Wéma, Mudaïtu, 'Adálí, Bukharto and Dinsarra; to the last of which the Vezír, and to the last but one, the Sultán of Tajurrah belongs. The Debenik Wéma and Mudaïtu are the most numerous, and the latter is perhaps the most powerful of these tribes (Kabáyil). They extend from Musawwa' to A'usá, which is their head-quarters. There a Naïb (Deputy) of the Sultán resides; and between that place and the valley of Kallu, the road from Tajurreh to Ankôber passes through their country. They were then at peace with the rest of the Danákil, though ill-inclined towards the Debenik Wéma, who some years ago got the mastery over them by the assistance of 400 Bedwíns brought over from 'Aden. This did not put a stop to the trade in salt between Shwá and Tajurreh; for the traders from that place went right to the Lake Asal, got a stock of salt, returned to Tajurreh, and proceeded through the Somálí country on the confines of Harrar to Shwá.

Friday, 3*rd*.—At 3 o'clock A.M. they continued their course, first westward, then for a short time to the N.W., then again W. and S.W., through barren valleys, till they emerged into a vast plain called Anderhadideba, which separates two ridges of mountains. The soil for the first ½ hour's march over this plain appeared to be good, but produced nothing, the ground being broken up; afterwards, however, it was fertile, overgrown with shrubs, especially juniper. Towards 7 A.M. they come to an open spot called Gagade, where they rested for the night. Near them was a Mudaïtu's tent, which was very low, and hedged round with brambles, according to the custom of the country.

Saturday, 4*th*.—They left Gagade at half-past 1 A.M. A part of their caravan had already separated from them, in order to go to A'usá, and the remainder travelled very slowly on account of the weakness of their camels: their servants and one of the missionaries always went on foot; the former because the camels could no longer carry them, the latter because he had no mule. Their direction was westwards till they reached the resting-place of Karautu, where the road to A'usá branches off towards the W. From Karautu their course lay southwards, between mountains showing signs of volcanic action, with scarcely any vegetation. In the valleys only were grass and brushwood seen, and even here the ground was covered with ashes. They soon afterwards entered a long glen, where they saw many date-trees, which seem to be quite neglected. The Bedwins cut off their tops, in order to collect the juice which flows from them, and is said to be intoxicating. At 8 P.M. they arrived at Dalalaï, a Dankalí settlement, where they rested, not having made in the last 6½ hours more than 3 hours' way. The Danákil, especially their women, when travelling, employ themselves chiefly in plaiting mats and baskets of palm-leaves, to hold salt and grain. Their women, who seem to be industrious, are very slovenly in their dress, and frequently wear nothing but a strip of blue or party-coloured cloth bound round their loins, and reaching down to their knees; they sometimes, also, wear a fancifully wrought belt or girdle. Notwithstanding this, they are vain and fond of wearing bracelets, anklets, ear and nose-rings, coral necklaces, and other finery.

Sunday, 5th.—They started at 3 o'clock A.M., and moved in a S.W. direction through the vale of Kurri to Saggadere, and thence to Little Marha, which they reached at 7 A.M., having passed over a distance of 2½ hours in 3½. Their two servants were suffering from illness and fatigue because they could not ride. At 3h. 45m. they quitted Little Marha, and after moving along the valley almost westwards, ascended a very stony hill about 300 feet high, and then took a more southerly direction to a halting-place on the table-land, which they reached at 6h. 45m. P.M., having travelled over a distance of about 1¾ hour (5 miles) in 3 hours. In the evening a hot wind blew, and the ground beneath them, as they lay stretched upon it, glowed almost like an oven.

Monday, 6th.—They set off at ½-past 3 A.M., and marched, stumbling over the stony table-land, till they descended, and passed through a ravine near the halting-place of Galamo, where they found a few Bedwín huts. Their general route was southerly, and having crossed a hill they came into another valley, where they would have halted, had they not been encouraged to pursue their journey by a cloudy sky, which sheltered them from the sun. From Adaïto they passed over a grassy plain in which there were deer and antelopes. After this the road soon led again over a hill commanding an extensive prospect. Further on they entered the vale of Ramudeli, where they encamped. They reached it at ½-past 8 A.M., having made scarcely 3 hours' way in 5 hours' march.

Tuesday, 7th.—At ½-past 3 A.M. they left Ramudeli; at 5 o'clock passed by Abú Yúsuf, and at about ½-past 8 A.M. reached Góbád. At 3 P.M. they left that place, passed through Sarkal and by a spot where there is a spring, and arrived at Arabdera at about 8 in the evening. The distance between Ramudeli and Góbád may be about 3 hours, and it is nearly the same between Góbád and Arabdera.

Wednesday, 8th.—They left Arabdera at 3 A.M. It is situated on a vast elevated plain, almost entirely covered with volcanic stones. Just before sunrise, they came to a low but extensive plain, where they saw some wild asses grazing, which took to their heels on their approach. At 10 A.M. they reached their resting-place, Daunileka, where their camel-drivers dressed a wild ass which they had killed. In these 7 hours they only made about 4 hours' way. There were this day some idle reports of a projected attack by a hostile tribe called Galeila, formerly repressed by the more powerful Wéma.

Thursday, 9th.—They started at sunrise ¼ before 6, and after a short march on the plain westwards, ascended a pretty high eminence called

Mari, southwards, and at ½-past 10 A.M. reached their encamping-place on the table-land. The air grew more and more cool and refreshing as they ascended, but they felt rather fatigued when they reached the plain.

At 3h. 20m. P.M. they set out from their halting-place Mount Mari, and descended a low terrace, then marched on a wide undulating but elevated plain, over loose stones, without any vestige of a path, their guides being at a great distance in front, till after sunset they reached a declivity, the descent of which was not a little dangerous. Several times the camels could hardly move forward, terrified by the dismal abyss on the right, while the darkness of the night rendered the path under their feet almost undiscernible. At length they reached a projection at the foot of the mountain on its western side, and there halted on a stony spot, where the Bedwíns frequently confine their herds between walls of loose stones, to guard them from beasts of prey; but they had neither fuel to light a fire nor water to drink.

Friday, 10*th*.—They started at a ¼ before 5 A.M., descended the remaining declivity, and came to Ahuli, where there are four or five hot springs, probably sulphureous; there they took in water, and after a halt of about an hour, continued their course through a large plain, extending S.E. and N.W.: their route lay S.W. across the plain. They afterwards crossed a low eminence called Lukki, which is nearly flat on the top, and covered with a volcanic stones, as are most mountains passed in this journey. After ½-past 9 A.M. they came to a tree, beneath which they rested. From this spot there is an extensive prospect towards S.W. and W.; the whole tract is nearly level, with the exception of some low hills in the neighbourhood, and two or three higher ones to the W. at a distance, which are the mountains of Argobba and perhaps of Shwá. At 3 P.M. they left Lukki, and having ascended in a south-western direction to a grassy plain, marched till nearly 7 o'clock P.M., when they rested on a level spot in the plain of Killele.

Saturday, 11*th*.—They started at 1 in the morning, in order to make a long journey this day; but had not proceeded far, when they sunk into the mud, rain having fallen the day before, and softened the clayey soil. However, on turning as far as practicable to the W., they soon reached a dry spot, and thence took a more northern course till they met with a new difficulty, and lay down to await day-break. At ½-past 7 A.M. they set out again, first northwards, then N.W., and afterwards W.; passed two large herds of fine cattle, found water of which they drank and filled their leathern bags, and after another hour's march, arrived at Barudega, where they rested under a tree till ½-past 3 P.M., when they left Barudega, and pursuing a S.W. course through the plain, drew near a low ridge of mountains stretching S.E. and N.W. Towards 8 P.M. they came to a place which had trees, brushwood, and water, where they halted to pass the night, and for the first time saw a hyæna. At about ½-past 5 P.M. they set off for Gáyel, the village of Mohammed 'Alí's uncle, which was only ¼ hour's distance from their halting-place. They were there informed that 'Adáyil is the Arabic plural of 'Adalí, the name of one of the Dankalí tribes. It is that to which the Sultán of Tajurrah belongs; and it generally encamps in the neighbourhood of Shwá; but the greater part is dispersed over the adjoining countries. It was formerly the most powerful of all, and gave its whole name to the whole territory it then occupied. The tribes apparently most powerful at present, are the Mudaïtus and the Debenik Wéma. The former have their chief seat in A'usá, and are sometimes at war with the

Wéma and the rest of the Danákil. They seem to be more numerous and powerful than any other Dankalí tribe: they call themselves 'Affár, the word Dankalí being Arabic.

Tuesday, 14*th*.—They set off from Gáyel, and ascended an eminence about 2 hours distant in a S.W. direction, where they encamped near the watering-place of Alibakele, which supplies the herds of cattle belonging to the neighbouring Bedwins.

Thursday, 16*th*.—At 3 P.M. they left Alibakele, where they had waited for the arrival of their guide's father, and ascending westwards, were in ½ an hour overtaken by a shower. After stopping till it was nearly over, they made their way with difficulty through the mud, and towards 7 P.M. arrived at a spot called Adaïto, where they passed the night. In the evening they saw the mountains of Harrar to the S.W. covered with clouds; the city of Harrar being distant only 2½ days' journey. They were then near the Alla Gállás, who had expelled Sheïkh 'Alí Jábí from Erer, and extended their ravages as far as that district.

Friday, 17*th*.—They started at about 7 A.M., and their course lay over a stony but grassy plain, where they saw many herds and singing-birds. At ½-past 8 A.M. they reached Hasnadera, their halting-place, the residence of Sheïkh 'Alí, their guide Mahommed 'Alí's father. The Wéma Danákil have 100 Somélí archers, originally prisoners taken in their wars with different Somálí tribes: though considered as incorporated with their masters, they still preserve their native language, and never intermarry with the Danákil, by whom they are employed, because that people is said to believe shooting with bows and arrows to be unlawful.

Saturday, 18*th*.—This morning at ½-past 6 they set off from Little Hasnadera, and continuing their course S.W. over the plain which rose gradually, reached Great Hasnadera at ½-past 10 A.M., where they halted, but quitted it in the evening at 10 minutes before 6 P.M., and moving westwards over very stony ground, reached Mullu at ½-past 8. This is nothing but a vast plain covered with stones, with a little verdure in patches, a few acacias, and hovels made of boughs here and there. As this was their guide's principal residence, they rested there till *Tuesday*, 21*st*, when they proceeded under the direction of Sheïkh Alí, their guide's father, who thought it necessary to take an escort of his people, as he was apprehensive of the Mudaïtus, through whose southern, and the Gállás, through whose northern, boundaries they must pass.

They left Mullu about sunrise, and moving S.W. over a plain, arrived at ½-past 9 A.M. at a place called Wadardarer, about 2½ hours distant from Mullu. There they rested till about ½-past 3 P.M., when they proceeded S.W., till 8 P.M., because Sheïkh 'Alí said they could not reach the nearest water at Kudaïti that night. But on the following morning, *Wednesday*, 22*nd*, they arrived there half an hour after they started, and took in a supply for themselves and their beasts. Proceeding onwards they soon reached the village of Kudaïti and alighted after they had passed it. In front of them to the N.W. were the Baadu and Ayalu Mountains, the latter being of a considerable height. To the S.W. was the Jebel Ahmar, or the mountains of the Gállás. Between them and that mountain was an undulating and nearly level country said to extend from the banks of the Hawásh as far as Berberah.

Thursday, 23*rd*.—They started about ¼-past 5 A.M., and descended gradually in a south-western direction through the valley till ½-past 9, then rested under a large acacia near Metta, by the dry bed of a small brook, on the banks of which were many of those trees. The air was very hazy, and they saw many whirling columns of dust-like smoke from manufactories. They quitted their halting-place at ½-past 3 P.M., and marching almost due W. over the plain, passed by the village of Metta. After 7 P.M. reached that of Kummi, and about an hour later encamped near a deserted and ruined village of Bedwíns, where there was no water, of which they were in want.

Friday, 24*th*.—They set off at ½-past 5 A.M., and pursuing their course over the same plain W.S.W., saw at a little distance to the left Mount Afraba, which is joined westwards by the small mountain of Fresiz, and to the N.W. of it by mount Asaboti, all inhabited by I'sas; to their right the high land of Shwá and I'fát was visible. The plain on which they travelled terminated in a valley overgrown with grass and trees: there they passed a village inhabited by Debeniks, and gained an eminence. At about ½-past 10 P.M. they reached the watering-place Gamnisa, whence the whole district takes its name, and there they found a caravan which left Tajurrah on the day of their arrival there, and had only reached this place the evening before.

Saturday, 25*th*.—They started this morning at 6 A.M., and moved nearly due W. over a fine plain full of grass and trees. At 9 they halted near the village of Little Mullu, surrounded by very luxuriant and gigantic grass, overtopping the head of a man on horseback. On this day they had a little elephant hunting. The country hereabouts swarms with wild beasts; and the hide of a zebra was sold to them for five needles and a few grains of pepper.

On *Sunday*, 26*th*, at 3 P.M., they left Little Mullu, and crossing a large plain, first covered with high grass, and afterwards with scattered bushes, where they occasionally saw an elephant; travelled till ½-past 8 P.M., and rested for the night at Berdudè, still in the same extensive plain. While they were there, some chiefs of another tribe of Danákil, the Taki'l, came to beg for tobacco: this alarmed their guide, and made him hurry them on. The other Danákil tribes in that part of Abyssinia are the Debenis, west of the Wémas, who extend very far into the district of Gamnisa, on the borders of which are the Masháïkh and Gasoba, among whom the Taki'ls are dispersed.

Monday, 27*th*.—They left Berdude at ½-past 5 A.M., and crossed the other part of the plain called Galakdiggi; saw much game, especially large deer, also two ostriches; and a little before 9 A.M. arrived at a watering-place called Ganni, where they rested.

Tuesday, 28*th*.—Having started at 10m. past 2 in the morning and marched westward over a barren part of the plain, they arrived at Great Galakdiggi; and then they crossed an eminence soon after sunrise, from whence the mountains of Shwá were clearly visible. From it they descended into the valley of Galakdiggi, and having crossed one of the hills which skirt the eastern side of the valley of the Hawásh, they descended into the deep and wide valley of that river, which they had already seen from the eminence above, whence some parts of the course of the river could be traced. At the foot of the mountain the road lay through a forest of acacias, from which the people of the caravan collected a good deal of gum-arabic. They then encamped on a spot called Debhille, from the trees near which, on one side of the village, there hung nests of small birds, sometimes as many as forty or fifty on one tree.

Wednesday, 29*th*.—At ¼-past 4 A.M. they started, and pursuing a south-western course, reached the Hawásh at ½-past 6 A.M., by a road winding through a fine forest abounding with plants and animals. The fresh tracks of elephants were often observed; the braying of a zebra, and the snorting of hippopotami were also heard near the Hawásh, but neither were seen. As they crossed the river, crowds of baboons were noticed on some of the trees, an animal not before seen in Habesh. This was near Melkukuyu; and although it was in the dry season, the water was from 2 to 4 feet deep. The breadth of the channel is about 60 feet, and the height of its banks, as far as they could judge, averaged from 15 to 20 feet. Both sides are covered with beautiful forests, the breadth of which, however, is not hereabouts considerable. The river runs N. and N.E. They could not learn whereabouts its source is. The right bank is inhabited by the Allas, Ittus, and Mudaïtus, and the left by the Danákil, who border on Shwá eastwards. From hence where it has the Argobbas on one side and the Mudaïtus on the other, it flows as far as Aussá, and there in an extensive plain forms a large lake, the water of which is said to be putrid, emitting an offensive smell, and being disagreeable to the taste: on digging near the lake, however, good water is said to be found. The Ittus, on the eastern, not the western bank of the Hawásh, as is marked erroneously in the maps, lay to the S. of the Missionaries' route, and further S. the Abarras join them, having the Allas and other Galla tribes still further southward. At noon they went to see a small lake W. of the Hawásh, which is about ½ a-mile long and ¼ mile broad. In it there were at least 100 hippopotami sporting about. They fired a few shots at them, after each of which they suddenly plunged into the water, and on coming up again blew a stream out of their nostrils like whales, and snorted like horses. There are also many crocodiles in this lake; one 9 feet long which lay in the water near the bank, was struck by their people: "a naturalist," they observe, "would have abundant employment in that neighbourhood."

Thursday, 30th.—At ½-past 4 A.M. they set off from Melkukuyu, and marched over a hilly track near a small lake, the water of which has a disagreeable taste and a sulphurous smell, and is believed by the natives to be peculiarly detergent. The tract of country through which they had lately passed is called Dofar. After passing through several woods abounding with game and enlivened by the notes of a great variety of birds, they reached a larger lake named Le-adu, at about 9 A.M., in which the hippopotamus is said to abound, but not one was then visible. Thence proceeding westwards they alighted at about 11 A.M. at Assaboti, in a large sandy plain full of acacias. Setting out again at ½-past 3 P.M., they left the caravan behind, and encamped in the evening at Atkonsi, having by the way seen several baezas, a fine animal of the size of a cow, but shaped like a deer, with upright, not branching, horns: their flesh is exquisite. This tract is like a garden of cactuses.

Friday, 31st.—They started at ½-past 4 A.M., and after sunrise entered a fine valley called Kokaï, with lofty trees, excellent water, abundance of cattle, and a great variety of birds. After crossing several hills, the outskirts of the Abyssinian high lands, which extend from the S. far northwards, at about 8 A.M. they reached Dinomali, the frontier town of Shwá, where soon after their arrival they were visited by Soleïmán Músa, collector of the customs, and Abbagaz Mohammed, governor of the confines, who came to inspect their persons and baggage. They were accompanied by Debtera Tekla Tsion, secretary for the salt trade. During this visit, Hájí Adam, whom the Rev. Mr. Krapf had seen the year before at Mokhá as a royal messenger, came in and said that he was on his way to Mokhá, and had a letter and a female slave for them;

as they could not conscientiously accept the slave, she was sent back to Ankóber. The letter expressed the king's wish for medicine, a gun, masons, &c., and if possible, the company of the Rev. Mr. Isenberg, to whom the letter was addressed. An answer was immediately returned to the king, and quarters were assigned to the Missionaries in the village of Farri, till the king's pleasure as a their further progress should be known.

On receiving the king's permission to proceed, they passed through a few villages, crossed the rivers Hachani and Welka Yebdu, in their way to a village called Aliu Amba, in a district so named, situated on a steep rock, where they met the first Christian governor, Yaunatu, who received them gladly as Christians. On the following day they took other porters and asses from that place (travelling at the king's expense), and ascended the high mountains, on one of the summits of which, Ankóbar, the capital of the country, is situated. They crossed over a ridge of the mountains, which commands an extensive view; on one side towards the vast plain they had lately crossed beyond the Hawásh, and westwards over Shwá to a great distance. They went round one side of the summit on which Ankóbar is placed, and passed through a part of the town: the houses are constructed chiefly of wood, with conical thatched roofs, and are generally surrounded by a garden. The upper part of the town is hedged in with long stakes interwoven with boughs as palisades; and on the summit is the king's palace, built of stone and mortar, with a thatched roof. The situation of this town with its rich vegetation and cool vernal, or rather autumnal atmosphere, threw them almost into an ecstasy. The king had given orders that they should be soon presented to him, and as he was at Angollola, a day's journey distant from Ankóber, they could not remain there. In their way onwards, they passed over stony roads along the side of some mountains, and crossed an elevated valley through which a crystal rivulet hurries along, and is to turn a mill begun by a Greek builder, named Demetrius, by order of the king, but not then completed. They here seemed to breathe Alpine air, and drink Alpine water. They then ascended another high mountain, where they saw many Alpine plants, camomile and penny-royal, densely covering the ground. The summit of this mountain was almost all covered with barley-fields, nearly ripe for the harvest. They put up at a poor little village called Metakwi, in a straw hut or rather stable, in which large and small cattle lay mixed together with men, and where the smoke of the burning cane and cow-dung was so offensive, that nothing but the cold out of doors could reconcile them to remain within.

On the following morning, the 7th of June, they left Metakwi, and pursuing their road eastwards over undulating table-land, halted at about one o'clock, P.M., in an elevated valley near Islám Amba, where the king's tent, of an oblong form and of black coarse woollen, was already pitched. He was expected to come that way, and to pass the night there in his journey from Angollola to Ankóber, to keep the annual Tezkar (anniversary) of his father Wusaï Saged's death, which occurred in 1811. They had not been long encamped before they saw a train of horsemen coming down the mountain westwards, and in the midst of them, the king, over whose head a scarlet canopy was carried. He sent for them immediately after his arrival. They had prepared their presents, and with palpitating hearts entered his tent, where he sat on a small low sofa covered with silk, and received them kindly. Their names were already known to his attendants; and a messenger whom he had once sent with Kídán Maryam to meet them at Góndar, inquired after M. Blumhardt. They first delivered Col. Campbell's letter, which had been translated by M. Isenberg into Amharic while on board ship; he perused it attentively. They then delivered their presents, among which the beautiful copies of the Amharic New Testament and Psalms particularly struck him; he seemed to intimate, however, that he would have preferred Ethiopic to Amharic books. He observed that with regard to their principal object, which they had distinctly explained to him, he would have further conversation with them at a future time, as it was a matter which required great consideration: for the present he wished only to see and receive them, and to say that he was very glad that they had come to his country. He desired them in the mean while to go back to their tent, and rest there till the following day, when they might proceed to Angollola, where he would again send for them, immediately after his return from Ankóbar. They were much pleased with their reception. The king's servants were ordered to treat them as his guests and friends, and to provide them with everything necessary. One of his attendants was appointed to wait on them, who had strict orders to keep off all troublesome people who might interrupt or annoy them in any way.

Very early in the morning of the 11th of June the king set off with his suite for Ankóber, and the missionaries proceeded to Angollola, which they reached at 2 P.M. Not long afterwards the king returned, and immediately appointed a house for their residence, and sent them a cow. In all these interviews with him it appeared that his great object was to obtain their aid in advancing the knowledge of arts and sciences in his dominions; but he did not betray any signs of displeasure when they reminded him that their business was of a spiritual and not of a secular character.

MR. COFFIN'S (MR. SALT'S COMPANION) JOURNEY FROM AMPHILLA BAY TO CHELICUT, NEAR ANTALO, IN JANUARY, 1810 (marked P on Map, page 39).

THE country round Madir, in Amphila Bay, yields neither water nor pasturage for the support of cattle, which had to be brought from Arena, in Howakil Bay.

Mr. Coffin started from Amphila on the 10th January, with a young chief named Alli Manda, who, notwithstanding that incessant rains fell for two days, kept him constantly on the march for 12 hours a day. Their road lay in nearly a westerly direction, over barren and rugged hills, where they met with occasionally a small village or encampment of the natives, who, out of respect to Alli Manda, treated the party in general with civility, though the provisions which they supplied were scanty, and by no means of the best quality.

On the 13th, in the afternoon, after having travelled nearly 50 miles, they reached a station on the edge of an extensive salt plain, where they stopped to refresh themselves under the shade of some acacias near some wells of fresh water. At this place they were provided by the natives with a sort of sandals, made of the leaves of a dwarf species of palm, which are invariably used by travellers for crossing the salt. The plain above mentioned lies perfectly flat, in a north-east and south-west direction, and is said to be four days' journey in extent. The first half mile, from not being firmly crusted, was slippery and dangerous to pass, the feet sinking every step into the mud, as is usual in crossing a salt marsh. After this the surface became strongly crusted, hard, and crystallised, resembling in appearance a rough coarse sheet of ice, which has been covered with snow, thawed, and frozen again; branches of pure salt, resembling pieces of madrepore, occasionally rose above the surface, and two small hills stood in the centre of the plain.

This plain took Mr. Coffin and his party about five hours to cross, when they reached the country of the Assa Durwa, which the Ras humorously terms his barbarian territory. On this side of the plain a number of Abyssinians were seen engaged in cutting out the salt, which they accomplished by means of a small adze, and the form of the pieces is similar to that of the whetstone used by our mowers in England. In some places it continues tolerably pure so deep as three feet, but in general not lower than two. From this plain the whole of Abyssinia is supplied with salt.

On the 14th, the party proceeded over some steep and rugged passes in the mountain, until they arrived at the village of Dafo, situated in an extensive and verdant plain, inhabited by the Hurtoo, a tribe of Danâkil, which was conquered at an early period by the Abyssinians, and has ever since been subject to the governor of Tigré.

The country beyond this was exceedingly beautiful, and game of various species appeared to be very plentiful. On the 15th, at night, they reached the foot of the mountain Senafé. At this station tribute is levied by order of the Ras. A camel carrying two hundred pieces pays eleven, a mule, whose load consists of eighty only, pays nine, and a loaded ass six.

On the 16th, the party ascended Senafé, which is said to be full as high, though not so difficult to pass over as Taranta. At the summit a complete change of seasons was experienced, and, instead of continued rain and tempestuous weather, the sky became unclouded, and they found the inhabitants busily engaged in gathering in their corn. Here they stopped at a village to refresh themselves, and at three proceeded through a rich and fertile country, which at six o'clock brought them to a large town called Hammee, where they stayed for the night.

On the 17th, they continued their journey to Dirbé; and on the 18th, they reached Chelicut.

MR. HAMILTON'S JOURNEY FROM SUAKIM TO KASSALA IN MARCH, 1854 (marked AA on Map, page 39).

THE port of Suakim is formed by a circular creek, with only a narrow inlet to the east, rendering it secure for the vessels of small burthen which can find anchorage in it. A single glance at the form of the harbour shews that it is a coral formation, with a small circular island in the centre, on which are the governor's house and some stone houses and huts. On the mainland is another group of huts. The inhabitants of the two places together probably exceeding 12,000 souls. Suakim contains a few merchants of some wealth, and a number of traders, with very small capitals. The distances in this journey are measured by hours of 2½ miles, being the rate of a laden camel. The camel men are Hadēndoa Arabs, a tribe which extends from Suakim the neighbourhood of Kassala.

Two roads exist to Tacca, the province in which Kassala is situated; the one by the Langeb is the most direct, the other, more southerly, skirts the frontier of Abyssinia. The first was chosen.

They started late on the 16th March, and only made two hours' march through the long plains, which extend north and south as far as the eye can reach, and which are covered with a thorny vegetation, affording excellent pasture for camels and goats, numerous flocks of the latter of which were scattered over it. The thermometer at 8 o'clock a.m. was 82°, a temperature which they found during those months pretty constant at that hour.

An hour and a half after starting next morning they were abreast of the Hamut, over the lower spurs of which the road passes into a wâdy, or valley, through which they travelled for four hours. From the rocks springs a meagre, thorny vegetation. At the end of the wâdy two high conical summits appeared to the S.S.E., and they came upon an oval basin surrounded by granite hills, covered with bright green grass, through which a streamlet flowed.

This stream flows down the Wâdy Goob, which they next ascended. The sides of the Wâdy were formed by low masses of granite, from the crevices of which sprung a multitude of plants. They reached the first water at 3 p.m., but went on until they had completed 7½ hours, when they stopped at the head of the valley, in a beautiful spot from which a lofty peak, Gebel Mikailôt, lay due N.W., Gebel Waraweh N. 80 E., and Gebel Nafait W. 19 N.

The next day, proceeding up the valley, they found themselves in a series of steep defiles. The Atabayat seems in many places utterly impassable; and looking round on the irregular masses of rock they had climbed over, it seemed difficult to understand how the camels had effected it. A shorter but not less precipitous descent brought them to a small plain, bordered on all sides by hills, and watered by a small spring, near which a herd of goats was browsing.

Riding W.S.W., they ascended the long but not steep pass of Haddameib, and, reaching its summit, found themselves, after an insignificant descent through Wâdy Gabut, on the edge of an immense plateau, covered with bulbous plants and shrubs. After travelling nine hours, they reached a grassy plateau surrounded by low hills, beautifully dotted with a variety of trees. This plain is called Subab. They then reached a long gentle ascent over ground, generally bare of vegetation, and covered with fragments of blackened granite; in places dotted with tufts of wiry, sharp-pointed grass, and with Mimosa trees. They had now reached a lofty plateau, to which the Sawakly come to pasture their cattle in the rainy season. Here they found some magnificent trees, a few neat tents of Hadendoa Arabs, and a flock of sheep, tended by a boy, who refused to sell any.

The remainder of the day's journey was continued up a gently sloped plateau, formed of sand covered with dark fragments of granite, and without vegetation. The camp was pitched in the plain Saballat, six hours from the starting place.

March 19th.—Traversed an immense series of gradually ascending plateaux, whose soil was for the most part barren gravel, with blocks of granite. Towards evening, eight hours from Sanballat, entered a pass were a few trees of a brighter tint attract the eye, and where there is water. This is the commencement of the country of Langeb, which is thickly peopled with Hadendoah Arabs, of whom, however, they saw little, as they fled at their approach, having been often the victims of Egyptian soldiers. Passed through a beautiful succession of gorges, the hills rising fantastically, their bases being clothed with a net fringe of doum-palms and other trees. Five hours from the entrance of Langeb, reached a well (Bir Walo), at which an immense number of sheep and kine were being watered.

Leaving the well, they continued up the wooded pass, following the watercourse, a strong north-east wind blowing. Four and a half hours through a thick wood of doums brought them to the watershed, where they encamped by a small spring called the Tkit.

Next day they descended the Wâdy Tkit, and then came to a new succession of short passes, barren of vegetation excepting a few thorns. Four hours and a half from Tkit they entered a lofty but not extensive plateau, entered by a defile, like a gigantic Cyclopean wall. They made only eight hours that day, and eight hours of the following were occupied in crossing a succession of almost treeless, featureless tracts of sand. They then came to a black basaltic defile, down which they rode, its steep turns opening into a vast plain, which stretches full of verdure to the east. They alighted at the Wâdy Araft, near a well.

March 23.—They ascended for a short time, and in half an hour reached a formation of white basalt and porphyry. Crossing this hill, the road lay for the rest of this march over the open plains of Wâdy Magwar, which are intersected by shallow ravines of sand, filled with trees. Towards evening they saw the long, apparently lofty, range of hills called Gebel Kuarit, under which the Tokka road to Kassala passes. They slept in the plain Wâdy Weidi, nine and a half hours from Hambulib.

The journey of the next day, over Wâdy Ambureib to a line of doums with a well, Bir Gadomeyb, though only four and a half hours, was one of the most fatiguing they had yet endured. There was no shade to ward off the sun, and the wind, a strong north-easter, blew alternately cool and burning blasts of great violence. The thermometer at midday was 99° in the shade of the palms; at two o'clock it was 101°. They then crossed an immense sand plain, covered with boulder stones, stretching on the west and south to the horizon. They slept two miles from the wells.

Next morning they traversed for 16 hours an immense Savannah, on which grew only one solitary tree. The ground was covered with long parched grass, and full of crevices, showing the abundance of the waters, which in the rainy season, about two months from that time, deluge the country, and convert it into a green prairie, like a rice field. The sun, added to the hot parched air, was very painful. The heat was so intense that the bridle hand was burned in white, transparent blisters, as if by the contact of fire, though the thermometer did not exceed 102°. Next morning they were fortunate in a fresh cool breeze to continue their ride across the treeless Atmur or desert. In less than four hours they reached the wells of Bu Shimah. The heat was intense; the wind not only blew burning blasts, from which they had to protect the face with as thick a muffler as from a northern ice wind, but also raised high columns of sand, of strange fiery colour, which swept slowly whirling across the plain.

Mr. Hamilton says "I had all this time resisted the temptation to drink of the small water jar suspended at my saddle, knowing how little it contained, and that the more one drinks, the more imperious become the calls of thirst. Hardly able to speak, I rode up to a thorn bush, round which the packsaddles and loads of the camels were littered, with their owners sleeping in the midst. One of them raised himself at my approach, and calling out to him for water, almost the only word of the Hadendoa language I possess, he pointed to a further group of trees on a low terrace on the other side of a small plain. Here I found a great number of wells, with a large herds of kine gathered round them; I dropped from my dromedary without taking time to make him kneel down, and a man reached me the round skin with which he was drawing water for his cattle, I literally buried my face in the not clear, but grateful fluid." This place contains a long line of wells, many of them within 20 feet of each other, which serve to water herds of many thousands of cattle.

From here for thee day's journey they crossed a country of the same nature, everywhere green, covered with trees and fields of pasturage, and presenting many groups of wells like these. "An immense sheet of water, probably nourished from the mountains of Abyssinia subtends the whole country." Two hours beyond Bu Sheinah, where their camels had been watered, they encamped in a plain thickly overgrown with large aromatic plants of absinth. To the left was an Arab village.

The next morning, an hour and a half after starting, they reached a village of straw huts called Fillik; this is probably rather the name of the whole tract of country they were entering on, as it means "populous," an epithet applicable to this tract in opposition to that of the Atmur, which they had just crossed. They then passed three groups of wells surrounded by innumerable herds of kine and sheep. The country is well wooded.

In 7½ hours they arrived at a large village called Elmit Kenab, the most populous in the country, and one of its chief markets. Lions, ostriches, giraffes, and elephants abound. Water very scarce. The thermometer at two o'clock was 108°, and they were oppressed with thirst as they had found no water since leaving the last well.

The next day they halted at midday at the last group of wells they were to find on the road, a beautiful scene, the ground studded with short thick grass. Gazelles, guinea fowls, and doves abounded. Leaving the wells, they rode through fields of durrah, a cereal which grows to the height of from 12 to 15 feet.

Towards evening they emerged from the fertile country and came on to a large gravel plain, abounding with antelopes and guinea fowl. After having made 7 hours they encamped.

The next day they reached Kassalah, after more than five hours' ride under a burning sun.

Kassalah, although only a small town enclosed by mud walls, contains a castle with a very large powder magazine, and barracks capable of containing at least 4,000 soldiers. It is the Egyptian frontier post towards Abyssinia. Though so recently built, it is already a market of considerable importance, and its vicinity to the chief marts in Sainaar and Abyssinia must render it, at no distant period, one of the best trading stations in the Soudan. The great gum market of Gadarif is only four days' journey from here; thence 8 days are sufficient to reach Matamah, the principal mart for exchange betwixt Abyssinia and Sennar, and much resorted to by traders from Dar, Fungi, and Fazogli. From Matamah there are two roads to Gondar; that usually followed, is both steep and in bad condition, but the other, which is closed to all but native Christians, is described as being half a day shorter and much better. Mr. Kotzika with great difficulty obtained permission to return by it from an excursion into Abyssinia made this winter. It took him $4\frac{1}{2}$ days in all, his first stage being from Gondar to Jendi, 4 hours, thence to Chankar, where the best coffee is grown, 4 hours; thence to the limits of the province of Takussa, one day; and $2\frac{1}{2}$ more to Matamah.

The hottest months in the country are March and April. The first days after their arrival the heat was intense, the thermometer at two o'clock being generally 103°, and at night seldom falling below 94°. One evening a little rain fell, and after that a cool wind at midday tempered the heat, and the night seemed almost cold, but Mr. Hamilton was informed that such early showers, for the rains would not begin for another month, are often productive of disease, however grateful they might seem.

Near Kassalah runs the Gash or Mareb. At that season its bed was dry, but when the rainy season arrives, it rolls a large body of water in its wide bed to swell the Atbara. Excellent water is found everywhere at a short distance from the surface, both at Kassalah and in the plains below, from which it may be inferred that there runs under the whole soil a vast sheet of water.

Industry might make this place an earthly paradise, but, when the summer rains have ceased, and the trees put forth their bright foliage, and the ground is covered with verdure and flowers, fevers spring up with them, and few in the entire population of town and country escape their attacks. These fevers are sometimes at once fatal, and often undermine the constitution.

Camels are wonderfully cheap in this country, being worth from 5 to 16 dollars, the latter, the price of the best dromedary.

BRUCE'S JOURNEY FROM GONDAR TO THE GWANGWE, NEAR METEMMAH (marked Z 1 on Map, page 39).

SET out on 26th December, 1771, by the west side of Debra-Tzai, having that mountain, which is close to Gondar, on the right hand. From the top of that ascent he saw the plain and flat country below, black, and in its appearance one thick wood, which some authors have called the Shumeta or Nubian forest.

He then set out down the steep side of the mountain, the course being nearly N.N.W., through very strong and rugged ground, torn up by the torrents that fall on every side from above. This is called the descent of Moura; and though they and their beasts were in great health and spirits, they could not, with their utmost endeavours, advance much more than one mile an hour. Three Greeks, one of whom was nearly blind, a janissary, and a copt were his only companions, besides some common men to take charge of his beasts, who were to go no further than Tcherkin.

At a quarter past four they came to the river Toom Aredo, which, arising in the country of the Kemmont, falls into the river Mahaanah. They crossed the river to the miserable village of Door-Maccary, which is on the east side of it, and there took up their quarters, after a short but very fatiguing days journey. They saw a high mountainous ridge, with a very rugged top, stretching from north to south, and towering up in the middle of the forest about five miles distance; it is called Badjena.

On the 28th, a little after midday, they passed Toom Aredo, and went, first east, then turned north into the great road. They soon after passed a number of villages; those on the high mountain Badjena on the east, and those belonging to the church of Koscam, on the west. Continuing still north, inclining a little to the west, they came to a steep and rugged descent, at the foot of which runs the Mogetch, in a course straight north; this descent is called the And. At a quarter past two, they crossed the Mogetch, the direction N.W. It is here a large swift running stream, perfectly clear, and they halted some time to refresh themselves upon its banks. At half-past three they resumed their journey. A sharp and pyramidal mountain, called Gutch, stood alone in the midst of the trees, presenting its sharp top about six miles due north. A few minutes after they passed a small stream called Agam Ohha, or the Brook of Jessamine, which is frequent here.

A few minutes past four, they entered a thick wood, winding round a hill in a south-east direction, to get into the plains below, where they were surrounded by a great multitude of men armed with lances, shields, slings, and large clubs or sticks. They showered stones on them, but did not hurt them, and eventually retired.

At some villages called Gimbaar, they procured some goats and jars of bouza, but could get no bread made, as the people all fled. They left Gimbaar on the 29th, at 10 A.M., Bruce putting on his coat of mail and exchanging his mule for a horse, in case of being attacked. In a few minutes they passed three small clear streams, in a very fertile country; the soil was a black loamy earth, the grass already parched or entirely burnt up by the sun. Though this country is finely watered and must be very fertile, yet is thinly inhabited, and, as they were informed, very unwholesome. At three-quarters past ten, they came to the river Mahaanah, its course nearly N.W.; it was even at that dry season of the year a considerable stream.

They rested here half an hour, and then pursued their journey straight north. They passed a large and deep valley, called Werk Meidan, or the Country of Gold, though there is no gold in it. It is full of wood and bushes.

At twelve o'clock, the Mahaanah was about a quarter of a mile to the N.E., and the sharp-pointed mountains of Gutch three miles distant S.E. by E. They were informed by some Abyssinians who passed them, that at a certain pass called Dao-Dohha, they would be attacked next day by above a thousand men, Christians, Pagans, and Mahometans. At half-past four P.M. they encamped in the market-place of Waalia.

Waalia is a collection of villages, each placed upon the top of a hill, and inclosing, as in a circle, an extensive piece of ground about three miles over, where a very well frequented market is kept. It lies due N.W. from Gondar.

On the 30th, at half-past six A.M. they left Waalia. In an hour arrived at the Mai Lumi river, and coasted it for some minutes, as it ran N.E. parallel to their course. The trees were loaded with fruit, which they found a great refreshment. At fourteen minutes past seven, continuing north-west, they crossed the river Mai Lumi, which here runs west, and continuing north-west, at eight o'clock they came to the mouth of the formidable pass Dao-Dohha. Before entering it, they rested five minutes to put themselves in order. They "found our appetites failing through excessive heat."

The pass of Dao-Dohha is a very narrow defile, full of strata of rocks, like steps of stairs, but so high, that without leaping or being pulled up, no horse or mule can ascend. Moreover, the descent, though short, is very steep, and almost choked up by huge stones, which the torrents, after washing the earth from about them, had rolled down from the mountains above. Both sides of the defile are covered thick with wood and brushes, especially that detestable thorn, the Kantussa.

Having extricated themselves successfully from this pass, they arrived at ¾ past 8 at Werkleva, a Mahometan village. Above this is Armatchiko, a famous hermitage. Having rested a few minutes at Tabaret Wimze, a wretched village composed of miserable huts, on the banks of a small brook, at a quarter-past two they passed the Coy, a large river which falls into the Mahaanah. From Mai Lumi to this place the country was but indifferent in appearance; the soil indeed exceedingly good, but a wildness and look of desolation marred the whole of it. The grass was growing high, the country extensive, and almost without habitation, whilst the few huts that were to be seen seemed more than ordinarily miserable, and was hid in recesses or in the edge of valleys overgrown with wood.

On the 31st December they left their station at the head of a difficult pass called Coy Gulgulet, at the foot of which runs the Coy, one of the largest rivers they had seen, but they did not discover any fish in it.

At half-past eight they came to the banks of the Germa, which winds along the valley and falls into the Augrat. After having continued some time by the side of the Germa and crossed it, going N.W., they passed at ten o'clock the small river Idola, and half-an-hour after came to Deber, a house of Ayto Confu, on the top of a mountain by the side of a small river of that name. The country here is partly in wood and partly in plantations of Dora. It is very well watered and seems to produce beautiful crops, but it is not beautiful; the soil is red earth, and the bottoms of all the rivers soft and earthy, the water heavy and generally ill-tasted.

They left Deber on the 1st January, 1772. At half-past ten they passed a small village called Dembic, and about mid-day came to the large river Tchema, which falls into the larger river Dwang below, to the westward. About an hour after they came to the Mogetch, a river not so large as the Tchema, but which, like it, joins the Dwang. Here they saw the steep mountain Magwena, where there is a monastery. Magwena, except one mountain, is a bare even ridge of rocks which seemingly bear nothing. In the rainy season, it is said every species of verdure is here in the greatest luxuriancy. All the plantations of corn are infested by small green monkeys.

Between three and four in the afternoon they encamped at Eggir Dembic, and in the evening passed along the side of a small river running west, which falls into the Mogetch.

The next day they passed through several small villages; at half-past eight they came to the mountain of Tchertrin, and at twenty minutes past ten pitched their tent in the market-place of Tchertrin, which resembles a beautiful lawn, shaded with fine trees and watered by a small but limpid brook. Here they found plenty of game, elephants, rhinocerosses, and buffaloes. There are immense forests in the neighbourhood.

On the 15th January they left Tchertrin, and entered immediately thick woods, but proceeded very slowly, the road being bad and unknown, if it could be called a road, and the camels overloaded. About an hour afterwards they passed a small village of elephant hunters, and then due north through thick dark woods overgrown with long grass. At half-past ten they came to another village and then turned N.W., passing several villages of Mahometan elephant hunters. At three-quarters past twelve they came to the small river which runs W.N.W., and falls into the Germa. At ten minutes past one they set out again through the thickest and most impenetrable woods, and at half-past four encamped on a small stream near the village of Amba Daid.

On the 16th they reached the Germa, a large river which runs N.N.W., and falls into the Augrab; at twelve they passed the Terkwa, which falls into the Augrab and then the Jibbel, Myrat and Woodo, which is full of small fish.

On the 17th, started before 7 A.M., going N. and N.W., and reached Sancaho at half-past one; which consists of about 300 houses, inhabited by Baasa. It is half surrounded by a river which was nearly dry.

On leaving Sancrato next morning, they passed through thick cane woods and crossed the Bedowi several times. They encamped on the Tokoor, famous for the number of buffaloes on its banks, which are shaded by large trees.

On the 19th they passed the village of Gilma and twice crossed the river of the same name, running northward. They then joined the Dabda road and crossed the Quartucca, a small river running north. The country here became more open, the thick woods having small plains between them. They crossed a stream and stopped at a Baasa village called Kantis.

On the 20th they only proceeded 1½ miles, being fatigued, to Guanjooh, where they found woods interspersed with lawns, cotton-fields, and a variety of game, especially guinea-fowls.

On the 22nd they started at three-quarters past six, crossed the Gumbacca and Tokoor, and then at half-past eleven the Guangue, the largest river they had seen except the Nile and Tacazze.

After passing this river the road entered the country called Ras-el-Feel, which, Bruce says, is the hottest country in the world. The thermometer was at one place 61° at sunrise, 114° at 3 P.M., in the shade, and 82° at sunset. The country is very destitute of water, and covered with thick jungle.

KRAPF'S JOURNEY FROM MAGDALA BY LAKE HAIK, ANTALO, ATEGERAT, AND TOHONDA (AVOIDING THE TARANTA PASS) TO MASSOWAH, IN MARCH AND APRIL, 1842 (marked R 1, O 1, N and E on Map, page 39).

MARCH 26, 1842.—As no message from our host beyond the River Bashilo had arrived, I judged it best to go from Gembarghie to Tanta, and inquire of the Imam what I should do in my perplexing circumstance. We kept close to the territory of Daood-Berille, whose capital is Saint, which we could see from a distance pretty well. This man, who had it in his power to forward me to Gondar, has the reputation of being a robber; and besides he was not on good terms with Imam Liban, without whose recommendation and protection it was impossible to proceed to Daood-Berille.

On our return from Gembarghie we had a pretty view of the course of the River Bashilo to the mount Samāda, in the north-east of Godjam. The high mountains of Begemeder were also presented to our view; and Debra-Tabor, the capital of Ras Ali, was pointed out to me by one of my servants who had formerly been there. As I would not return to Imam Liban in too great haste, I pitched my tent near the stronghold of Magdala before mentioned, and sent a messenger to the Imam, to ask his advice in my critical situation. In the meantime, I inquired whether there was any other road to Gondar except by way of Daunt and Saint; but my inquiry was in vain. My messenger returned without having seen the Imam, his whole court being in confusion and preparing for war. My man was like to be plundered and deprived of his cloth and mule by the Imam's own people. Under such circumstances, and the way being obstructed on all sides, the best plan appeared to return to Shoa through the territory of Adara Bille, on whose friendship and kindness I thought I could rely.

Under these circumstances I compared the road from Ankobar to Tadjurra with that to Massowah, and was led to the following conclusions:—

1. Although the climate from Ankobar to Massowah is superior to that of the Danākil country; and although there is everywhere plenty of water, and a cool and healthy air on the Massowah road, yet the Tadjurra road is more preferred by the traveller.

2. It is true, that the difficulties arising from want of water and excessive heat in the Danākil country are very great; but you do not meet with the disturbances which almost continually happen on the road through northern Abyssinia, and which either delay or considerably endanger your route.

3. On the Tadjurra road you have only to agree with one guide and proprietor of camels, which will carry your baggage as far as Efat; while on the Massowah road you pass from the hands of one Chieftain into the hands of another, each of whom wants a present for the assistance which he gives you. Besides, as there is no road for camels, you are obliged to procure your own beasts of burden, which cannot carry the same quantity of baggage which a camel can.

4. These beasts of burden cannot stand a journey, which is almost three times farther from Ankobar, than that from Tadjurra. Your animals will die, and you will be exposed to many difficulties till you have procured others.

These and other reasons led me to the conclusion, that the Tadjurra road is, notwithstanding its inconveniences and difficulties, preferable to the Massowah road; and that therefore the road from Tadjurra must be kept open and secured.

March 27, 1842.—We returned this morning to Imam Liban, who appeared to have been in great sorrow and apprehensions regarding myself. When he saw me, he said, "You have done very well in returning to me, as you cannot go to Gondar under present circumstances. If you like you can take your refuge with my Governor Joossoof on the stronghold of Hoait,* which my enemies will not be able to conquer.

* This is a high hill on the junction of the river Bashilo with another, the name of which I have forgotten. The hill is situated in the north of Tanta.

In the course of a month you will be able to see whether you can again attempt your journey." I replied, that I could not take the part of any of the combatants, and that I would prefer taking any other route which he would recommend to me; or if not, that I would return to Shoa. He said, "Just as you like; but I cannot send you to Gondar, as all the roads will be closed for some time." It appeared that he wished to send me to Hoait, in order that my gunners might assist in the defence of the place; but I would never have consented to this, except under most perplexing circumstances. I learned afterward, that the stronghold had been attacked by Berroo Aligas and his brother Faris, who joined him at the time of my return to Adara Bille, and that many men had been killed on both sides.

I took leave of Imam Liban, and returned to Tartar-Amba, where Abba Gooalit the Governor of Adara Bille's territory in Worra Himano, received me well, and provided me with provisions, which had been very scanty for several days.

March 28, 1842.—Abba Gooalit, our host, treated us kindly. He is a Christian. In general, there are many Christians in Worra Himano.

We left Tartar Amba about sunrise, accompanied by a servant of Abba Gooalit. We took great care to avoid going towards the territory of Ensenne, the famous robber in the tribe Charso, which I have mentioned before. We kept our route in the territory of Worra Himano, which is bounded on the north by Wadela and Yechoo, on the east by Tehooladere, on the south by Berroo Loobo's and Adara Bille's countries, and on the west by Begemeder. The people knowing that I came from Shoa, frequently asked me how many ounces of gold I had received from the King of Shoa, it being the general opinion of the Abyssinians in the north, that there is much gold in Shoa; and that its king gives this metal to all strangers, who leave his country. In some instances this report is true, as the king has given gold to some strangers; but Shoa is not the country where gold is found. Occasionally some may be found; but the gold which comes to Shoa, is brought from Guragne and beyond, where it is found in the bed of rivers after rain. But no Shoan subject is allowed to possess gold, which is only in the hands of the king, who would severely punish any of his subjects who had any, except the king himself had given it.

About ten o'clock we passed Fala, where a celebrated market is held. It is situated on a hill, with steep and high banks in the east and west. In this direction a wall of about three or four feet in thickness has been built to close the road against an invading army. This difficult passage secures from the south the access to the interior of the possessions of Imam Liban. In the west of Fala is the mount Amŏragadel, which is a natural stronghold against the inroads of the Galla tribes in the south-west. In the east we saw a high hill, called Kemmer Dengai, which was produced by a former Imam, according to a tradition, which states that when the Imam was resting on a stone, he ordered his servant to lift it up; and that when the servant did so, the stone became a large hill.

About three o'clock we passed close to the market-place of Totola in Berroo Loobo's country. This is one of the most celebrated markets of Abyssinia. We saw immense flocks of people coming from all quarters, as the market was to be held the next day. Even the Boranna Gallas, of the western Wollo tribes, visit this market. Merchants come from Gondar, Tigre, and Shoa. Whatever Abyssinia produces, is sold in this market, particularly horses, skins, clothes, and slaves. The duties which Berroo Loobo levies on this market are said to be very little; but, notwithstanding, he receives weekly about 6,000 or 8,000 pieces of salt. It must be remarked that a dollar is changed for thirty pieces of salt in Loobo's country. The people here are as scrupulous as the Shoans in selecting a certain kind of dollar. The dollar must not only have seven points distinctly expressed above the star in the middle, and s. f. below; but it must also look very white, and must not appear dirty, as they believe that filth has been applied to the dollar for the purpose of covering the tin, of which it had been composed by impostors. I am sorry to say, that they are not so particular in having their faces cleaned, or their clothes washed, as they are in selecting this sort of dollar.

There are several other important market places in Berroo Loobo's country, and I have often heard that Berroo encourages trade, and in general has great order in his government. The Danākils like him much, and his people trade to Tadjurra. In this respect he must be superior to the King of Shoa, who did not allow his subjects till hitherto to go to the coast, probably from motives of superstition or narrow ideas, as if the entrance to his kingdom would become known to strangers, and his subjects having been acquainted with the Danākils, might run over to them when they are male-content with him.

About five o'clock P.M. we again reached the territory of Adara Bille, and intended to pass the night in the house of a Governor called Edris; but on arriving in his village we learnt that in consequence of a quarrel which arose between him and his subjects, he had been compelled yesterday to take flight. The whole village was still in confusion, a circumstance which was extremely unpleasant to us, as we had believed that as soon as we had returned to the territory of our great friend and kind host, Adara Bille, our difficulties and privations would be at end. The behaviour of the villagers was rude and daring, and every appeal to Adara Bille, to whom we represented their proceedings, was in vain. Our guns, however, frightened and prevented them from falling upon our baggage like a vulture on his defenceless prey. I found it necessary to put on a sentry; and as my people were very tired from the fatigues of the day, I watched in my turn.

From the village where we had pitched the tent, I had a majestic view over almost all the territories of the Wollo Gallas. Ranges of mountains run from south or south-east to north and north-west. Each range is separated from the other by a plain, a river, or a torrent. Each range is inhabited by another Wollo tribe, just as I have observed in the country of the Gallas in the south of Shoa. The river or torrent serves the inhabitants of the mountain to defend their territory against another tribe. The rivers run chiefly to the Bashilo, which has the same destination as the river Adabai in Shoa; viz., to collect the tributes of water of a few hundred miles around and to carry this tribute to the great lord Abai or Nile. I must confess, that the system of the mountains and rivers of Abyssinia always replenishes my mind with astonishment at the wisdom of Him who has created all things with the best order and organization.

March 29, 1842.—When the man who had accompanied me from Tartar Amba had left, we started from the village where we had been treated very rudely. As our animals were tired from the continual fatigues, we had great difficulties in giving them their loads. Several mules were sore and could not be mounted. I thought that if I should undertake this journey another time I would pack up all my baggage on horseback, but with a very light load. I would be mounted myself on horseback, and my servants also. A guide would shew me the road. I would take such a quantity of provisions that I should not be obliged to halt at places where there is any danger, and should I accidentally fall in with dangerous people, I would mount my horse and escape. This is the only way of traversing these hostile regions.

We arrived at Gatira, the capital of Adara Bille, at three o'clock. I immediately sent my compliments, and explained the reasons of my speedy and unexpected return. He sent word, that I had done exceedingly well in returning to him, and that God had delivered me from being plundered and murdered on the road to Gondar. At the same time, he sent some refreshments, and promised to give all that I wanted, as he wished to make me very comfortable. Can you fancy this to have been the language of a man how himself was going to plunder or to kill me in his own house? After an hour's rest, I was called to see him; and when I appeared, he used the same expressions as before, and appeared to be extremely sorry at my disappointment in the prosecution of my journey. How could I suppose that Adara Bille, whose house I considered as my own—who always pretended to be the most sincere friend of Sahela Selossieh—who assumed the greatest friendliness—who sent every moment to inquire after my wants—and who, in one word, treated me with the utmost attention—how could I suppose that this man was the very worst man whom I had ever seen in my life?

My people, as well as myself, hoped that we should in a few days be within the boundaries of Shoa; but our Almighty Guide had intended to lead us by an opposite road, and to try me with indescribable privations, hardships, dangers, and difficulties.

March 30, 1842.—When I intended to leave Gatira after sun-rise, I was ordered by Adara Bille to stay with him, till he had informed the Governor of Dair, and through him the King of Shoa, whether I should be permitted to return to Shoa or not, as he had only received orders to conduct me to the road of Gondar, and not that he should assist or allow my return.

Thinking that Adara Bille intended to detain me for the purpose of obtaining from me some presents in addition to those which I had given him on my first stay in his house, I gave him several valuable things, hoping he would allow me to depart. But of course after he had once made up his mind to plunder my whole baggage, he was not content with these. His head wife Fatima, the daughter of Berroo Loobo, Chieftain of Worra Kallo, sent for a looking-glass, which she received.

They were detained some days at Gatira, and eventually not allowed to proceed to Shoa, but obliged to return by the north.

April 5, 1842.—We were still uncertain what would become of us. Reports were spread this morning, that Adara Bille would keep my servants as his slaves, and send me off alone to a road of which nobody could give any information. This report drove my people almost to despair, and made my own heart ache so much, that I could not refrain from weeping with them. They said that they would rather die than be separated from me. However, the Lord gave me strength to console my heart and that of my afflicted people.

About nine o'clock a servant of Adara Bille appeared, with an order that we should leave the house, and follow the six soldiers, who were to conduct us beyond the territory of Adara Bille. He did not tell us which way we should be conducted, and I could not venture to ask, as Adara Bille might have become angry. Silent, and defenceless, we followed the soldiers, who went before us with spears, shields, and swords. Almost the whole population of Gatira was assembled; most of them wept, others wished us a happy journey; none praised their Chief; and many expected a punishment from Heaven would be inflicted upon the country in consequence of the injustice shown towards strangers.

As well as I could ascertain, from the position of the sun, my compass having been taken by the robber, we marched north-east-east. It was now evident that Adara Bille intended to send us to the road of Tehooladere. I was quite indifferent regarding the way, as I could not lose anything more; and indeed I could only profit from being conducted to a road hitherto untrodden by Europeans.

Our road led us continually over a level country, which however was but little cultivated. In general, nature seems to have refused to the Wollo Gallas that fertile country, and that state of wealth, which the Gallas enjoy in the south of Shoa. This is perhaps the reason of the thievish character for which the Wollo Gallas are truly blamed. We saw very few villages, and the population cannot be considerable in this part of Adara Bille's territory. We crossed several rivulets, which presented to us their cool and delicious water.

In the afternoon we were joined on the road by the robber's chief priest, who was returning to his village, not having obtaedin any new revelation concerning my party, as all our property was lost. His name is Tahir. On meeting me, he gave his compliments with a smiling face, and said, "If you will come with me, I will give you something to eat and drink; but your servants may look out for themselves by begging in the village." I put very little confidence in the cunning man; but he did more than I expected; for he gave us a house, lighted a fire, it being cold and rainy, and gave me to eat and to drink as well as my servants, who did not find anything in the village. I ate with the greatest appetite, being rather hungry from the confinement in the prison.

April 6, 1842.—Early this morning we left the village of our host Tahir. When saying good bye, I expressed my thanks for his hospitality, which I could not now reward, as he was well aware. He said, "Never mind: it does not signify. I have my share in the property which Adara at my advice on the Wodacha has taken from you." He laughed, and walked off. This is something of the character of the Wollo Gallas, namely, friendly cunningness and rapacity.

About eight o'clock we crossed a rivulet, and about ten o'clock we left for ever the territory of Adara Bille, having entered into that of Berroo Loobo. We first passed Totola, the celebrated market-place of which I have spoken before. Totola means, properly speaking, the whole beautiful valley and district into which we had entered, having left the country of Adara. It is intersected in the middle by the river Gherādo, which runs from south to north-west to the river Bashilo. On both sides of the valley is a range of hills more or less elevated, and covered with juniper-trees. These hills are covered with hamlets and villages. The whole scenery is so beautiful, that I cannot recollect ever having seen such a fine sight in Abyssinia. You can scarcely imagine that you are in Africa. The cool climate—the fresh and healthy air—the green plain, watered artificially by aqueducts from the river—the activity of the inhabitants in cultivation—the quantity of cattle grazing—and the multitude of travelling merchants whom you meet on the road with their goods—all these and many other things give the place an European appearance. It is great pity that such a magnificent district of ten or fifteen miles is not in the hands of a better people and government. I waited several times to rest on the wayside to see more of this pretty scene; but our soldiers drove us on, repeatedly saying, "Are you not our cattle, with which we can do as we please?" The principal market places of Worra Kallo are, Totola, Ancharro, Regghe, Dawe, Kallo, and Fellano.

Our guardians said, that they were ordered to accompany us as far as the river Millē, where there is a wood-like wilderness, in which they evidently intended to plunder the rest of our clothes, and thus leave us to certain death. But Providence watched our lives. About twelve o'clock we crossed the river Berkona, and entered into the territory of Tehooladere, which is governed by Amade, or Abba Shaol. The latter is the name of the Chieftain's favourite horse, which has given him the same name. The Berkona was not more than twenty feet in breadth at the part where we crossed. Its sources were pointed out to me as rising at the foot of a hill called Boroo, about six miles from the place where we crossed the river. Near the hill Boroo is a village called Kombolcha; therefore the people generally say that the sources are at Kombolcha, where there is a marsh ground. The Berkona was on our passage at a very low height of water, being about a span in depth. It runs first to the south, then turns round to the east, near Ayn Amba, and finally joins the Hawash in the country of the Adals. Not far from our passage, the Berkona forms a cataract. Most of the waters of Worra Kallo join the Berkona, a very important fact, which shews that we had passed this forenoon the watershed, being between the river Gherñdo, which runs to the river Bashilo, which goes to the Nile, while the Berkona goes to the east to the Hawash. The continuation of the range of mountains observed in the east of Shoa is consequently the range which runs through Worra Kallo toward Ambassel, leaving the Berkona in the east, and the Bashilo in the west. This most important fact throws a great light upon our maps of Abyssinia, because the watershed of a country, if it is once correctly known, throws a light upon many other subjects which are in question.

We approached Mofa, the capital of Amade, which is built on a steep and high hill, from which there is a pretty view of the lake Haik, of which I shall speak afterward. The soldiers of Adara Bille observing that we were marching toward Mofa strongly objected to it, and a quarrel arose between us. We declared that we had nothing to do with Ali Gongool, who was not the lord of the country; but they replied, that they had received orders from Adara Bille to deliver us into the hands of Ali Gongool, who would send us with a large escort to the river Millē and to the wilderness between Tehooladere and Yechoo.

They carried their point, and proceeded to Amade, who delivered them from the soldiers of Adara Bille.

It appeared to me from observation, that the eastern tribes of the Wollo Gallas consist of a better set of people and government, and possess greater wealth, than the tribes of the west. The western Gallas are continually lurking on the way sides, till they observe a caravan or a single traveller. They frequently run after you to the distance of a mile, in order to inquire after the state of things of other tribes, or to learn who you are, and whither you are going. Their curiosity is then converted into robbery, if they think themselves strong enough to overcome the travelling party. This clearly shows a trait in their character, which is truly blamed with committing robberies and hostilities against each other. The people of the eastern tribes may be less blameable in this respect, as they have more intercourse with the other parts of Abyssinia, being more concerned in carrying on some trade, for which a considerable number of market places have been selected, as I have before mentioned. Travellers have always been more protected among them; but whether European travellers would be treated like the Abyssinians is another question. The eastern tribes also assume a greater show of dependency on the ruler of Gondar than the western tribes do. Their rulers are principally invested with a lineal succession, and endeavour on this account to keep up their country in better order, and their people submit themselves more, having been accustomed to obedience for a long time to the descendants of one ruling family. This is evidently the case in the tribes of Worra Himano and Tehooladere.

The nature of the territory of Tehooladere is most conspicuous and excellent, and gave me the appearance of those Galla countries which I have traversed in the south of Shoa. The soil of Tehooladere is excellent for cultivation, if there were only hands enough to cultivate the black fallow ground. I was told that the population of this tribe was very considerable six years ago; but that it was considerably thinned, first by the cholera, which raged six years ago almost over the whole of Abyssinia and the countries beyond; secondly, by a famine which laid waste so many tracts of Abyssinian provinces; and finally, by a war, in which Ali Marie, the former Chieftain of this tribe, was engaged with the Chiefs of Worra Kallo, Lagga Ghora, and Worra Himano, who assisted the present Chief of Tehooladere against Ali Marie, his relation, who was entitled to the government by right.

Tehooladere is rich in wood, and grass for cattle. The climate is finer, as the country is lower than that of the western tribes, although there are some high mountains. In geographical and historical respect, it has a certain celebrity, which I will presently mention. I have already mentioned that the river Berkona rises in the territory of Tehooladere. The lake Haik is also situated in this territory. This lake is one of the most important lakes of Abyssinia. Its Christian population gives it still more importance. A former great king of Abyssinia had established his seat in this country as I shall mention hereafter. Before I enter, however, into a description of this lake, I must mention another called Ardibbo, which I have never seen marked on the maps. This lake is in the tribe of Imam · Faris, whose capital is in Gherfa. This tribe is situated between the country of the Danākil in the east, and Worra Kallo and Tehooladere in the west. Imam Faris is said to be frequently engaged in war with Berroo Loobo. He is in the possession of a few field-pieces, which he has bought from merchants trading to Mocha. He is on good terms with the Danākil, and his territory extends as far as a journey of four days from Aussa. If a traveller could succeed in penetrating to Abyssinia by way of Aussa, the former capital of the Kings of Adel, he might be able to obtain most valuable information of the countries between the Danākil and those Wollo tribes through which I have travelled. He might be able to throw much light on the geography of these countries of old, and by this means he might make us better understand the accounts which we have of the annals of Abyssinia regarding the wars of its Christian rulers with the Mahomedan Kings of Adel.

The lake Ardibbo, near Gherfa, is said to be not much less in circumference than Haik; but there is no island in the Ardibbo. I must strongly recommend travellers attempting a journey to Abyssinia, to endeavour to the utmost to get in by way of Aussa, although I cannot conceal that this journey would be attended with many dangers. The traveller having arrived at Aussa from Tadjurra, could probably proceed either to Berroo Loobo of Worra Kallo, or to Imam Faris of Gherfa.

As the head servant of Amade, Chief of Tehooladere, had told us that there was a Christian village at the foot of Mofa, we resolved to pass the night at that place. From the capital of the Chief we had a very steep and long descent to the village; but our feelings of joy and cheerfulness at having been delivered from the hands of Adara Bille's servants, made us forget every difficulty and fatigue. It was dark when we arrived in the village. We applied to a merchant of Gondar, who kindly received us into his house, and provided us with food sufficient for our party.

April 7, 1842.—This morning the merchant with whom I had passed the night, started early from his home to visit the market of Ancharro, which I have mentioned before. He promised to send some intelligence to Shoa through merchants of Alio Amba in Efat, whom he would see at Ancharro. I regretted that I was unable to give him a copy of the Holy Scriptures, as he had expressed a great desire for it; but I will send some copies through people going from Shoa to the lake Haik. The village, the name of which I have unfortunately forgotten, consists chiefly of trading inhabitants, who are all Christians. Their trade is carried on from Gondar through Worra Himano to the country of Berroo Loobo and to the territory of the Yechoos, with articles which are found at Gondar.

Having taken leave of our friend, we directed our course to the lake Haik. The road led us through a most beautiful and fertile valley, being rich of trees, grass, and rivulets. The soil was chiefly black; but it is scantily cultivated, for the reasons which I have before mentioned. They principally cultivate maize of different kinds.

My joy on arriving near the shores of the lake was indeed great, as I had been desirous several years of visiting the Christians on the lake, and as the large mass of water reminded me again of the water-stock of the Red Sea, to which I had so often committed myself in former times, and to which the end of my journey would bring me again.

The Alaca of the Convent of Haik had been already informed of my occurrences with Adara Bille, with whom he is personally acquainted. But when I arrived, I did not find him at home, as he was gone out on the principal road to the Yechoo country, believing that Adara's soldiers would not allow me to see him in lake Haik. He came to this conclusion from the circumstance that I did not arrive yesterday evening, when he had expected me immediately after the arrival of my previous messenger. Not being permitted to cross over to the island in the lake without the Alaca's special orders, I was obliged to wait on shore till he returned. In the meantime I was engaged in contemplating the shores and the very interesting country around, and in inquiring after the state of things on the island. The multitude of people also, who assembled soon after my arrival, gave me an opportunity of speaking on many topics; so that my long waiting for the Alaca was no lost time.

The shores of the lake in the west and north are not high, nor steep; but those of the south and east are surrounded by high and steep mountains. The circumference of the lake may amount to forty-five English miles. Several bays are observed extending inland a few hundred yards. The greatest extent of water is from east to west. The lake is full of water birds of different plumage. I was also told that it is rich in fish of a large size. The water is sweet, as may be expected from being a land sea. The island, called Debra-Nagoodguad, (hill of thunder), is distant from the north-western main land about 260 yards, and might easily be battered by riflemen. The anchoring-place is called Mad-gebata, and the village, where you must halt before crossing over to the island, is called Debra-Mariam (hill of Mary). This village is chiefly inhabited by the wives of such priests who are married, as by an ancient law no female is allowed to enter on the island. All the inhabitants around are Mahomedans, who are not prevented however from visiting the island; but their wives are under the same restrictions as those of the Christians. A number of acacia-trees are observed near the anchoring place, between the village and the lake. These trees afford a pleasant shade to those who must wait for the rafts taking them over to the island. The eastern mountains of the lake are inhabited by the tribe Worra-Babbo, which is governed by the Chief Ali Adam, who is dependent on Imam Liban. There is but little wood around the lake, except in the south-east, which is far off from the island; but the inhabitants of the island cross the lake on rafts to fetch wood. Beyond the tribe Worra-Babbo is another tribe in the east, called Chaffat, and is independent. In the east of Chaffat is the country of the Danākil.

The old Alaca at last returned. I was delighted at seeing him again. I had made his acquaintance a year ago, when he called upon me at Ankobar. I sent at that time a copy of the Amharic New Testament to the church of the island. I also gave him a copy of the Æthiopic New Testament when I met him at Dair. Thus my name was pretty well known in Haik, as well as the object of my stay in Shoa. The Alaca took me over to the island on a raft, composed of a thick stratum of reeds. The raft was about twelve or fifteen feet in length, and about three or four feet in breadth. The whole stratum of reeds is tied together with ropes at both ends, and in the middle. Two rowers moved this curious machine, which carries about six men over to the island. The depth of the water increases with the distance from the shore. About one hundred yards from the main land the water is very deep till almost close to the island. I was told that on most places of the lake the bottom cannot be found; but although I would not object to this, I doubt whether they have ever taken the trouble to examine the depth of their lake, especially as they are unacquainted with the plummet.

The western and northern winds raise high waves on the lake; while the winds blowing from east and south are prevented by the high mountains from displaying their full power over the water-heaps of the lake. As to the rise and formation of the lake, I am at a loss how to explain as I could not learn whether there is any volcanic action in the neighbourhood, nor could I discover volcanic traces from the nature of the country around. I could not however examine the eastern and southern shores, being too far off from the island; and I do not venture to judge from rocks scattered around the village Debra Mariam, as these may be ascribed to the destructive power of the violent annual rains. In my opinion, an observer should be careful in drawing a conclusion for the existence of former volcanoes from his perceiving stones scattered around, as it is well known, which I could prove by facts, that the rains have demolished considerable hills. A traveller of late, who has not however been in Abyssinia during the rainy season, seems to me to have been greatly mistaken when he seemed to observe nothing but volcanic traces in Shoa.

It must be remarked, that the name "Haik" is a general expression, and means in Æthiopic "sea," or rather "shore." I should think that this lake is in a straight line from Ankobar, perhaps a little more to the east. I did not observe that there were any shells on the shore, nor did I hear that there were on other parts of the coasts. There is plenty of grass in the water, where it is not of considerable depth; and this is the place where the water birds are gathered in immense numbers, so that one shot would afford a great booty to the sportsman, if the prejudices of the inhabitants of the island would allow you to fire a gun. Their conviction of the sanctity of the island, in consequence of Tecla Haimanot's having resided on it, and blessed the water, seems to have produced this prejudice. The same prejudice would be in your way if you attempted to kill a bird on the island, though I saw there several trees, on the branches of which was such a multitude of vultures, that I wondered the branches were not broken.

The population of the Island amounts, as I was told by the Alaca, to 350 souls, consisting of monks, priests, scholars, and servants. Before the Gallas abridged them of the ancient benefits, the population amounted to upwards of 1,000 souls.

April 8th.—As this day was Friday, I was obliged to cross over from the island to the main land, the rafts not moving on Saturday and Sunday on account of the sanctity of these days.

We prosecuted our journey in the direction of north-east; but being already too late to go a considerable distance, we went to the village Bora, in the district Wordai, distant about five miles from the lake. Here we found a Debtera, who received us kindly and provided us with whatever his circumstances would admit. He is the only Christian inhabitant of the place, all the others being Mahomedans. I shall not forget to reward him if he comes to Ankobar, as he has done his duty towards his fellow-creature in affliction and poverty. He stated that he had seen me at Ankobar, and that he was glad of having been enabled to render me a little service, and of making his personal acquaintance with me.

Several high mountains were visible in the neighbourhood of Bora to the north and north-west, the highest of which is Sagarat, on the northern foot of which the river Bashilo was said to rise. This would not be far from the sources of the Berkona. Sagarat belongs to the territory of Imam Liban, and the sources were to be placed between the territory of Imon Liban and that of Yechoo. I have no doubt that the high mountain of Sagarat, and the whole ridge of hills branching to south and north, form the watershed in this part of Eastern Abyssinia, and is evidently the continuation of the famous range of mountains which I have frequently mentioned in my Journal.

April 9, 1842.—This morning, about six o'clock, we left Bora and our kind host, Debtera Atkoo. He gave us some provisions for the road. From Bora we had to descend a little into the pretty valley of the river Millē, which rises on the northern end of the mountain Mofa, and runs toward the country of the Adals. This river separates the territory of Tehooladere from that of Ambassel and Yechoo. The course of the river Mille to the east, and that of the river Bashilo to the west, shows that the mountains of Mofa and Sagarat form the watershed, and are the continuation of the famous range which surrounds Eastern Abyssinia like a girdle. In the east of this range you enjoy a milder-climate, which gets hot the more you descend toward the country of the Danākil. This descent takes place over little hills and valleys almost impassable on account of thorns and other kinds of wood.

We crossed the river Millē about seven o'clock. It runs through a most beautiful valley, being rich in trees and grass, and a good soil for cultivation; but notwithstanding this, the valley is neither cultivated nor inhabited, but left a complete wilderness. I have never seen such a variety of birds, of the most beautiful plumage, as in this valley, and I am sure that a good collection could be made for zoology. The bed of the river is of considerable breadth; but its real breadth, where there was water and where we crossed, was only fifteen feet and a quarter in depth; but it must be remarked, that this was the hot season of the year. The river runs north-east-east, and we followed its course for a distance of a few miles, till we took a more northern direction in the vicinity of the mountain Ambassel, from which the tribe and the whole country around has its name. The height and steepness of this mountain raises the greatest astonishment. It is one of the most important strongholds of Abyssinia, which, if well guarded, would be able to check a large army for a considerable time, as there is only one road, which is steep and dangerous, leading to the top, where there is a plain with water and good ground for cultivation. This mountain was for some time the state-prison of the former Emperors of Abyssinia. The royal princes were frequently confined on this stronghold, which is not far from that of mount Geshano (not Geshen) which is in the north-west from Ambassel, as well as I could ascertain without the compass. Besides these mountains, those of Damo in Tigre, and Weihne in the west of Abyssinia, were selected for the imprisonment of the royal issue.

The mountain Ambassel has several high and prominent peaks, and extends from south to north with a little east. It is about nine or twelve miles in extent from south to north. Its banks in many places resemble walls of an immense height, and I doubt whether the ball of a cannon of the best calibre would reach the top of the mount. This stronghold would be of the most decided importance in a better military system of Abyssinia, in order to secure its eastern frontiers against the Gallas and Danākil, who could be conquered with the greatest ease by a small detachment of regular troops starting from Ambassel. In general, my road from Shoa to Tigre has convinced me that Eastern Abyssinia is almost unconquerable, and would be so if its rulers once adopted the European military system.

Having crossed the river Millē, we entered into the districts of Seeba and Goombisa, through which the Millē runs, whereupon it is lost in the sandy deserts of Adel. Both districts belong to the tribe of Ambassel. Having passed the district of Seeba, we traversed the district of Woochāle, in which we travelled through a village called by the strange name Sekdat-teherk. On enquiring after its meaning, I learned that the inhabitants formerly used and manufactured clothes from the wool of black sheep, which is called in Amharic Sekdat; but having become acquainted with cotton and the manufacturing of it, they relinquished the use of black clothes, which they then considered as Teherk, *i. e.*, rags; thus dishonouring the improved state of the skill of their countrymen. Having traversed Woochale we came to the district of Worra Kallo in Yechoo, which must be well distinguished from Worra Kallo in the Wollo country, which is governed by Berroo Loobo, as I have stated above.

We halted a little in Worra Kallo in order to beg for some provisions, as we were very hungry.

As it was late when we arrived in the village, and having been overtaken by a violent rain, we took the liberty of entering into the nearest house on the way side, and asked the proprietor for a night's lodging.

April 10, 1842.—Early this morning we departed from Leebso, moving towards Mersa, a celebrated village, inhabited chiefly by merchants of the Yechoo country, into which we entered yesterday afternoon on arriving at Worra Kallo. Having proceeded on our way for about half an hour we were overtaken by a heavy rain, which compelled us to seek for shelter under trees, no house being visible in the whole neighbourhood. The second rainy reason—between February and April—appears to be heavier in these regions than in Shoa. Perhaps the mountainous country, which must be always clouded, contributes to this phenomenon. It is a fact, that where there is high land in Shoa, the rains are more frequent and heavier. We were in a large valley, a complete wilderness, though it might nourish many thousand of inhabitants. The acacia-trees and bushes were in such abundance that we lost our road several times, and were entirely at a loss how to extricate ourselves. The mountains around were quite clouded, so that we were unable to find and correct our direction, which was pointed out to us by our kind host at Leebso. We did not know whether we should not fall into the hands of the Gallas and Danākils, who dwell on the eastern end of the wilderness; or whether we should be attacked by ferocious beasts, against whom we had no weapons of defence. Fortunately, however, the rain ceased, and the clouds were dispelled, and with these our embarrassments were dispersed, as we could now distinguish the mountain which we should pursue. However, the violent rain had made the slight soil so slippery, that I frequently fell down. The vapourous air besides and the thorns made our walking very inconvenient.

About nine o'clock we crossed the river Ergebbo, which runs to the country of Adel, as is the case with all the rivers rising in the east of the famous range of mountains in Eastern Abyssinia. Probably there is a large river down below toward the country of Adel, a river which may take up all the rivers, brooks, and rills, of which we passed several since we passed the river Millē yesterday. This river, which probably receives the waters of Yechoo, Lasta, and Agow, is most likely the upper course of the river Anazo marked on the maps. It may be the general conductor of the mighty reservoir of water which is contained in the mountain range so frequently mentioned. The Hawash takes up all the waters coming from the east of the watershed of Shoa and Worra Kalloa; why should we not therefore be allowed to suppose, that a companion of the Hawash takes up the numerous water tributes of Yechoo and Lasta, collects these tributes to one common stock, and conveys them to the coast? but that the long journey through the sand of Adel prevents it from reaching the Red or Indian Sea, as is also the case with the Hawash. Had I been able to take my route through the country of the ferocious Raia Gallas, as I intended to do, I should have obtained more particulars for or against this opinion.

The Gallas have intruded themselves around the whole eastern girdle of Abyssinia, between the Danākil and Abyssinians. They live at enmity with both these nations, although they have adopted the Mahomedan religion. In the east of the great plain which we traversed, there are several tribes which pay tribute to the Governor of Yechoo; namely, the Chorrē, Logana, and Boora tribes. It must be observed, that the Yechoos are not Gallas nor Pagans, as it would appear from

Mr. Bruce's work. At least, at present, they are Christians, and speak Amharic; and I did not find that their features are the same with other Gallas. Probably Mr. Bruce, who although the best writer on Abyssinia, yet is sometimes greatly mistaken, took those tribes which are dependent on Yechoo for Yechoos themselves. In the north of these tribes towards Lasta and Agau are the Ana and Raia Gallas, who could not be subjected by the Abyssinians on account of their mountains, which appeared from a distance to extend to the very sky. The Raia Gallas, of whom I shall speak frequently hereafter, are the most ferocious set of people, plundering and murdering for the sake of pleasure. They are divided into several small tribes, which dwell in the higher and lower countries of their mountains. The mountain ridge which they inhabit, probably extends a hundred miles from the south to north-east. There they watch the opportunity of carrying terror and death against the lower countries in the east and west. If the traveller had not to fear this inhuman set of people, we would be able to reach Tigre in a much shorter time; and the route between Shoa and Massowah would be considerably abridged. But thus the traveller is compelled to take a long and tiresome route through the country of Lasta and Wag, on account of the Raia Gallas lurking like lions at the foot of their mountains.

On the banks of the river Ergebbo I saw the coffee tree. It was about fourteen feet in height. The leaves were very long, and the husk of the fruit, which was not yet ripe, red and sweet. Coffee is not dear here, as the Mahomedans plant as much as they want for themselves, the Christians refusing to drink from religious motives.

Mersa is the point where Christians begin to become frequent, and their number increases to the foot of the Yechoo mountains, where the Mahomedan power was seldom felt. The Christians and the people of Yechoo in general are said to be good, simple, and hospitable. This testimony appears to be true to a certain extent. Since I had left Shoa, and been without means, I had not been so well treated as in Yechoo. It appears that they have kept up much of the ancient Abyssinian manners. Their mountainous country separated them from the intercourse and political movements of other Abyssinian provinces, and this circumstance contributed much to the preservation of their former character. Their hospitality may be partly ascribed to the great wealth which they enjoy. They have everything that an Abyssinian wants in abundance. They have a beautiful soil for cultivation, a soil which will produce all that they want.

Moving toward Mersa, we met multitudes of people going to the market of Goobhāra, a village through which we had passed yesterday. I observed a very strange custom of the Yechoo women whom we met on the road. They either turned backward, or turned their faces to the ground, standing still on the way-side. Believing that this arose from the fear which they had at seeing a white man, or that it was a trace of modesty customary in their tribe, I inquired the reason; and I learned that in doing so, they request a blessing from the traveller, who has to address them with the words: "May God have mercy upon you;" or, "May He bless and preserve you." I observed afterward almost the same custom in the Wag country, though only in the male sex.

I have already mentioned, that the immense plains of the Yechoo country would admit a more numerous population; but on examining this matter more fully, I found that they leave them uninhabited on purpose. These plains, which are complete wildernesses, are narrow in the west toward the foot of the mountains, but very considerable in breadth toward the east and the Galla country. Thorns and other kinds of wood grow up on these plains in such abundance, that you can scarcely find your road through this thorny wilderness, which is dreaded on this account by large wild beasts. Thus naturally fortified against the inroads of the Gallas of the east on points where the only entrance is presented to these savages, the Yechoo people do not feel inclined to deprive themselves of this thorny stronghold by means of cultivation, for which they have room enough in other places. Besides, the cultivation of such a wilderness would require great exertions, which the laziness of the Abyssinians will not attempt, unless the utmost necessity compels them. The climate in these plains is beautiful, neither too hot nor too cold; the air being always refreshed by the winds blowing from the mountains. There is plenty of water poured out from the veins of the neighbouring mountains.

About twelve o'clock we crossed the river Mersa, which carried in its narrow bed such a mass of water that we had great difficulty in passing the river. The heavy rain which fell this morning had caused this swelling of the river, which at other times cannot have much water. Much cotton is planted on the banks of the river. But I was particularly struck with the manner in which the natives plant their red pepper. They dig small pieces of ground near the river, which they surround by a fence. In this the young pepper plant is placed, and covered with reeds, which, however, do not touch the top of the plants, as they stand very close together. These reeds are frequently sprinkled with water, which drop down on the plants gradually. This treatment evidently contributes to the speedy and luxuriant growth of the plant. When it has grown about a foot in height, it is transplanted into another tract of ground. I was told, that a pepper-plantation of only about ten or twelve feet in circumference, will bring in to the proprietor a revenue of two or four dollars, as he is enabled to plant a large field with the previous produce of but a small garden.

Having crossed the river Mersa, we immediately saw the village of the same name before us, and entered into a little house close to the wayside. The people of the house proved to be Mahomedans. Upon entering and saluting them, an old sickly looking woman returned our salutation, and bid us walk in and sit down on a skin, which she spread out before us. She then ordered her daughter to make some coffee, and to bake a few cakes. In the mean time, she gave us some hog's beans, till the coffee was ready.

April 11, 1842—We left our kind hostess about seven o'clock, A.M. We travelled toward Woldăia, the capital of Dejasmadj Faris, Governor of Yechoo. On our road we met a number of priests coming from Gondar by way of Begemeder and Wadela. They told us that the robbers near the river Chĕchĕho had deprived them of their clothes and provisions. They had nothing on their bodies except the skins of bullocks, which some merciful people had given them to cover their nakedness. This fact is a further proof of what we might have experienced if we had been able to prosecute our road to Gondar by passing the Chĕchĕho. The river Chĕchĕho has its source in the mountains of Lasta, and runs between Begemeder and Daunt into the Nile.

Our road led us through plain land as yesterday, but it was less woody. As it was already evening, and a shower of rain approaching, we would not enter the capital of Woldăia, but preferred seeking for a lodging for the night in a village called Shelte, a few miles distant from Woldăia. Our intention was to move to-morrow to Woldăia, and if possible to rest there for a day or two, as our daily journeys had tired us considerably. Besides, we thought that we might be able to collect a stock of provisions for our journey through Lasta, which, we had learned, was a poor country, abandoned by the inhabitants. Having arrived in the village of Shelte during the rain, we entered a house to beg for shelter and a lodging for the night.

April 12, 1842—As we wanted to pass the day in Woldäia, we were in no great hurry to leave the village of Shelte, Woldäia being not very distant. On our road we met a great many people, who were going to the market which is held this day at Woldäia. They came from all quarters. We saw many hundreds of donkeys and mules loaded with salt-pieces, barley, cloths, &c. A dollar is exchanged at Woldäia for thirty-six or forty pieces of salt, consequently double as much as in Shoa. I observed that the Yechoo language varies in many things from the Shoan Amharic, which differs in many things from the dialect of Gondar, which is considered the purest Amharic. As to the rest, I could understand the people of Yechoo as well as the Shoans.

Woldäia is a considerably large town, situated in a plain with slight hills. It may contain a few thousand inhabitants. Probably Faris has chosen the place, in order to be at hand against the inroads of the eastern Gallas. The houses differ but little in construction from those in Shoa.

On account of the insecurity of the road, we had been advised by some people to join a caffila going to Lasta and Wag. As we did not know the day of its departure, we were told to apply to Atkoo the Negad Ras (head of the merchants) in Woldäia, and to ask him about this matter. Trusting that he would give us the best information, and would perhaps allow us to stay in his house a day or two, we went to him; but we were immediately refused admittance into the house. He was sitting in the house-yard; but probably thinking that we did not know him; he said, "The Negad Ras is on the market: he is not here," though the neighbours had told us that he was at home. Upon endeavouring again to enter the gates, he cried out and said, "I have told you once that the Negad Ras is not here." At the same time he ordered his servants and many ferocious dogs to drive us out of his sight. We went away very sadly indeed, and grieved at the man's uncouthness. We resolved, however, to prosecute our way without caring any more about the departure of the caffila or the insecurity of the road. I must confess that the rude behaviour of this man made my heart weep: but at the same time it led me to cast myself upon Him who is a merciful Father to all those whom the world turns out, and who was my only friend and protector in an unknown country, where I had neither friends nor funds.

Thus the plan which we had schemed yesterday for collecting a store of provisions at Woldäia was entirely frustrated; but we entertained the hope that we should find what we wanted at other places, and at a time when we should absolutely require it.

Upon leaving Woldäia in a north-easterly direction, we had to descend a great deal from the plain of this town. We had no guide with us; but we proceeded on our way, continually inquiring after provinces and places which I knew from the maps of Abyssinia. A narrow path from Woldäia led us down into a small valley, through which the river runs, called in Amharic, the Black River. It had much water from the rain of last night. It runs to the country of Adel.

About nine o'clock we halted in a village called Gooddo, where my people wanted to go and beg, as the village had the appearance of a wealthy population. But this was not only an optic delusion when we tried to obtain something from the apparently rich people. With great difficulty, and after long supplications, my people brought back a small quantity of hog's beans from their begging excursion. A Mahomedan woman allowed us to boil the beans on the fire in her house. She also allowed us to make a little black coffee, which had been given us the day before yesterday by our host at Mersa. We could never prevail on Christians to allow us to make coffee in their houses, as they instantly took us for Mahomedans and sent us out of their houses; nor would they by any means give us a vessel for making the coffee, because it would make the vessel unclean.

After we had left Woldäia, we seldom met Mahomedans, who are not very numerous in the Christian country of Yechoo. They are still less in the country of Lasta and Wag.

Starting from the village of Gooddo, I made the acquaintance of a man from the village of Shal, near the district of Angot. He came from the market of Woldäia, and was on horseback. He inquired after the country from which we came, and where we were going. On learning that we came from Shoa, he said, "The Shoans are the best Christains of Abyssinia, and their king is the best ruler." This remark was made by many people of Lasta, Wag, and Tigre. Both the king and the people are in favourable reputation with the rest of Abyssinia. The king's generosity is known every where; therefore they flock from all quarters to Shoa, principally monks and priests.

Our road led us over a very fine country, extremely adapted for cultivation, the soil being that of our European gardens. In the west we had always the sight of high mountains, ranged from south to north and north-east. About twelve o'clock we crossed the river Ala, which rises in these mountains, and runs toward the country of Adel. It carries a considerable quantity of water in its narrow bed, and during the great rainy season must be impassable. Being late, and the clouded peaks of the mountains menacing the approach of rain, we thought it best to look out for shelter in due time. We beheld the village of Shal, the name of which we have heard previously from the man whom we had met on the road. He had left us before we crossed the river Ala. We did not know his name, nor did he invite us to pass the night with him; nor had we asked him for any favour of this kind. On entering the village, which consists of single houses scattered over a considerable distance, it happened that we directed our course to the very house belonging to the man whose acquaintance we had made before.

April 13, 1842—I got up very early this morning, as the fleas and other insects would not allow me to take rest for a moment during the whole night. The great number of cattle in the stable in which we were quartered, gave an attractive power to these little tyrants, who vexed us at night, after we had been pained by the people during the day. We left early the village of Shal, which is in the district Sanka, belonging to Yechoo. From Shal we had to ascend a long time. Many rills intersected our road and refreshed us with their delicious water, coming from sources which we could observe, at a distance of a few hundred yards, gushing from the rifts of the rocks.

About ten o'clock we finished our tiresome work of ascending to the higher country. We rested a little on a spot, where two highways request the traveller to decide which he will choose for his journey. The north-western highway leads to Lalibala and Gondar; while the north-eastern road will bring you to Sokota and Antalo. Had I been furnished with proper means, I would have changed my mind and taken the route to Lalibala and Gondar, as I had more than one motive to see the latter town; but my misery and affliction compelled me to prosecute the north-eastern route toward Tigre, as this would lead me quicker to Massowah, the end of my journey.

They were here detained on a charge of slave-dealing, from which they cleared themselves.

Disagreeable and annoying as this occurrence was to us, yet it turned to our great advantage; for had we not been detained, we should have traversed the district of Angot, and then we should not have found a village on the road before night, in a cold and dangerous wilderness. Thus frequently many circumstances are insignificant and disagreeable, but in course of time are found to be very providential indeed. O that my heart were more thankful to Him, whose gracious hand was to be seen so manifestly during the indescribable misery and distress of my journey!

We were now in Angot, which appears to be a large district. It begins with the point of the separation of highways mentioned above, and extends as far as Lasta, to which it is considered an additional part. It is at present dependent on the Governor of Yechoo, to whom Lasta is also subjected. This is evidently the province of Angot marked on the maps; but it must have been formerly much larger than it is now. It must have extended more to the east, where there is at present a part of the Raia Gallas. I had frequently asked such people in Shoa as I thought would be able to tell me something about the province of Angot; but I was left in ignorance till I asked a native this afternoon about the name of the district which we were traversing. The same was the case with the districts of Bugna and Wolaka, which are mentioned by Mr. Ludolph and by Mr. Bruce. According to the latter (Vol. II. p. 441) the daughter of the Jewish King Gideon was married to the Governor of Bugna in Lasta. Bugna is still to this day a district near Lalibala in Lasta. Wolaka is another district, through which I shall pass to-morrow. I am convinced that many names of the ancient geography of Abyssinia would be again discovered, if travellers would go over the whole of the country. In some distance in the east of Angot is the high mount Sobel, inhabited by a part of the Raia Gallas. The climate of Angot is very cold, as it is high land. On the eastern frontier of the mountains of Angot I saw a large plain, situated very low between Angot and the mountains of the Raia Gallas. The beauty of the prospect which I had of this plain, and the high mountains of the Raias beyond, is truly indescribable. The plain must be very considerable in breadth, and a river runs through it from what I could see and learn from the natives. If this be true, and I believe it is, it must be the river Millē mentioned above. This, I suppose, takes up all the waters of the Yechoo mountains, and runs between Angot and the Raia mountains north-east-east, where it receives the waters of Angot, Lasta, and Wag, and perhaps also the waters of Wofila and a part of Tigre; whereupon it attempts to reach the coast, but it is prevented by the sand and the rising country toward the coast. I inquired much about this plain; but people told me that they did not go over to the Raia Gallas, and therefore did not know whether there was a large river; but that there was water running through the plain. This information compelled me to suspend my judgment of the subject, till other travellers shall throw more light on the matter. It frequently happens that travellers form their own idea of a subject, and turn their observations or information according to these their preconceits, which is rather a loss than an advantage to geography.

Having reached the village of Saragadel, we learned that there was no other village on the road for a distance of about fifteen or eighteen miles. As it was late, the rain approaching, and we were tired, we resolved to pass the night in this village. We entered into a house; but the inhabitants immediately set their dogs at us. I withdrew a little, and sat down on a rising ground, where the rocks afforded me a little shelter from the cold rain which began to fall. My servants went through the village to seek for a night's lodging. Pensive and grieved at the hardness of man toward his fellow-creature, I sighed after the assistance of Him who had not hitherto forsaken me on my pilgrimage. My servants went from house to house; but all their endeavours were in vain, till at last a sick old man offered his cow stable if we would be content with it, which of course we thankfully accepted. The old man introduced us to the stable and ordered his children to light a fire, as we were trembling with cold. He then had some bread prepared for us. There was nobody in the room except ourselves and the cattle, which did us no harm, except that they attracted those disagreeable tyrants of which I have spoken before, and which would frequently have rendered our nights entirely restless, had not the fatigues of the day produced such an overwhelming sleep that we did not feel the tormentors. I sometimes checked them by leaving the room and staying outside in the cold for a few minutes.

April 14, 1842—We left Saragadel about seven o'clock, and moved toward the mountains of Lasta, still ascending till about nine o'clock. Our road led us to a complete wilderness, very different from those we had passed a few days ago in the lower country of Yechoo. There we had plenty of water, a warm climate, and could always find the road when we had deviated. But this was not the case on the high land of Angot and Lasta. Coldness, want of water, and difficulty in finding our true direction, was painfully felt by our whole party. There was not one large tree, and nothing but grass, called gooassa in Amharic. With this grass they cover the roofs of their houses. A country where there is this sort of grass frightens the Abyssinians, as the name reminds them of a country being cold. The country where you find the gooassa, requires a height of 8000 or 10,000 feet above the sea.

The sky was clouded when we traversed the wilderness, a circumstance which rendered our situation still worse, as we could not distinguish and make out our direction from the peaks of the mountains. However, we went on, being convinced that the road must lead us to some place or other. We saw no village, no cultivated land, no cattle, no beast, except some foxes; no travellers, in fact nothing but desolation, and we ourselves seemed abandoned. Few places ever gave me such a melancholy impression as this wilderness, an impression which I can scarcely forget. After a walk of three or four miles, on a sudden we observed at a distance through the mist covering the wilderness a number of people, who were sitting on the ground on the side of the way which we had blindly taken. Their appearance was not agreeable to us, as we took them for lurking robbers, of whom we had been warned yesterday at Saragadel. To our great joy, however, they proved to be merchants of Woldäia coming from the market held at that place. They were just eating their breakfast, of which they kindly gave us a share after they had heard of our misery. They also provided us with some meal for our use on the road. One of their party also accompanied us for some distance, and showed us the road so plainly that we could not go astray. I took both the food and the guidance as coming from the gracious hands of Him who always helped when help was necessary.

About one o'clock P.M., we reached a few houses on the road, where we halted and had our flour which the merchants had given us made into bread.

We left the hamlet about two o'clock P.M., continually descending on our route, which led us again through a tract of country entirely abandoned by inhabitants. I must remark, that we began to descend after we had left the merchants mentioned above.

The wilderness through which we now travelled had a very different appearance from what I observed this morning. We now found more water; we had fine scenery for our eyes; juniper-trees, kolquall acacia were in abundance; and sometimes we found it difficult to extricate ourselves from the abundance of thorns. But we saw no inhabitants; we met no travellers; nor did we see any wild beasts, but beautiful birds of the finest plumage. Fortunately we could find our road easier than had been the case this morning, when the grass and mist prevented us from keeping up the direction pointed pointed out by our host in Saragadel.

The present population of Lasta seems to be almost nothing, having been destroyed by famine, war, and sickness, as I was told by the natives whom I asked about this subject. Ras Ali was blamed for having ravaged the country several years ago in the most barbarous manner. There would be much room for the maintenance of a numerous population; but it would require an active hand, till the thorny ground could be made arable. A single farmer might now possess himself of as much ground as he likes. I shall never forget the refreshing water which I drank out of the rivulets which run to the north-west in small but deep beds under the shadow of a thicket of wood, so that the sunbeams can never touch the water, and which is therefore agreeably cool. Their course is north-west to the river Tacazze; a circumstance which shows that we had this morning passed the watershed as soon as we had passed the cold wilderness. The country of Lasta is high and hilly in the east and west; and therefore the running of the waters must force their way to the north-north-west. From the point where we travelled to-day we saw no more a rivulet running to the east till we reached the frontier of Tigre to Massowah. Having left the country of Angot, we crossed only such waters as belong to the waterstock of the river Tacazze. But I have no doubt that the high mountains in the east of Lasta, Wofila, and Enderta, pour out many rivulets toward the country of Adel, as is the case with the eastern mountains of Yechoo, Worrakallo, and Shoa. The space of a journal does not allow me to dwell upon a subject which would give occasion for writing a volume about the system of waters and mountains of Eastern Abyssinia.

About five o'clock in the evening we reached a village, called Deldei, which means in Amharic "bridge." In many respects there is some truth in this name, as this village really presents the passage you must take either in going to the country of Wag in the north, or of Yechoo in the south. It leads you in both cases to uninhabited tracts of country. It is therefore the general assembling place of merchants going from Sokota and Wofila to Woldai, or vice versa. In Deldei, the market people join together, in order to frighten the robbers of the road with an imposing party. The robbers especially lurk on such days when the merchants return from Woldaia or Sokota. Wo met a company of merchants; but our plan was now positively against joining their party, who wished us to take the road to Sokota, the capital of the Wag country, which we endeavoured by all means to avoid, having heard of the rapacious character of the Governor of Wag.

We entered the first house which we saw in Deldei on the way side.

April 15, 1842.—Early this morning we departed from Deldei, taking an easterly direction toward Wofila and the lake of Ashanghē. We did not like to go to Sokota, having heard of the bad character of the Governor of the Agaus. Last year he robbed a French gentleman, who intended to go to Shoa. This gentleman had a fine sword, which the Governor wanted to buy; but as he would not bargain with the Governor, he was angry, and sent his servant on the road to rob and kill him on his way from Sokota. They wounded him with a lance; whereupon he fell to the ground; and the robbers taking him for dead, took his luggage and clothes, and returned to their master. This fact is true, and was afterward related to me by the servants of the Governor of Wag. About six o'clock we halted on the banks of the river Terāri, where we finished the remainder of the bread which our host had given us yesterday evening.

On our road to Wofila was the convent of Shamado Mariam, which is in great reputation with the Abyssinians. We did not visit the convent. The principal convents distinguished for sanctity are in Axum Tzion, Lalibala, and Debra Libanos. Our road led us through countries quite destitute of inhabitants, although the good soil would admit a considerable degree of cultivation. The ground was overgrown with grass and thorns, and intersected with rills and brooks. The road was hilly, but not rocky. We could see in the north-north-west of Lasta the high mountains of Semien, the peaks of which presented to us the appearance of large towers. The hilly country of Lasta and Wag, as far as we could see, had exactly the appearance of a raging and stormy sea, presenting numerous hills of waves, with a large space between each wave.

We observed only a few hamlets on our road; namely, Ahio, Tartara, and Atemie Galla. The ground is full of grass, thorns, and bushes; but this is exactly the country which suits the purpose of the gangs of robbers.

We travelled to-day almost in an easterly direction; but on arriving at Atemie Galla, we deviated from our road to north-east-east, having learned that our eastern direction would lead us to the country of the Raia Gallas, who would certainly kill us if we fell into their hands. Having no reason to doubt the correctness of this statement, which was given to us by an old man of the hamlet Atemie Galla, we turned off immediately to north-east-east toward the village Enalka, which we could see from a distance, and where we intended to pass the night. Marching over a thorny field, we saw two men running after us with large sticks. When they came up to us, they were silent for some time as to their object in coming after us. Upon asking them what they wanted, they said, that they wanted medicine. Their behaviour, however, clearly showed that this was not their real intention of coming to us. Unquestionably they contrived this falsehood to make us believe that they did not intend to plunder us, when they saw that they could not manage us, our party being too strong for them.

We reached the village of Enalka about four o'clock.

The village of Enalka belongs to the district of Wofila, being dependent on the Governor of Wag. I have forgotten to remark in my notes of yesterday, that the village of Deldei is the frontier of the government of Dejasmadj Faris. All the country in the north of that village is governed by the Governor of Wag, which is the country of the Agaus. Lasta was formerly in the hands of the Governor of Wag; but Faris conquered it, and has been confirmed in his government by Ras Ali. Lasta is bordered in the south by Angot, Yechoo, and Wadela; in the west by Begedemer; in the north by Wag; and in the east by Angot and the Raia tribes. The country of Wag is dependent on Ras Ali; but this dependency appears to be very loose. The capital of the Governor of Wag is Sokota. Wofila is dependent on him, as already mentioned. The principal places in the south-east of Wofila, are Zelga, Bora, and the lake Ashanghe. The language spoken in Wofila is that of Tigre, by which it is bordered in the north-east; while the language of Wag is totally different from any language in Abyssinia, so that I could not understand a word of it. It has neither affinity to the Æthiopic and Amharic, nor to the Galla language. It is totally a different tongue. I have been informed that the other tribes of Wag, which reside towards the sources of the Nile, have a language which is not understood by those Agaus whose country I have traversed. They told me, that the whole Wag country is divided into seven houses or tribes; but they could not tell me their names, nor could they inform me of their former histories. I have collected a number of words of the Agau language; but unfortunately they were effaced by the rain, as they were written on reeds, in consequence of the scantiness of paper which was left me by the robber Anara Bille.

The Agaus differ as much from that of the rest of Abyssinia in their features, manners, and customs, as in their language. In one great thing, however, they agree with the other Abyssinians; namely, the Christian religion, and which has certainly tempered a little the character of savageness, spirit of independency, bravery in warfare, irascibleness, revengefulness, and rapacity, which is ascribed to them by the other Abyssinians, and which, I think, is pretty correct.

April 16, 1842.—We left Enalka at sunrise. The priest whom I mentioned yesterday accompanied us for some distance, and showed us the road to Lat. In consequence of his advice we gave up the plan of taking the shortest road to the lake Ashanghe, and thought it better first to proceed to Lat, and there to make further inquiries respecting the security of our way. We ascended till about ten o'clock. On the whole of our road, we saw only one hamlet, called Dafat. Having arrived on the mountain, which we had been ascending since we left Enalka, we had a pretty view of the mountains of the Raia Gallas in the east. They pointed out the position of the lake Ashanghe; but it being surrounded by mountains, I could not see the water. From what I heard, however, I must conclude that it is not so large as lake Haik; at all events there is no island in it. I was told that there are many villages around the lake, where there is a weekly market held. If I am not mistaken, I heard that the largest village, where the market is held, is called Wofila, close to Ashanghe. This is at the same time the name of the whole district or province. On the eastern shores of the lake are Gallas, and therefore great care must be taken which road you go in these hostile regions. I afterward very much regretted that I allowed the people on the road to discourage me with their statements of the insecurity of the access to the lake from having seen this interesting part of the country, as I was not more than eight or ten miles from the lake. But the desire of getting rid of his miseries and hardships frequently prevails on a traveller to let many opportunities escape, which, if he availed himself of them, would afterward afford him the greatest pleasure from the favourable success which might have crowned his scientific endeavours. I was told by a native, that there is another small lake at some distance from the large one: but I have forgotten its name.

About twelve o'clock we arrived in the village of Lat, which is of considerable extent.. I do not recollect having seen such a large village since I left the country of Yechoo. Probably the name of Wofilat is to be derived from the Amharic Wof-Lat, which means *a fat bird*. But I do not know to what this origin of the name refers.

We only intended to rest a little from the fatigues of our road, to inquire after our route to the lake Ashanghe, and then to go further; but the Alaca of the Church of St. George, who pretended to have seen me at Ankobar, begged me to stop. He delighted us with a cake of bread and a quantity of hog's beans. I learned from him that the Governor Wolda Medhen had encamped at Wofilat close to the lake Ashanghe, in order to collect the annual tribute of the people, consisting of sheep, cows, barley, hog's beans, &c. From the description which the Alaca gave me of the soldiers of Wolda Medhen, I was not induced to go to the lake under present circumstances, although our route to Antalo would have been thereby shortened. A compass would have been useless, as the very route which we had now taken to avoid going to Ashanghe, afterward took us to the Governor of whose soldiers we were apprehensive.

April 17, 1842.—We travelled for some time in the bed of a river which flows to the Tacazze, from which we were distant only a few days' journey. We were obliged to halt in the river till after daybreak, in order to be sure of our direction to Antalo in Enderta. After daylight we saw a village at some distance, and people coming up to us. But they could not tell us anything about the route to Antalo, which was still far off. We then asked whether this was the route to Bella Georgis, to which they answered in the affirmative. On asking about the residence of the Governor, we learned that he had not yet moved from the villages of Ashanghe.

About ten o'clock we crossed another river, the name of which I could not ascertain. Its course was north-north-west, and it carried down a considerable quantity of water. Before we reached this river, we could scarcely find our way through the thorns and bushes, which caused us many difficulties in advancing toward the river. Our clothes, which we were obliged to preserve as well and as long as possible, as we had no others, were considerably damaged in this thorny jungle.

About eleven o'clock we reached another river. We halted a little, and collected a quantity of ripe fruit of the wanza-tree, which appeased our appetites a little. From thence we passed by a village situated on the foot of a high mountain, which he had now to ascend. The country of Wofila appears to be better inhabited and cultivated than that of Angot and Lasta. Since we had left Lat, we observed many villages and tracts of land well cultivated; but the reason is, that the destruction of Ras Ali's war had not extended so far.

We reached the top of the mountain after mid-day. Our passage was sometimes extremely difficult and narrow. The banks of the mountain had sometimes the appearance of high walls of rocks, a slip from whence would cause certain death. Toward the end of our ascent we observed several houses close to the way-side. We understood that they belonged to a Governor who is charged with watching the road. Nobody troubled us. as we had nothing that attracted their attention;

but should a traveller pass by with much luggage, he would certainly be detained by this Governor.

Having reached the top of the mountain, we learned that the Governor Wolda Medhen with his troops had moved this morning from Zelga, and that he was expected in Bella Georgis this afternoon. This was bad news to us: however we hoped that we should be able to pass by before his arrival. We marched as quickly as possible, although we were so tired, that we could scarcely move our legs, having commenced our march before daybreak. We had two roads before us; one leading east, and the other north-east. The position of Zelga, where the Governor was said to be, appeared to me precisely east. I therefore proposed to take the route of north-east, thinking that the distance from Zelga might be so considerable that we should not meet the Governor. But in this I was perfectly mistaken.

We went on as quickly as we could; but unfortunately we met no one who could give us better information of the Governor's movements. We at last saw a large village, to which we directed our steps; but on a sudden we were stopped by the deep and wall-like banks of a torrent. We had then to turn eastward; but having travelled about three miles more, we reached the banks of another steep hill, from which we could see down into a little valley, where a part of the Governor's troops were encamped. Escape was now impossible, as they had seen us on the top of the hill.

April 18, 1842.—The Governor treated them well, and they started the next day. When we had crossed the river Ghebia, we were overtaken by violent rain. The wind and rain rendered the air rather cold, and the hard work of ascending and descending the mountains and hills on foot produced a continual perspiration. Fortunately we reached in due time the village of Karanghē, where a man kindly received us into his house, lighted a fire, and made us a litte comfortable. Falling upon my knees, I offered the sacrifice of humble thanksgiving to Him who had graciously brought me a further step on my tiresome pilgrimage. Thus it is with the life of a Christian, who is now in sorrow, and then in joy; who weeps in the evening, and rejoices in the morning, till his earthly journey is over, and he enjoys eternal and immutable happiness in heaven.

April 19, 1842.—We left Karanghē before sunrise in a north-east-east direction. As this day was the anniversary of St. Michael the archangel, our host was gone to church before we could take leave of him. Although we had yesterday ascended considerably, yet we had to ascend this morning still more through a country full of thorns and grass. We saw, however, no village, nor did we meet any inhabitants. I was struck at the great number of partridges, which I have nowhere seen in such abundance as on this mountain. One charge would have provided us with food for several days; but our weapons were in the hands of Adara Bille. The country around was extremely hilly, and reminded me of Geshe in northern Shoa. Torrents run between the high and steep mountains, which were full of thorns and trees of various kinds of wood. Having reached the top of the mountain, we had a pretty view of the provinces of Wag and Semien. The mountains of Semien appeared to be elevated to the sky, till the clouds withdrew their tops from our contemplation. One of the highest mountains of Wag is Biala, on the eastern foot of which Sokota, the capital of Wag, was said to be. Here resides the present Governor, Taferri, who sends from hence his officers at certain times over the whole country to collect tribute.

About ten o'clock we entered the district of Bora, the name of which is derived from the white stripes which mark all the hills around. Each stratum of rocks presents a white and somewhat grey appearance to the eye. There are several large caves, which might give shelter to several hundred men.

In the first village of Bora we met the Governor, Woldaa Michael.

About twelve o'clock we passed the river Shemsheho, which runs to the Tacazze. It carries much water in its bed, which is surrounded on the banks with beautiful trees. The river is full of fish; and we saw about thirty naked men busy in catching the fish with their hands. It is well known that fish is a substitute also in Abyssinia for all other kinds of meat during the time of fasting. I understand that the new Abuna has forbidden the eating of fish during the time of fasting.

From the river Shemsheho, which comes from the east, we had to ascend through a wilderness, the thorns of which again damaged our clothes, of which we took care as of gold or silver.

About three o'clock we were overtaken by thunder and lightning, and subsequently rain, which troubled us in general every afternoon till we had reached the province of Tigre, where we had very little rain. Fortunately we found shelter in a cave of rocks close to the wayside. After the rain had ceased we continued our march, having resolved not to pass the night in the Agau country, but in the first village of Enderta, which we were told we could reach before night. We were unable, however, to accomplish this, as the rain again overtook us. At a distance we saw a village on a small hill, and we thought it better to go on in the rain, than to be overtaken on the road by the approaching night. But when we came near the village, we learned to our great astonishment, that the Governor of the district had encamped in this place, and soon afterward we saw some soldiers coming towards us.

Thus we had arrived on the frontier of the Wag country, which is decidedly one of the most important and interesting provinces of Eastern Abyssinia. It would admit a larger population and a high degree of cultivation of the soil, if a better government ruled this country. It would be necessary, however, for such a government to do away with the system of annually plundering their own subjects, as this is the very means to destroy commerce, order, cultivation of the ground, and every improvement of human society. At present the Governor comes annually with his troops and takes away what he pleases; and the consequence is, that the inhabitants conceal their treasures, and take flight to the mountains; whereupon the Governor destroys their houses and fields. As Wag is a country intersected by deep dales, torrents, and steep hills, which only allow certain passages to their tops, and as the rivers of Wag would be defended, and their storehouses well preserved on the top of their almost impregnable hills, the inhabitants would be able to check the strongest invading army, as has been the case frequently in the annals of Abyssinia. Ras Ali on his last expedition, which he several years ago attempted against Wag, is a remarkable instance of this. He invaded the country with a considerable army; but finding the natural bulwark of the country too strong, he returned, being content with devastating those districts which admitted of access. This natural fortification is the reason why the people of Wag yield only a very loose allegiance to the rulers of Gondar; why they throw off their yoke whenever they please; and why they use a haughty language toward the rest of Abyssinia. The chief Governor of Wag is said to be in possession of several thousand matchlock-guns, a report which may be true, as he can obtain everything that he wants from Massowah.

The principal market-place of Wag is Sokota, the capital of Wag's Governors. Its merchants carry their goods to Woldaia, the capital of Yechoo, and go and fetch other goods from Antalo, or even from Massowah. They are principally engaged in carrying salt-pieces to the south of Wag, a business which proves of great profit to them, as the value of salt-pieces increases in the southern countries.

The Agaus chiefly cultivate barley, wheat, red pepper, and maize. Their houses are of the same construction as those of other Abyssinians.

I was told that gold is found in the country of Wag; but I cannot say whether this report is true or not. I was frequently asked whether I knew how gold was to be discovered in mountains, and how the works in mines are managed. They entertain, like all other Abyssinians, the idea that white people only come to their country in quest of gold; and that a white man knows the places where there is gold.

April 20, 1842.—Early this morning I went to the Governor to thank him for his kindness, and to take leave of him. On being admitted to his presence, he appeared still more civil to me than yesterday. I did not venture, however, to ask him for provisions, and he did not offer any thing of his own accord. Having at his request given him a blessing, I started from the camp, and moved toward the river Tzana, which separates the country of Wag from that of Enderta and Tigre. This river rises in the east of the mountains of Wag and runs to the Tacazze. It is a fine river, and carries much water in its bed. Its banks are surrounded with trees and bushes, and the inhabitants of Wag and the people of Enderta on the other side cultivate every spot of soil which they can gain from the river. We arrived at this river at eleven o'clock A.M. The last part of the Wag country, through which we had travelled this morning, was well cultivated and inhabited. This is probably owing to the inhabitants being close to the frontier of another country, to which in time of war they can take refuge. I left the Wag country with feelings of great satisfaction, although I had experienced in it a great many difficulties. Its costly water—its healthy air—its pretty scenery—the manner of its inhabitants in asking for a blessing—and its interesting system of mountains and rivers, will be ever remembered by me.

To our great sorrow we learned from people coming from Antalo, the capital of Enderta, that the whole of Tigre was in a state of confusion and rebellion.

The man who had apprised us of the state of things in Tigre, likewise informed us that the expelled Governor of Silloa had collected a new force, and would probably come to an engagement with Guebra Medhen. He therefore advised us to reach Antalo, if possible, before the road should be disturbed and rendered insecure by the fighting parties, and before starvation, which is always the consequence of such disturbances, would render our situation still more precarious.

Having crossed the river Tzana, which runs in a deep bed between a range of mountains on both its banks, we had a long and difficult ascent before us. The cry of the Agau ploughmen resounded strongly in the dale of the Tzana, and made us sometimes believe that there was a body of troops engaged in fighting, or a brisk quarrel between some parties. I observed the same custom in many parts of Enderta. When ploughing, they make as much noise as possible, in order to driveon their bullocks, with which they converse as if they were rational companions. A traveller unacquainted with the custom, or not understanding the language, would imagine that there was a quarrel or a plundering party at hand, and give way to unnecessary apprehensions.

Having arrived on the mountain which we had been ascending since we left the river Tzana, we halted in the village of Bora, being the first village in the province of Enderta. Here we learned that Guebra Medhen had moved his camp to the east, toward the village of Shebrāra. We resolved to go and see the Governor, having found from experience that it is better to travel under the protection of the Governor of a district or province. Our road led us over rocky hills, dales, and torrents, which we cared little for, as we wished to reach the camp of Guebra Medhen before night. I can understand why the Governors of Wag and Enderta ordinarily live in peace with each other: the frontier of both provinces being of such a nature, that the difficulties of making inroads or entering these countries for the purpose of conquest are almost insurmountable.

Having arrived in Shebrāra, we learned that Guebra Medhen had moved still further to the east. We were therefore compelled to give up our intention of reaching him to-day; besides, it was already evening. We then looked for a lodging for the night; but every body in the village made an excuse, by saying that the Governor had taken their property, and rendered the people unable to receive strangers, as they had scarcely food for their own wants. With sorrowful hearts, and sometimes weeping, we went from house to house, till at last we found a host who gave us shelter and food for the night, which was a rainy and cold one.

April 21, 1842.—Soon after day-break we departed from Shebrāra. We did not, however, take our direction to the camp of Guebra Medhen, as we had been warned by our kind host not to go there, intelligence having been received that the Raia soldiers had killed some people travelling from Antalo to the Wag country. Our host accompanied us for a considerable distance through a by-way, by which we were able to avoid the Governor's camp. But this by-way soon led us into a wilderness, where we entirely lost our road.

About ten o'clock we traversed a large woody plain, where we found several ruined villages, but no inhabitants whom we could ask about our road. We went on in this wilderness till about midday, when we found another village, likewise ruined, but not a single person able to show us the road. We still proceeded on in an easterly direction; but we saw nothing but thorns and ruined villages on some hills. Fatigue, thirst, and still more our apprehensions of the neighbouring Raia Gallas, began to vex us to a considerable degree. Being always afraid of these Gallas, I proposed to turn round to north-east and north, as this direction could not lead us to their country, whatever might be the consequence. But this new route led us to a complete wilderness, where we could not observe the least trace of a human foot. The grass was so high, that we could not see one another, and one of our party was several times left behind; a circumstance which caused the most painful feelings, as we could not venture to make much noise on account of the Gallas, and as the approach of night precluded us from halting, and compelled us by all means to find our right road. However, we fortunately joined our man, whom the fatigues of the day, the height of the grass, and thicket of thorns, had prevented from keeping pace with us. But the worst part of our difficulty now began. Driven in on both sides by the steepness of mountains, which we are unable to ascend, not knowing the path, we were compelled to jump from rock to rock, the space between which was sometimes overgrown with grass. Having fortunately finished this manœuvre, we were received again by our enemies, the thorns, through which we had to wind with the utmost precaution. Profuse perspiration from this exercise—thirst and weariness—fear of the Gallas—apprehensions as to the road and approach of the night—and sorrow for my poor starving people, exhausted me so much, that I laid down on the ground to rest, whatever might befall us in this precarious situation.

All around being as still as night, and seeming destitute and lost, we heard the purling of some water in the neighbourhood. Creeping up and down through high grass and thorns, we reached a small rivulet. This discovery produced a greater cheerfulness in our minds, than the discovery of the sources of the Nile would have given, as we now were in hope of getting out of this dreadful and endless wilderness. We first refreshed ourselves, by drinking of the delightful water of the rivulet, and then followed its northern course, in the cheerful conviction that it must lead us to our lost road. About five o'clock we had the unutterable joy to find a road crossing the rivulet and leading up to a mountain, from which we thought we should be able to see or to learn something of the position of Antalo. The road was pretty large, and trodden by the paces of men and animals, and we had no more doubt of this being the way to Antalo. Having walked about half-way up the mountain, we were met by a small company of people, who proved to be some priests and soldiers of the Governor Guebra Medhen. They had set out from Antalo for the purpose of joining their master in the camp. They were astonished at finding us quite alone without a guide or a caravan in the present state of disturbances. They regretted that we had not visited their master, who, they said, would probably have given me a mule. They then confirmed us in the truth of our way, which we then prosecuted as cheerfully and as quickly as our tired legs would allow. On arriving at the top of the mountain, we saw a large valley, and many villages in it. We accordingly directed our course toward the valley, and took shelter for the night in the Church of St. Michael, in the village of Mawoini, as the villagers would not receive us.

April 22, 1842.—The terrible noise of the singing priests, and still more the fleas, which are always the greatest torture to those who pass the night in churches, had annoyed me so much, that I waited for the break of day with the most ardent desire. As soon as it dawned, we got up and departed, having taken leave of the priests last evening. On our road, which was exactly west, we saw plenty of villages ruined by Oubea, of whose barbarity the whole country bears witness.

About nine o'clock we crossed a rivulet called Gumalo. It runs through a dale of steep banks. On the western banks we saw a village, which we took for Antalo, as the priests of Mawoini had told us that Antalo was quite close, and that we should reach it in the forenoon; but I frequently found that priests and monks in Abyssinia have not the least idea of distances. Having crossed the rivulet Gumalo, we had to ascend considerally till we reached the village, likewise called Gumalo. The banks of the bed of the rivulet are well cultivated, as the soil can be watered at all times. From Gumalo we marched southwest, and were led to an immense plain with some slight elevations. On arriving at this plain, we got a sight of Antalo, situated at the foot of a mountain, the soil of which presented a red appearance. The plain, however, was considerably lower than the situation of Antalo; and the nearer we approached the town, the more we had to ascend.

When we first saw it, we thought it nearer than it really was, as is frequently the case when a traveller calculates distances from a plain. He is often greatly disappointed, and it requires more practice in calculating distances, than one would commonly think.

We arrived at Antalo about four o'clock. Being unable to find a lodging for the night, we repaired to the Church of St. George, which was splendidly built by Ras Wolda Selassieh, who is well known by Mr. Salt's Mission to this ruler.

I had intended, for many important reasons, to proceed from Antalo to Adowah; but the distance of three or four days deviation from my route, the way being through starving and disturbed countries, and my ardent desire to reach Massowah, and the end of my miseries and hardships, would not allow me to make this deviation from my road, although I knew that I should deprive myself of much valuable information respecting the state of the country, which I might have been able to collect at Adowah.

Antalo to Ategerat.

April 23, 1842.—At a very early hour this morning we set out for Chelicut, about six or eight miles from Antalo. A body of soldiers, who were going to join Balgadaraia on his march to Adowah, accompanied us for a considerable distance. They were in a very good humour, and had great hopes of a favourable result of their master's expedition. Our road was pretty plain. Not far from Antalo we saw an immense flock of baboons, called in Amharic Ratchie, of a somewhat white colour. They were close to the wayside. I was surprised at the good order in which they marched, some large ones walking before and behind each line, which they formed. After a few moments they halted a little and gazed at us, as if they were about to make an attack upon us. They marched round a small elevated spot, and then crossed our road, where they again halted a little, till they walked up to a larger hill in as perfect order as if they had been ranged in a square. The noise which their movements produced exactly resembled the bustle of a small body of horse marching over a grass plot. I thought it would be well were the Abyssinian soldiers in their military movements to imitate these brutes in the regularity of their march and continual circumspection and reconnoitring. This kind of monkey is generally acknowledged by the Abyssinians as a sort of more ferocious ape.

Chelicut is situated in a little dale intersected by a rivulet which provides the town with water. This situation in Europe would give rise to the establishment of numerous manufactories; but in Abyssinia nobody thinks or exerts himself to make use of the benefits arising from the nature of the country. The inhabitants are ready to accept with the greatest eagerness pleasing things as presents; but they have little desire to manufacture them by hard labour. Having heard that there was in Chelicut an Armenian who worked in leather, I called upon him; but I found him suffering in his eyes. I did not venture to ask him for a supply of anything, as I was aware of his being a poor man; but even the offer of a piece of bread, or a horn of Abyssinian beer, which at all events he possessed, would have delighted me more than anything else; but he appeared during his stay in Abyssinia to have adopted the same unkind and inhospitable manners which the greater part of the Abyssinians observe towards an unfortunate traveller.

From Chelicut we took our direction to Adigrate. Our road was pretty plain. Sometimes we had to ascend a slight hill. But although we had now a better road compared with that in Lasta and Wag, yet we were considerably inconvenienced from not having plenty of water, which we had found in abundance in those countries. Besides this, the heat of the valleys of Tigre was an addition to those inconveniences, with which our journey through that country abounded, from the inhospitable reception of the natives, and from the rumours of war and dissoluteness of the soldiers.

The approach of the evening reminded us of our unpleasant business of begging the people for a night's lodging. For this purpose we halted in a village called Arena Mariam, at some distance from the wayside. Having waited for some time on a place where most of the villagers could see us and guess our demands, we were invited by a man who was bleeding a sick bullock before his house. He sent us to a wretched cottage close to his house. This cottage, which served as a stable for his cattle at night, was surrounded by a stone wall, and only a part of the roof was slightly covered with grass. As the proprietor had invited us of his own accord, his duty was to treat us with some attention, but we had ourselves to look out for our supplies for the night. My servants went out begging, and returned with some flour, which we begged the wife of our host to make into bread; but she positively refused to do it, except she had a share of the flour, which was scarcely sufficient to provide half a ration for each of our party. The other materials, wood and water, we of course had to fetch ourselves. After a long dispute she gave us the vessels in order that we might prepare the bread. On giving us the vessels, she said, "I am giving them for my soul's sake."

We had rain at night; but fortunately not much, as our stable would not take off the water. I felt great consolation in thinking of the approaching termination of our journey, and I troubled myself little about the increase or decrease of difficulties.

April 24, 1842.—The unkind treatment of our host hurried our departure from Arena Mariam. Probably on our arrival he had expected that I could cure his bullock with charms, and with this expectation perhaps invited us to pass the night with him; but as soon as he found himself disappointed, he took no further notice of us.

The road to-day and the nature of the country was almost the same as we observed yesterday, namely, plains, with slight hills. The want of water was again very perceptible, and we seldom saw a village or a hamlet. The soil was stony, and not particularly fit for cultivation. We found no water till we reached the river Haikamesal, which runs to the Tacazze. Before we crossed this river, we met a man, who hearing of our distress for want of food, gave us a little basso. This is the flour of barley, which is first roasted on the fire in a vessel of clay, and then ground. When mixed up with water, it is no bad food; and it is quickly prepared when you are on a journey. We accepted this present from the man with cordial thanks, as our scanty repast of yesterday evening had left us nothing for this day, and the villages were far off from the wayside; so that we should have tasted nothing at all to-day, if Providence had not inclined this man to provide us with as much food as was requisite for our starving bodies.

Having refreshed ourselves with the water of Haikamesal and with the basso, we continued our march, in the cheerful confidence that He who feeds the birds, would also feed us in the evening, the approach of which always caused painful feelings in our minds, since we had experienced the inhospitality of the Tigrians. Had we possessed some property, we should not have been at a loss; for if the Tigrians see property, they will seldom be inhospitable. Such was the case with those travellers who overvalued Abyssinian hospitality. But let these persons travel without money, or without articles passing for money, and you will find that they will give another, but more correct idea and description of Abyssinian hospitality.

About midday we were met by a man, who informed us, with tears in his eyes, that he had been deprived by soldiers on the road of his provisions, his sword, and his cloth. We were sorry that we could not help him. At the same time our sorrow and apprehensions of perhaps sharing the same fate in the course of the day, caused us to look with anxiety and timidity after those places which we thought dangerous, and where we might fall in with a roving party of robbers. But our heavenly Father, who well knew that our sufferings had already been considerable, graciously preserved and protected us, and brought us to a village called Maberka, where, although a stable full of fleas was given as our lodging, and a sour look and other unkind treatment wounded our hearts, yet we were content with a few horns of beer and some paste called tello, being confident that we were daily advancing nearer to the sea coast.

Tello is prepared from the flour of barley. The barley is first boiled in a little hot water, and then roasted in a vessel of clay. It is then ground, and the flour is mixed up with water, or oil, or the Abyssinian pepper-soup. It is indeed a very miserable and disgusting dish; but necessity had taught us to despise nothing.

April 25, 1842.—This morning at a very early hour we left Maberka, and halted a few hours afterward in the village of Atzbie, where I learned that a European had been there a few days ago.

Having left Atzbie, the attack of fever, which had commenced this morning, increased. It was occasioned most likely in consequence of my sleeping last night outside the house on the wet ground in the cold, and without sufficient bedding. The fleas had tormented me so much, that I was compelled to risk sleeping outside the house. Happily, we were met by the same man who had yesterday given us some basso. As his village was close to the wayside, he observed us after we had passed by the village, and ran after us, and called us to his house. This invitation was certainly most providentially, not only because we had taken a wrong route toward the Shoho country, but still more as our halting in his house relieved me from fever. I had first hot water prepared for a foot bath, and then took a little coffee and went to sleep. On getting up I felt considerably better. We then ate and drank whatever our kind host placed before us; and having refreshed ourselves, we resumed our march. May God in His mercy give the man his reward in heavenly and eternal goods and blessings for the kind assistance which he gave me when help was needed! The meeting with this man yesterday and this morning, when we were in great distress, was indeed one of my most remarkable experiences on the whole journey, and clearly showed me the Divine interposition.

I was told by my host that we were only a distance of five days journey to the village of Borre on the coast of the Red Sea; but that the road was extremely dangerous, and that nobody would guide us there for fear of being murdered by the Danākil. Last year some Christian merchants endeavoured to open that route; but they were all killed by the Danākil. However, the Danākil bring merchandize from Borre to Tigre; but they will not allow other people to go down. As Borre is not far from Mocha, I should have liked to have gone this road; but who would guide me, how could I afford the means for the guide, and where had I other means of defence? Travellers, however, well provided and armed, might succeed in opening a road, which would be of the greatest importance for the intercourse with Tigre.

The white appearance of the mountains of this district struck me much, having nowhere observed it before. Our road was generally plain, and I should think camels might be able to go from Atzbic to Antalo.

In the evening we halted in the village of Masaot. As the villagers would not receive us, we went to the Church of St. Michael, the priests of which proved to be very light-minded, and frivolous in their conversation and manners.

April 26, 1842.—As there was a priest in Masaot, who promised to leave the place in a few days on a journey to Shoa, I availed myself of this opportunity to write a few lines to Capt. Harris, Her Majesty's representative in Shoa, to inform him of my safe arrival in Tigre. The priest promised to take charge of my letter.

We intended to take our direction toward Senafe to the Shoho country; but we were advised not to do so in the present circumstances of the unsettled state of things in Tigre. We therefore marched toward Adigrate. At nine o'clock we reached the market Gooila, and halted at midday in the village of Agoddi. Here we went into a house to beg for some food, as we were exceedingly hungry. Tello, mixed up with oil and red pepper, was again offered to us; but it had a better taste than formerly. The man who received us into his house, bitterly complained of Oubea's tyranny last year against the district, when he plundered and burnt their houses.

To our right we saw the high mountain Haramat, where Cassai made his defence before he was imprisoned by an artifice of Oubea, who swore eight times before ninety priests that he would do no harm to Cassai, if he would surrender himself and his stronghold. Cassai did so; but was immediately put in chains; and the priests, who reminded Oubea of his solemn oaths, were also imprisoned.

April 27, 1842.—We departed early from the village of Mashagheria-Mariam, where we had rested last night. About nine o'clock we arrived in Adigrate. I went to the Church of St. Chirkos, which, I was told, had been built by Mr. Eichinger, the German carpenter who was with Mr. Gobat in Tigre. The form of the interior does not differ from other churches, though there is more regularity observed, and is richly embellished with paintings and pictures, representing lions, elephants, hyænas, &c. It is a square building; but in other respects it is like the rest of the Abyssinian churches. The priests recollected Mr. Eichinger, as well as Mr. Gobat. One of them gave me some flour, and his wife also baked a few cakes for our use on the road. I could not help thinking in my mind that probably this priest had experienced much kindness from Mr. Gobat, which he was now desirous of returning to me.

Adigrate is situated in a large plain, surrounded by mountains. The village is not very large, and at present almost the whole is in ruins. It was built by Sabagadis, if I am not mistaken. The present Governor is Ayto Beraki, who however was not at home, having set out with Balgadarai on the expedition above mentioned.

Adigrate to Degonta or Tekunda.

Before we entered the village we crossed a small rivulet, which supplies the villagers with water. We did not stop long; but continued our march. On the left we saw a large village, which they called Kersaber, being much larger than Adigrate.

Our road this afternoon was not so plain as it had been for several days. Near the village of Mamberot we received some beer and basso from a man, whom the narrative of our journey had affected. He was ready to sell some grain; but what could we give in return? My head-servant, Atkoo, agreed to sell his belt, saying, it was better to do so than to starve. We then bought some barley and hogs' beans. Afterwards we went to the church of Mamberot; but as there was no priest, we could not pass the night in the church. We then endeavoured to beg for a lodging in the village; but all our petitions were in vain, till a man offered a stable, which was sufficient to shelter us against the coldness of the night and the wild beasts. Our situation daily became worse with the increase of the inhospitality of the Tigrians. We had travelled through a very hilly country, which was not much cultivated. probably for want of water.

April 28, 1842.—We left Mamberot with sunrise. Our direction was north-east. We marched through a very rocky territory, and but little cultivated and peopled. Here and there we saw a hamlet on the wayside. In the village of Dagadi we got some bread and beer. From thence our direction was east. We arrived at Behat about three o'clock P.M. We intended to stay here till the next day, in order to inquire after our road, and to collect some provisions for our journey through the Shoho country; but when we applied to the principal priest, whom we met on our way, we got the rude answer—"There is the road: do not stop here: rest somewhere else. We have no room for you." This roughness made us so sad, that we resolved to use our utmost endeavours to leave Tigre as quick as possible.

Behat is a large village, situated in a plain. Before we entered this plain, we had a very difficult and steep descent. Having passed the village, we had to ascend again and to march toward Teltal, a part of the Shoho country. Having reached the summit of the mountain, which we had been ascending since we left Behat, we came to another extensive plain, which was much cultivated and peopled, being surrounded by villages. This plain is the eastern boundary of Tigre, beyond which plain in the east there are no more Christians.

On our road we had heard from a merchant that the people of Senafe were very bad and fanatic Mahomedans; and that we could not travel close to the Shoho country without a guide or some other kind of protection. He therefore advised us to go to one of the numerous villages around, and to wait for the market-people, who, coming from a market in the neighbourhood, would pass Senafe to-morrow; and with them we should go to Tekunda, where the Governor, Ayto Habta Michael, would send us to Massowah. This intelligence was most providential, and indeed saved our lives, because, had we not known this, we should have continued our march, and should either have been entirely stript by the people of Senafe, or killed by the Shohos, as we did not know the road, and should have been led to the Shoho villages, which we should have taken for villages of Tigre.

Having obtained this important information, we went to the church of St. George, in the village of Shemasana.

April 29, 1842.—At daybreak this morning we set out from Meshaikh, the market people being desirous of departing before the heat would be too powerful. They came from the market of the village of Tchegnara, which was held yesterday. I was most fortunate in having joined this party, as the Governor of Senafe, a very ferocious looking Shoho, stood in the way, and declared that he would not allow the Egyptian to pass before he had paid him a dollar.

Having settled the business with the Shoho of Senafe, we proceeded on our road with the caffila through a stony and woody wilderness, being continually apprehensive of a body of Shohos attacking us from the east of our route. This woody wilderness, which was full of juniper-trees, is exactly the place for a gang of robbers, as they could break forth on all sides from the wood and catch you up in the narrow rocky road. About nine o'clock we found a well of good water. Here we halted till the whole party had assembled, and then the caffila separated, each individual taking the direction to his own village. Some took the direction to Halai, in the north; while we proceeded with some people to Tekunda, which is at present the usual starting place from Tigre to Arkeeko. I understood that an arrangement had lately been made between the new Naib of Arkeeko, whose name is Hassan, and Ayto Habta Michael, the Governor of Tekunda, that the Halai route should be given up, and that of Tekunda substituted. It appears that the Governor of Halai had a quarrel with the Naib, who opened another, and I must add, a much better route to Tigre.

We arrived at Tekunda after ten o'clock.

April 30, 1842.—The Governor promised this morning that he would send to the next Shoho village for a guide to take us to Arkeeko, as the Shohos would kill us if we travelled through their country without having a guide from them; but that this guide would require a dollar for his trip to Dohono or Harkeeko—not Arkeeko, as we always pronounce according to the maps. I said, that I had no money; but that I should be able to procure some at Massowah. He said, "It does not signify: the Shoho must wait, and I will wait also, till you have got money." It is customary for the Shoho guide to receive half a dollar, and the other half is given to the Governor; but I believe the Abyssinians pay only half a dollar to the guide, and nothing to the Governor. However, I would not bargain about this, as the Governor showed me much kindness in my distress.

May 1, 1842.—This morning the Governor settled the matter with the Shoho guide and wished us to depart; but the Shoho wanted his money to be paid before moving from Tekundo. To this I would not consent. Then the Shoho requested me to make oath that I would not leave him in the lurch at Dohono. I replied, that whether I made oath or not, it would be the same thing, as it would depend on the man who would lend me money at Arkeeko or Massowah; that if any one would lend me money, I would pay him without taking an oath; and that if no one would lend me money, the oath would be useless, as I could not pay him. It would therefore be better for him not to press this matter, but to go with me to Dohono, believing on my word.

Tekunda *to* Arkeeko.

At length the Shoho gave in, and we immediately left Tekunda. After walking about a mile we reached a well, which is the spring of a river running to the Samhar. The people of Tekunda must go thus far to fetch their water. Tegunda is a small hamlet, situated on a hill; but it is now important on account of the communication with the sea, and I believe I am the first European who went this new route, which however falls into the old road after you have travelled about ten miles. I was delighted with having seen the beginning of a river, which runs from this point as far as the sea, some distance from Arkeeko; but the river is dry during the hot season, and the traveller finds water only at certain places. As far as I could ascertain without a compass, it runs first from south to north, then to north-east-east. We had a very good and plain road through a woody wilderness. It is much superior to the road of Halai, which leads over the difficult mount of Shumfaito. On the Tekunda road you descend by degrees, and the road might be trodden even by camels, if it could be improved a little by removing some rocks in the way.

We saw several burial grounds on our road. On passing by, our guide always recited certain prayers, the nature of which I could not make out, as he did not understand either Amharic or Arabic. The appearance of graves in the wilderness also produced a deep impression on my own mind.

About eleven o'clock we arrived at that part of the Tekunda road which joins that of Halaia. I immediately knew this route, and we afterward came to a few high trees, where the camels are usually discharged and sent back to the coast. I recollected the spot well, where, four years ago, I had three days quarrel with the Shohos for the hiring of bullocks to carry my luggage up the mount Shumfaito.

May 2, 1842.—With ardent desire we saluted the approach of this day, which should bring us almost to the end of our miserable journey We arose about four o'clock; and although my legs would scarcely move, in consequence of the exertions of yesterday over the stony bed of the torrent, through which the road led us, yet the anticipation of my journey's end gave me, I might say, more than human strength. We travelled almost the whole day, although the heat in the narrow valley was very excessive. About two o'clock P.M., we arrived at a place, where the guide told us that we were to take water, as at this season. no water would be found till we reached Arkeeko. We then filled a large water-bag, which the Governor of Tekunda had given us for the road, the guide being ordered to take it back to Tekunda. We travelled on till after night, when my legs were so tired and sleep overtook me, that I frequently fell down on the ground. Thirst and hunger also tormented me to the utmost. The more I drank, the more I got thirsty; and the few hogs' beans which my people still carried with them, did not satisfy my appetite. After night-fall I had requested from the guide to let us have a rest anywhere in the wilderness, as it was almost impossible for me to move any further; but he declared that he could not stop on account of the Bedouins and the wild beasts, and that we should go further on till he would point out a proper spot; I said, "All this does not signify, let us only rest a few hours." However, I continued marching for some time; but as the guide would still not listen to me, I laid down on the ground, and said, "You may now do as you like: you may stop or go; for my part I will not move from here till I have rested a few hours." He then gave in, and we rested on the sand only a few yards from the wayside. In this situation we were most fortunate. I did not think that the Bedouins strolled over the wilderness at such an early hour—about two o'clock in the morning. But a party of Shohos passed the very road where we were sleeping. Probably they were of Waia, a Shoho place a few miles distant from us. Happily they did not observe us, probably taking our figures for stones. Besides, they were talking so loud, and making so much noise, that they did not hear the snoring of my people. Only the guide and myself awoke from their noisy conversation; but we kept silence as much as possible, till they had passed altogether. I was in a very melancholy and apprehensive situation at this moment, as I did not know whether they were friends or enemies. I could not however expect that a Shoho, who is an enemy in the day, would be a friend at night. As soon as they had passed by without observing us, we resolved to depart. The moon just rose, being in her wane. The idea that we were so close to the coast, and that any unfortunate occurrence should happen, gave us courage and strength to reach the coast as quickly as possible. Besides, hunger, thirst, and weariness, drove us on. The immense plain of Arkeeko annoyed us very much. We saw the place from a distance; but although we exerted ourselves to the utmost, it still appeared distant and unattainable.

About nine o'clock A.M. we reached the wells of water close to Arkeeko. We first paid a hearty welcome to this water, as ours was all gone. We then moved, almost half-lamed, toward the house of the Governor, whom we happened to meet on the road. He ordered his servants to give us a room and to make us comfortable.

DR. BEKE'S JOURNEY FROM DEBRA TABOR BY SOCOTA AND ANTALO TO ADOWA, IN MARCH AND APRIL, 1842 (marked Q and L in Map, page 39).

March 12th.—I remained at Máhdera Mariam over the Saturday's weekly market, in order to inquire as to the state of the country beyond Antálo, it being reported unsafe to travel through, but could obtain no definite information on the subject. Nevertheless, I this morning left for Débra Tábor on my way north-eastward, it being my determination not to go to the so oft trodden road of Dembea and Sémien, if it could by any means be avoided. The road is very irregular, winding up the mountains, and crossing the rivers Sébat-Wódem-Gumára and Sénsaho-Gumára, the latter being the principal stream of those bearing the common name of Gumára, which have their rise in the high range of Mount Gúna, bounding the basin of Lake Tzána to the E., or perhaps more correctly the S.E. The country, as far as I have seen it, is but scantily peopled, and little cultivated; to the westward, however, the low flat country towards Lake Tzána is said to be extremely fertile. Débra Tábor, where we arrived soon after noon, was founded by Ras Gúksa, the grandfather of the present Ras Ali, who made the title hereditary in his family, and whose remains lie buried in the church of "Yesus," situate on an elevation of the ridge between the basins of the Gumára and Rĕb, which elevation is properly Débra Tábor, the capital of that name being placed on its skirt to the north-eastward. This town was last year burned down by Dejach Biru, shortly after my arrival in Gojam, and is not yet rebuilt, it consisting for the greater part of a rude assemblage of mere huts. On announcing myself at the palace, and stating, in answer to numerous inquiries, from whence I came, where I was going, &c., &c., I was not allowed the honour of an interview, the Ras (as I was told) being apprehensive that I might have brought "medicine" from his enemy Biru Goshu. But his curiosity led him, notwithstanding, to come out and view me from a distance, and he even went so far as to send for my cap to look at. He readily consented, however, to my journey through his country, and gave me a guide to Ebenat, the residence of Dejach A'bbic, ordering me a supply of food until my departure. From Débra Tábor, Lalíbala was said to bear N. 80° E.; the road thither, gentle travelling, is as follows:—Mókeria, Cháat Waha, Dibúkko, then cross the Tákkazie to Lalíbala. Beyond this one day's journey farther to Imerehá. Mount Gúna bore from hence S.E. Beyond it are Wádela and Daunt, between which and Amhára the River Báshilo is the boundary. The head of the Báshilo said to bear E.S.E. from Débra Tábor. Isti, S. 15° E., about seven miles; Yefag, N. 60° W., one day's journey; Mount Melza, N. 20° E.

14th.—This morning I left Débra Tábor without holding any further communication with the Ras. We now descended into the valley of the

Rĕb, crossing on the way the Gibúda, a tributary of the former, nearly of the same size as the principal river of the name of Gumára. The Rĕb itself is of larger size than any of the Gumáras, being at the ford at least 10 yards in width, but shallow—that is, less than a foot in depth. Lower down, however, when all the latter rivers have united and formed one stream, the Gumára is said to become a more considerable river than the other. On crossing the Rĕb we entered the district of Ebenat, and began gently ascending the mountain towards the residence of the Governor, which bears the same name. In about three hours we reached the summit, when we came to a plain on the N. side of Mount Melza, a cross ridge of the range of mountains forming the watershed between Lake Tzána and the river Tákkazie. The elevation of this plain I cannot estimate at much, if anything, above 7,000 feet. Crossing it north-westward, we, towards evening, approached Ebenat; but it not being considered proper for me to make my appearance before the Governor so late in the day, we turned off to a small village called Gúltoch, situate close to the first break in the ground towards the Tákkazie, and near to a small brook named Sánkisa, the waters of which descend towards that river. Beyond Débra Tábor the country improves somewhat in its character and appearance, being, although still mountainous, of more even surface, and possessing more population and cultivation. From Gúltoch Mount Melza bore S. 35° E.

15th.—Ebenat, whither we went up this morning, is a place of no size, consisting merely of the residences of the Dejazmach and his family, with a few huts for their attendants. The market is, however, very considerable—lasting two days, Friday and Saturday—it being the point where the merchants of Gojam, &c., meet those from Sókota bringing salt, for which they give cloths, coffee, and a large number of cattle, oxen, and heifers brought from Gúdera. Farther from Ebenat they do not go eastward, this being a cross-road. The two grand mercantile lines from N. to S. through Abyssinia are, the one by Adowa and Góndar (or Dérita), and the other by Antálo and Sókota to Warrahémano, besides which, as I have learned since leaving Máhdera Mariam, there exists a middle road from thence by the way of Hádesha in Beléssa (Belessen) and Témbien, which is sometimes taken by the Baso merchants going to the coast. The Governor of Ebenat is a Mohammedan; and many of the inhabitants of this country are of the same religion. Islamism is, in fact, making strides over Abyssinia (as it is through the Galla countries in the S.), numbers of the Christians continually passing over to it. I am now writing (at Ebenat), whilst in my hut chatting with my servants are two of their relations, natives of Gojam, settled here, both of whom have become Moslems. Mount Guna bore S.S.E.; Mount Melza, S. 55° E.; Mount Débra Tábor, S. 10° W.

21st.—I remained at Ebenat over the market, intending to leave for Sókota with the merchants returning thither on Saturday evening. But the two servants whom Wálda Georgis had engaged, and who had thus far conducted themselves to my entire satisfaction, having here fallen in with several friends and acquaintances, merchants from Yaush, were by them ill-advised and frightened, and on the Saturday morning, when preparations were being commenced for our journey, they "struck." The one, having obtained from me a dollar "to send home to his family," left without hinting his intention to any one, and was not to be found; the other, after refusing all day to accompany me farther, was at length in the evening induced by Wálda Georgis (whose relation he is) to go on with him. But it was now too late, for the merchants for Sókota had left, so that we were compelled to wait till we could find another opportunity of proceeding on our journey. This occurred to-day, when the Dejazmach (who had previously refused to furnish me with a guide) being on the point of proceeding to Débra Tábor, whither he had been summoned by the Ras, in consequence of reports of the hostile approach of Dejach Biru Goshu, I was forwarded by him to Nikwára, the residence of Fitaurari Síyum, through the petty *shums* of the intervening country, in the same way that I was "passed" through Damot in December and January last. On leaving Ebenat this morning, the road lay at first for a short distance N. by E., through a rich cultivated valley, when it began ascending the mountains north-eastward, crossing a ridge of the name of Jírzu to the N. of the river Tékken, the ridge of Mélza (already mentioned) being to the S., and both being portions of the range bounding the Tákkazie to the westward, of which Amba Hai and the mountains of Sémien are also part. The summit of the ridge crossed by me, which I rate at probably 8,000 feet above the sea, is covered with olive-trees in such numbers as to form complete woods. This tree is common throughout Abyssinia, but I never met with it in such abundance as here. After continuing over the tolerably level summit for some time, we reached the village of Zibákwaha, overlooking the river Tékken, where we put up for the night. From Zibákwaha, Zoz-amba, a remarkable flat-topped isolated mountain, bore N. 20° E. beyond the river Níli. Lalíbala said to bear from hence S. 60° E.

22*nd*.—We did not leave Zibákwaha till the afternoon, as we were told that we could not reach Nikwára to-day. The real reason was, however, that our host of last night was unwilling to see me farther than the neighbouring village of A'derseg, distant less than a couple of hours' ride, its bearing being about N.E. of our last station.

23*rd*.—Notwithstanding what we had been told of the distance of Nikwára, we reached it this morning before noon, our course being about E.S.E. Shortly after leaving A'derseg we crossed the river Tékken, forming the boundary of the district of Ebenat. Its bed is about ten yards in width, but in the present season it is for the most part dry, there being merely two shallow streams, of two yards and one yard in breadth, running through it. The country of Biégemider, from the rise of the basin of Lake Tzána eastward, is a mass of mountains intersected by deep valleys, getting more barren and desolate the farther one proceeds. In Ebenat they said that the country was formerly peopled, and at Zibákwaha they pointed out to me the sites of several villages, the inhabitants of which, they said, have either died off or removed in consequence of the oppression of the present governor. Whatever may be the truth of this statement, it is quite certain that this is not the sole nor even the principal reason, since after passing the limits of the Dejazmach's jurisdiction, so far from finding improvement, the country appeared to me to get worse. It is the scarcity of water that is the main cause of its deserted state. At one of the villages on the road, where we stopped to change our guide, the distance that the women had to go to fetch water for domestic purposes was quite appalling. In Europe, Biégemider is said to be noted for its fine flocks of sheep, and the name of the country has been derived from the word *bag*, meaning "sheep." But it is only necessary to see the country to be satisfied that it never was and never can bé a sheep country. The high Wollo plains to the S.E. are, on the contrary, justly celebrated for their fine breed of long-wooled sheep, the fleeces of which are prized throughout the whole of Abyssinia, being worn over the shoulders by persons even of the highest rank. From Nikwára, Mount Melza bore S. 45° W.; and Lalíbala was said to bear S. 75° E.

24*th.*—Fitaurari Síyum being at A'zwi, a short distance to the E. of Nikwára, we left this morning for that place, which we reached by noon. The barren mountainous country is so far from improving, that during a ride of nearly five hours I did not pass a single village, or cross or come within sight of a single brook or spring. From A'zwi, Zoz-amba bore N. 45° E.; and the valley of the Merri, between Lasta and Waag, N. 85° E.; Mount Melza, S. 65° W.; Mount Biála, a high mountain mass above Sókota, N. 75° E.; Lalíbala said to bear S. 65° E., four days' journey, or three quick travelling.

27*th.*—Yesterday, on Fitaurari Síyum's leaving for Nikwára, he recommended me to the care of his brother Aito Háilu, whom I was to accompany across the Tákkaẓie. The latter accompanied the Fitaurari a part of his way, but returned this morning; and by his direction I removed to the village of Dugaláiba, about a mile lower down, he having taken up there his quarters for the day. At Dugaláiba I was informed that Tzelásferri and Mékkina are two A'mbas in Lasta, on the opposite sides of the river Tákkaẓie, the former being beyond the river towards Wádela; the latter on this side towards Lalíbala.* The head of the Tákkaẓie is less than a day's journey beyond Lalíbaḷa. To Lalíbala from Dugaláiba is three days' slow travelling; quick, it may be reached in two days. The first day Búgana in Lásta is reached; the second day Lalíbala. The head of the Tákkaẓie is in the district of Gedán, beyond which is A'ngot. All the opposite E. side of the Tákkaẓie, from Waag to Gedán, is Lasta.

28*th.*—This morning early we left Dugaláiba: our road kept descending north-eastward,* at the end very steep, till we again came to the river Tékken, down the bed of which we continued, going now more eastward, for nearly two hours, incessantly crossing the small stream: if I counted right, it was 36 times in all. On reaching the junction of this river with the Tákkaẓie, we stopped to rest during the heat of the day, Aito Háilu at the same time superintending the collection of duty from a numerous party of salt-merchants passing into Biégemider from Sókota. In the afternoon we continued our journey, crossing the Tákkaẓie and entering Lasta. Our course was for about half-an-hour N.E. down the bed of the river, the stream of which we crossed three times before we left it. From this fact alone it will be evident that the river is of no great size here. At the places where we entered it it was some 20 yards wide, with a depth of about a foot, running briskly, but by no means with violence: at the deeper part of the stream it was less than ten yards in width, whilst the current seemed scarcely to possess motion. The elevation here of the bed of the river above the ocean must, I think, be about 4,000 feet. Since the destruction of my thermometer I have no means of deciding as to heights beyond my personal judgment, grounded on past experience: no certain dependence is therefore to be placed on my estimates; nevertheless they may not be altogether without value as rough approximations. We continued down the Tákkaẓie till we reached the river Mérri, an affluent of the former, of about the same size as the Tékken, along the side of which we ascended, and then crossing it we began a steep ascent to the high land of Lasta. It was near sunset when we reached Zélesa, the residence of Aito Háilu, who is the Shum of A'bam, a subdivision of Dáhana, as

* This is totally irreconcilable with the information given by Mr. Salt in his "Voyage to Abyssinia," p. 279, on the authority of Pearce.

the country to the N. of the Mérri is called: beyond that river to the southward is Búguna. These two districts, together with those of Sókota or Waag Proper, Wóffla, and Bóra, form the dominions of the Waag Shum—to call him Dejazmach would be an insult—who is one of the most important chiefs of Abyssinia, and whose territories are far more extensive than one can have any idea of from the existing maps.

Even at the present day, when the empire of Abyssinia exists only in name, the Waag Shum, although in a great measure subjected by the Ras, is not looked upon as a dependent chief bound to pay tribute; whilst the King of Shoa, although virtually an independent sovereign, is considered *the governor of a province*, who does not render tribute simply because the Negús (or his representative the Ras) has not the power to enforce its payment, but who, if the empire were reinstated, would do so as formerly, which the Waag Shum never did and never would. It is to be remarked that neither Sáhela Salássie, Negús of Shoa, Goshu Záudie, Dejazmach of Gojam and Damot, Gehanécho, Thato of Kaffa, nor any other of the numerous descendants of the imperial family *in the female line*, has the slightest pretensions to succeed, now or hereafter, to the imperial throne, the right to which is subject to the same law as as that regulating the descent of the crown of Israel—the power of selection among the *males* of the royal family; but with total exclusion of all claiming descent only through females. From Zélesa, the head of the river Mérri bore S. 75° E.; Nágala (high country), S. 20° W.; Aunánur, an Amba, with a convent of that name thereon, in Mékiet, S. 5° W.; the high country of Wádela, stretching from S. to S. 30° E.; the river Bérberi-wans is between Aunánur and Wádola, and joins the Tákkazie. Lalíbala, said to bear S. 50° E., three days' journey slow, or two days' quick travelling.

30*th*.—I remained a day at Zélesa, and this morning proceeded on my journey eastward towards Sókota, ascending the mountains between the valley of the Mérri and that of the Sábbaha, a smaller affluent of the Tákkazie, the summit bearing the name of Nárbilu; and in about three hours we reached the high level country—the elevation of which I estimate at from 6,500 to 7,000 feet—over which we continued E. a couple of hours longer, when we came to Gor-ámba, the residence of a relative of my late host. The language of the inhabitants of the valley of the Tákkazie and of Waag generally is the A'gau—the native name is Hhámera—in a dialect not near so harsh as that of A'gaumider, and to which (remarkably enough) the language of the Faláshas of the latter country bears a closer resemblance than it does to that of the people among whom they are now residing. At Gor-ámba the position of Lalíbala was visible, (which, at all other stations, it had not been,) bearing S. 40° E., two days' journey, or perhaps thirty miles. Tzelásferri is beyond Lalíbala, about half a day's journey; Mékkina nearer than Tzelásferri to Lalíbala, and the Tákkazie runs between them. Mékkina is the same as the head of the river. The road from Débra Tábor to Lalíbala is good, being *plain*. The head of the Bashilo is a long way off, said to bear about S.S.E. of Gor-ámba. At this place were further visible Mount Guna, bearing S. 35° W., and Mount Biála, N. 80° E.

31*st*.—Beyond Gor-ámba the plain country ceases, and the road continues winding, on a general bearing of E.N.E., along the summit of the ridge—sometimes only a few yards in width—between the valleys of the Sábbaha and the Mízrib, a tributary of the Mérri. After about two hours there is another flat, but of no very great extent, and then again a narrow ridge between the Mízrib and the A'rri, a tributary of the Tákkazie. The mountain-ridge thus traversed bears the name of Amdawárk; and I am told that the whole of Lasta is a succession of ridges and valleys like this. The flat summits of the mountain, which, being nearly the same throughout, give one at a distance the idea of an uninterrupted plain like Gojam, may be estimated to be throughout somewhere about the height last mentioned. In six hours after leaving Gor-ámba we stopped at Síkuna, formerly said to have been a convent of celebrity, but now reduced to a few huts, situate under the precipitous face of the rocks bounding the basin of the Arri. From Síkuna the summit of Mount Biála bears N. 85° E., and Mount Gazgíbla, the head of the river Mérri, S. 70° E.

April 1st.—It was my desire to reach this evening Máskalo, the residence of the Dejazmach of Dáhana (or Dáhana Shum, as he is called here), and therefore I was off early from Síkuna. But on reaching Kitára, the frontier-post of Waag Proper, distant about five miles E. by N., I was stopped by the "custom-house officers," and, in spite of all I could say or do, detained until the Dáhana Shum's pleasure was known, for which purpose Wálda Georgis and a servant of Aito Háilu, who had accompanied us, were despatched to Máskalo. I in the meanwhile remained encamped in a fine level plain, forming a portion of the ridge along which our road still continued, whilst my detainers were employed in collecting the duty from the salt-merchants arrived to-day from Sókota; and in the evening I was taken by the Shum to the small village of Katzemán, there to wait the return of our messengers.

3rd. Yesterday morning Wálda Georgis returned with orders from the Dáhana Shum that I should instantly be forwarded with my luggage, &c., without molestation; however, it was not till this morning that I started. The road continued eastward over the mountains, which now form the southern boundary of the basin, first of the Arri, and afterwards of the Télla, another tributary of the Tákkazie, till in rather more than six hours we reached Máskalo—properly Máskala Krístos—a town with a convent at the foot of Mount Biála, which the present Dáhana Shum has chosen for his residence, although not within his government, in order that he may be near Sókota, the capital of his brother the Waag Shum, Mount Biála (which is sometimes called Mount Máskalo) is a prominent portion of a large mountain-range which apparently descends from the system in which the Tákkazie has its source, and which range, running northward or north-westward, separates the basin of the upper course of that river from that of the Tzelári, a river of which I shall have occasion to speak in the sequel. The summit of Mount Biála bears from Máskalo S. 75° E.

4th.—Leaving Máskalo this morning in company with the Dáhana Shum, who fortunately was going to Sókota on business, a sharp ride of about five hours N.E. by N. brought us to the latter place. The road skirts the foot of Mount Biála, crossing spurs from it, and then traversing a tolerably level country it descends to the capital. Sókota is a place of considerable size, but is so very straggling that it is not easy to form a definite idea on the subject. It has a large market, held on Tuesday and Wednesday weekly, which is frequented by the merchants of the south and west, this place being the great centre of the salt-trade, the Tigre merchants coming thus far only, and then returning. By walking about half a mile eastward from the town to the summit of a low ridge of mountains, I obtained a view of the country to the E. beyond the river Tzelári, some portions of which were pointed out to me by a person who accompanied me. A'shangi (this being the name of a district), distant two days' journey, bears from Sókota about S.

80° E. The first day's journey is up the mountains to Záffa in Wóffla, bearing S. 60° E. The lake of A'shangi is between A'shangi and Wóffla; and Lake Máchakh (the smaller lake of A'shangi) is in Wóffla, being to the S. of the large one. The rivers of A'shangi go to Ráia, which is below the former; beyond it being Zóbul, and then Adál. A'shangi belongs to Tigre, and Wóffla to Waag, but the language of both is that of Tigre: the inhabitants are Mohammedans. The Ráia Gallas are generally pagans, but there are Mohammedans among them. Further, from this spot Mount Adamahónni, in the district of Ebálli in Tigre, bore N. 70° E., below it to the left being Bóra. The valley of the river Sássela, coming from Wóffla and joining the Tzelári, S. 75° E. Mount Biála, S. 15° W.; Warrahémano, S. 15° E.

6th.—I had fully calculated on a detention at Sókota of several days. On applying, however, to the Waag Shum for a guide to Tigre, he readily gave me one, and I left this morning early. My road of to-day was travelled for the most part in company with a numerous party of Tigre merchants, who had brought *amols* to Sókota, and were now returning with cloths, mules, and oxen. They had left over-night, encamping by the small river Mai Lomi, a tributary (I believe) of the Tzelári, along which the road descended N. for about an hour, and then leaving it, it ascended, keeping on N. during the whole morning along what appeared to be a ridge of the mountains bounding the Tzelári to the W. Soon after noon we began a sharp descent from this ridge, and came into the dry bed of the river Shagálu, down which we continued till evening, when we came to the Tzelári, which we crossed, and then ascending from it a little way, encamped for the night in the open air. The Tzelári is a river of considerable size, which rising to the eastward of the head of the Tákkazie, receives a large proportion of the waters of Lasta as far as Wóffla, and being afterwards joined by the Zámra, a river of smaller size rising in Wójjerat, the united stream falls into the Tákkazie, which by this accession must be fully doubled in size. The country traversed by these rivers is very different from the rest of Abyssinia, being for the most part an almost uncultivated, desert, sandy tract, of much inferior elevation, and almost entirely devoid of water; the so-called rivers being mere *wadies*, which are filled with water only during the rainy season. The Shagálu, down the bed of which we descended to-day, has a great resemblance to the Wady Kéllu on the road from Tajúrrah to Shoa; with this difference, however, that the latter contained good water along its whole extent, whilst during two and a half hours' march down the bed of the former we came to only one spring and well, the water of which was filthy. The later season of the year has perhaps something to do with this. The elevation of the bed of the Tzelári can certainly not exceed 4,000 feet. The desert tract thus traversed by this river forms the natural boundary between Lasta and Tigre, as it does between the Agau and Tigre languages. In order to prevent misunderstanding it is proper to remark here that Waag is the name of the northern portion of Lasta, the southern portion being known as Lasta Proper. In Waag itself they repudiate the designation of Lasta, but throughout the rest of Abyssinia both Waag and Lasta Proper pass under the general appellation. It is the same with Gojam. The inhabitants distinguish between Gojam, Damot, Enábsie, &c.; but on this side the Abáï the whole peninsula is styled Gojam. So too, in fact, with Tigre, Amhara, and Shoa.

7th.—Leaving the Tzelári, the road began ascending gradually northward for more than an hour to Sákka, from whence Mount Biála was just visible, bearing S. 5° E., and then continued still N. over a level barren country till, in about three hours, we reached the Zámra, which forms the frontier between the province of Bóra in Waag and that of Sálowa in Tigre, by the side of which river we encamped during the heat of the day. The district between the two rivers through which our route lay is not properly Bóra, but Zebád, a dependency of the former, which lies farther to the E. In the afternoon we went on N. for about a couple of hours more, to the village of Fenárwa, where we stopped for the night. At Fenárwa, Mount Adamahónni bore S. 55° E.

8th.—From Fenárwa we continued gradually ascending E.N.E., and then E. by N. towards the high land of Tigre, the country improving by slow degrees, the sandy soil and want of water being opposed to vegetation. In near six hours we reached the town of Sámrie, the residence of the governor of Sálowa, where we stopped. From hence Mount Biála bore S. 15° W., and Mount A'mbera, an isolated peak rising from the *Kolla* (low country) of Tembien, N. 35° W. Sámrie is the salt-market of Tigre, in direct correspondence with Sókota in Lasta, and the difference in the value of *ámole* on the opposite sides of the desert is very marked; here 50 or more going to the dollar, and there only about 40. At Ebenat the number has already decreased to about 30; and on crossing the Abáï only 25 or 20 are obtained, the rate falling as the distance increases. It is not to be imagined that the difference all falls to the merchant: in fact, I have in the course of these pages given more than one instance of the levying of duty on these *ámole* in their passage across the country, which duty forms no unimportant item of the revenue of the native princes and governors.

9th.—Dejach Ubie being in the field against Baalgáda Aráia, a grandson of Ras Wálda Salássie, and being encamped a little way from Antálo, I this morning proceeded to his camp, which I reached after a ride of about five hours N.E. by N., the ground continuing to rise gradually till the table-land of Tigre is reached. I had been told that when we arrived here we should find it like Gojam: as yet, however, the difference between the two countries is very marked both to my eyes and to those of my servants, who constantly exclaim (and with perfect truth) that they have not yet seen a country like their own. The grand cause of the superiority of the peninsula of Gojam appears to be, that the elevated mountains in the centre of it collect the waters of the heavens more or less at all seasons of the year, so that the numerous rivers descending from them, although they decrease considerably in the dry season, are never quite dry. The Dejazmach's camp was by the village of Asta, about four or five miles almost due E. of Antálo, which town is however not visible, it lying, as I was told, just on the other side of some low hills. On arriving in camp I went up to the gate of the Dejazmach's court to announce myself, where I waited some time till the Dejazmach's eunuch came out. After inquiring as to my country, my journey, &c., he refused to announce me to the Dejazmach, notwithstanding my repeated request, and although I said I required no more from him than I had obtained from the Ras and the other chiefs through whose countries I had passed—a guide. I had therefore no alternative but to return to my baggage, which I had left below; and finding some unoccupied huts in the camp, I took possession of them with my people, and passed the night there. From hence Mount Alájji in Wójjerat bore S. 20° E.; Mount Adamahónni, said to be beyond it, was not visible.

13th.—The following morning I went up to the Dejazmach's, but with no better success than before; so that I returned, leaving Wálda Georgis to negotiate with the eunuch, who on the promise of a dollar—one of my remaining two—engaged to procure from the Dejazmach a guide to Tembien, the direct road to Massówah through Agámie, as also that to A'dowa by Girálta, being rendered impassable by the rebels. To this arrangement I could make no objection; on the contrary, I was rather pleased than otherwise, as the circuitous route enabled me to visit an additional, though small, tract of *new* country. As to my approaching the Dejazmach, this was utterly denied me, for what reason I am not able to say. But notwithstanding the arrangement made with the eunuch, it was not till this morning that the guide was given me, and in the meanwhile I had to remain in my miserable quarters, and to shift for myself as well as I could. On leaving the camp this morning, the road led north-westward, over an almost level country, with the exception of a ridge bounding to the eastward the valley of the small river Durgebáa, a tributary of the Arékwa, which latter stream (I was informed) has its course *separately* to the Tákkạzie. After a ride of about four hours and a half we reached the village of Kabári, where the guide, as directed, found us quarters for the night.

14th.—Our journey of to-day was, if anything, less than that of yesterday, owing to the delay and difficulty caused by the constant changing of guides. In about two hours and a half after leaving our resting-place, our course being about N.W., we came to the brink of the valley of the river Gebáa (Gibba), over which we had an extensive view, and down the side of which we now began descending. In about a couple of hours more we were compelled to stop at the village of A'dega Músie, the persons who were now called on to see us across the river into Tembien refusing to take charge of us till the morning. As far as the Gebáa the whole country is Sahárto, a subdivision of Endérta.

15th.—This morning we started long before daylight, and, descending a little way farther, began crossing N. the level plain through which the Gebáa winds. This river, which, like the Arékwa, has its course separately to the Tákkạzie, is not more than about six or eight yards broad at the ford, and quite shallow. In rather more than five hours N.N.W. we reached the town of A'biyad, where we stopped. We were here in view of the high mountains of Semien, stretching from W.S.W. to W., which were, however, but imperfectly seen from their being enveloped in clouds and mist. Mount Ambera in Tembien bore N. 85° W. A'biyad is the principal place of Tembien, and a large market town.

16th.—The Shum of A'biyad sent a man on with me this morning to the Mohammedan village of Tạkirákira, about an hour's ride distant to the N., with orders to the Shum there to see me across the river Woréi, the boundary between Tembien and Tigre Proper; but the valley of that river being said to be infested by the Baalgáda's people, he refused to take charge of me. I sent back to A'biyad, but could get no redress; and consequently had only to sit quietly down in Tạkirákira till I could make some other arrangement. From this place Mount Ambera bore S. 80° W. The mountains of Semien were not visible.

19th.—The Shum of Tạkirákira had refused to take charge of me, but he was very civil in advising me as to the road, &c., and this morning some of his people leaving for A'dowa and the neighbourhood, he willingly consented that I should accompany them on my own responsibility. Descending north-westward towards the river Woréi, we, in about five hours and a half, crossed that stream, a good-sized brook, forming at the ford an almost stagnant pool, and then began ascending the valley of the Gwáhero, a tributary of the former. From the Woréi, a ride of about three hours, in general bearing about N., brought us to the village of Nạráa, where we alighted.

20th.—From this place in about eight hours I arrived in safety at A'dowa, our course being N. by W. Here I received a most hearty welcome from the European residents, consisting of Mr. Schimper, the German naturalist, and the Abbate de Jacobis, with the other members of the Roman Catholic mission to this country. Nothing could equal the kindness of these gentlemen in supplying me with everything necessary after the privations to which, in consequence of my limited funds, I had unavoidably been subjected, especially during the latter portion of my journey; and they laid me under still further obligations by furnishing me with the means of continuing my journey to the coast. Having thus, through God's mercy, arrived at a place so well known from the number of Europeans who have visited it, my further journey can present nothing of novelty; and I therefore here close my journal.

RÜPPEL'S JOURNEY FROM HALAI TO ADDIGRAHT, MAY 1832 (marked N in Map, page 39).

10*th May.*—Travelled over an undulating plain with low shrubs. After three-quarters of an hour passed on the right a village on a hill, called Sambi. Trees then became scarcer, and after another half-hour, he reached a large village, called Dera; then through a deep ravine to a valley with meadows. Passed Burhanet village on a steep hill on the right, and encamped at the village of Woklune, where there is a pool and meadows. The whole march occupied two and a-half hours (an hour's march is about two and a-half miles).

11*th May.*—Ascended a steep narrow path, and reached a high plateau, near a large village, called Auhenne. Soon afterwards he passed Artegeh; the land barren and uncultivated. After three and a-half hours' march he reached a steep slope, which he descended into a narrow valley with meadows, and encamped about half a-mile south-west of Dogonta. Grass was found only in the valleys, and isolated shrubs of acacias, roses, &c., on the plateau.

13*th May.*—Went for a quarter of an hour across a table-land, sloping to the east. At the end of half an hour from starting ascended a hill of sandstone breccia, to the foot of a steep terrace, with shrubs. At the end of an hour, he reached the summit by a difficult road. The plateau was undulating, with loose pieces of gravel, and low shrubs, and uninhabited. The district is called Kaskassé. The road then led between steep heights of lime marl, and after a three and a-half hours' march, he encamped near Sanafé.

14*th May.*—He remained at Sanafé.

15*th May.*—He advanced first over a table-land, then somewhat down hill through a valley with meadows. After an hour's march, he saw on the left the large village of Drogaro, and much cultivation. The district is called Hakerai, and is well populated. He then descended by a steep road, apparently hewn in slate rocks, to a broad, fine, grassy valley, bounded by sandstone terraces, and encamped about a mile north-east of the large village of Barakit, where there is a rivulet, but no wood in the neighbourhood. The day's march was two hours.

He stayed six days at Barakit. One hour to the north is the plateau of Edatra-Akhfesie. The village of Gunna Kuma is reached by a valley in an hour to the south, and then by another valley half an-hour to the west.

22nd May.—Travelled for one and a-quarter hours down a grassy valley. Then passed the valley Mai Muma, in which is the village Gunna Kuma to the right. A rivulet flows from it in an easterly direction, which joins subsequently the Oha Bead Meriam river, and then descends towards Amphilla Bay, but does not reach the sea. The road along this rivulet is said to be easier than that by the Taranta Pass. Went along the banks of the Mai Muna Valley for half-an-hour. He quitted the river as the valley narrowed, went for three-quarters of an hour direct south, and encamped near the Chapel Getana Maherat.

23rd May.—Ascended hills formed of clay-slate and sandstone. No cultivation. Low shrubs. The road mounted at first a steep hill, then turned to south-west, along a hill forming the boundary between Agame and Akalo-Kassai, and then descended by grassy vales, bounded by high hills. After one and a-half hour's march, encamped below large village of Soruxo, where there are fine pastures.

24th May.—After two hours' journey, he encamped near a pool in the valley of Omfeito, near a group of villages. From here there is a road by a mountain torrent, which leads to Amphilla Bay. Luxuriant pastures, and fine groups of trees.

25th and 26th May.—Journeyed over rugged, steep, sandstone hills, and desert table-lands, with no trace of agriculture. Low thorn shrubs on heights, and ferns in the rocky ravines. Many springs in low places. After four and a-half hours' journey, he reached Addigraht.

RUPPELL'S JOURNEY FROM ADDIGRAHT TO TACKERAGGIRO, MAY AND JUNE 1832 (marked O 2 on Map, page 39).

ATEGERAT lies along the foot of a sandstone ridge, extending from north-west to south. Three-quarters of an hour to the W.S.W., Mount Alegua rises above the chain which forms the watershed between the Red Sea and the Tacazzy. Rivulets flow in grassy vales to the east, and soon form the considerable rivulet Oueret, which flows to Amphilla Bay. Mount Gondegonta, ten hours E.N.E. of Ategerat, rises to 1,000 feet above the plain.

28th May.—Travelled one hour, and encamped near the sandstone hill, Nogel Emeni.

29th May.—Ascended for one hour and three-quarters, up the south-east slope of the isolated volcanic cone, called Alegua, to a pass, above which the mountain rises 300 feet. He then descended a steep hill to a rocky "bench" of basalt, three-quarters of an hour wide, on which were low shrubs. Thence reached a fine alluvial plain, cultivated by the inhabitants of Gaba, which lies to the west. He then descended some sandstone cliffs, by a dangerous path to the valley of Saheta, where he encamped. In the valley there were luxuriant grass, and fine trees, and a rivulet flowing to the south-west. About one hour and a quarter from the top of the pass.

30th May.—Travelled along the valley, which was about half an hour wide, well cultivated, with but little wood, with sandstone terraces 300 feet high. Observed many villages on the terraces and along their foot; this was part of the Agame Province. The day's journey was half an hour.

31st May.—The eastern terrace receded gradually. Encamped after an hour and a-half's march, close to the village of Ambalul, on the western terrace. Much rain.

1st June.—Continued along the valley. On the right the sandstone terrace forms a steep promontory, called Amba, 1,200 feet above the plain. The soil barren. Sandstone, lava, and granite. Encamped after two and a-half hours' march, at a rivulet flowing to the south-west, through a grassy plain, called Welled. The country to the west was open. Thunderstorm in the afternoon.

2nd June.—The country sloped towards the west. Sandstone and granite formation. After one hour's march he reached the foot of a hill, on the top of which are the ruins of Mai Quarar. Descended a steep hill to the bed of a rivulet, nearly dry. After two and a-half hour's march from the starting place, encamped on an undulating plain, strewed with granite boulders, half an hour to the east of the northern end of a steep sandstone terrace, extending five hours to the south-east, on the northern extremity of which was the town of Magab, west of the camp. The town of Mugga was two and a-half hours' to the south.

6th June.—After half-an-hour's march, left Magab on the left. Crossed immediately afterwards a deep river bed, which was dry, sloping to the north-west. To the left of the road and parallel to it were steep sandstone hills. Half an hour further on the road separates, one branch leading to Adowa (two days distant). He passed for seven hours over a wide, uninhabited plain, the soil rocky sandstone, with low, thorny shrubs. On a hilly range to the south were several villages, including the church Jemata. Three hours and a-half after leaving Magab, they passed the village of Gorara, at the foot of some sandstone hills to the left of the road. Four hours after leaving Magab, they crossed the permanent River Gedjeda, which rises three hours to the south-east, and where it was crossed, flows to the west. The town of Gelibetta is situated on its banks, at one and a-half hour's distance from the route. He then went for one hour and a-half along a rivulet, and crossed the road from Adowa, *viâ* Mugga, to Antalo in Enderta. No cultivation. The soil dry and rocky, with bushes; the country then improved. He travelled along a fine, grassy plain, with groups of trees, and observed several villages at the foot of sandstone hills towards the south. Eight hours after leaving Magab he crossed the River Warie, which comes from the E.S.E., and at a distance of seven hours from the route unites with the Geba. Where crossed it flows north-west, forming the boundary between Giraldâ and Temben. Nine hours from Magab, they encamped in a fine meadow.

7th June.—For half an hour the road ascended slightly, and then traversed a hornblende plateau. One hour and three-quarters after starting, he reached Tackeraggiro.

DR. BEKE'S LETTER ON A ROUTE FROM ANSLEY BAY TO TOHONDA (marked E and F on Map, page 39).

"Bekesbourne, August 14, 1867.

"Though so many suggestions have been made as to the preferable road into Abyssinia for an invading army, I have not seen any special notice from any one but myself of that which unquestionably affords the shortest, readiest, and best means of access; namely, the caravan road of Arrian's *Periplus of the Red Sea*, which was explored by myself and my wife last year, in order to verify the opinion I have long entertained that it is the key of Abyssinia.

"I allude to the valley of the Hadás, which river, rising near the village of Tohónda (Tekunda), about fifty geographical miles south of

Massowah, flows between lofty mountains for about twenty miles in a course a little to the east of north, then turns to the east, and, as I was so fortunate as to discover last year, after running in that direction about twenty miles more, enters the Gulf of Adulis, or Annesley Bay, near its south-western extremity.

"The natural importance of this road is manifest from the fact that on the left or northern bank of the Hadás, on the sea-shore, stood the famous emporium of Adule, or Adulis, described in the *Periplus* as lying in a deep bay, and having in front of it an island called Orine (Oreine, the "rocky" or "mountainous"), and as being the place whence the caravan road passed to Coloë, whose modern representative, Hálai, retains its ancient name in a corrupted form, and thence to Axum, and so across the Takkazye, the "Nile" of the Ethiopians. The representative of Adule at the present day is Zulla, a poor village on the opposite bank of the Hadás; and Orine is marked in our charts as Dissee Island (properly *Dissyet*, meaning in Amharic "*the* island"), to which in 1805 Lord Valentia gave his own name.

"The distance of Adulis from the sea is given in the *Periplus* as twenty stadia; by Cosmas Indicopleustes it is said to be two miles; but its ruins are now about four miles from the coast, owing to the gradual uprising of the land, of which evidences exist along both shores of the Red Sea.

"During the dry season, the Hadás has now no water in a considerable portion of its channel, which is doubtless the reason why it has not more attracted the attention of modern travellers; but down one-half of its course, from its head at Tohónda as far south as Hamhámmo, a well-known camping ground of the caravans, situate a little to the south of where the river turns towards the sea, water is met with at certain spots all the year round; and, even when at the driest, wells dug in the sandy bed of the river afford a constant and copious supply of that necessary fluid. During the rains in the upper country, the floods of the Hadás and of its large tributary the Aligaddi (with respect to which river I shall have to trouble you with a few remarks on some other occasion), find their way down into the sea, and often render the river itself impassable.

"In February, 1866, my wife and I found the dry bed of the river between Adule and Zulla to be 25 or 30 yards broad; and at about a mile nearer the sea we came to wells sunk in the sandy soil, at which numerous horned cattle were being watered; small, well-shaped, fat beasts, giving a delicious rich milk, the pasture at that time of the year being plentiful almost down to the sea-side.

"From Zulla we went five or six miles further inland, and struck the dry watercourse at three different points. Had the natives been well disposed, we should have gone on to Hamhámmo, which was not more than six miles from our extreme point, and we should then have had the satisfaction of coming to water actually in the bed of the river, within one march from the sea-shore. But, though disappointed in this, we succeeded in finding the road from Zulla to Hamhámmo, which, instead of following the circuitous course of the Hadás, goes directly across the country, making the distance of Hamhámmo from the sea-coast not more than 16 geographical miles: from the wells near Zulla it is only 13 miles.

"In the following month of March we went from Massowah into Abyssinia by the modern caravan road taken by Bruce, Salt, Rüppell, Krapf and numerous other Europeans, and in May we returned to the coast by the same road. In doing so we had to traverse 26 miles of low and barren country between Arkiko and Hamhámmo, where no regular supply of water is to be had; and instead of continuing up the bed of the Hadás to its source, we stopped at about 10 miles below Tohónda, and turned up the steep side of the valley by the pass of Shumfáito, of a similar character to the adjacent and better known pass of Taránta.

"We took seven days' slow travelling between Arkiko and Hálai on the road up, including the stoppage at the foot of Shumfáito, whence we had to send to Hálai for bullocks to replace the camels which had brought our baggage thus far, and we were four days in returning. The actual time we were on our mules' backs (from which we had to alight at only two or three difficult spots), was twenty-five hours going from the sea up into Abyssinia, and twenty hours and three-quarters coming down. Of these intervals, respectively, five hours were spent in ascending, and three hours and a half in descending Shumfáito. Had we continued straight up the valley of the Hadás to its source near Tohónda, we should have avoided the steep mountain, and the time of our journey would still have been about the same, by a road gradually rising and practicable for camels almost all the way to the summit.

"At Hálai, at an elevation of upwards of 8,400 feet, or a mile and a half perpendicular height above the ocean, we had reached the table-land of Abyssinia; and yet we were so close to the coast—from the head of Annesley Bay it is little more than twenty geographical miles as the crow flies!—that we could perceive the sea beyond Arkiko to the north of us, and hear the firing of cannon at Massowah; while to the south and south-west, for hundreds of miles, extended the Abyssinian table-land, of which Amba Mágdala is a detached spur, at a lower elevation than the table-land itself, approachable by a practicable road through Agame, Enderta, Bora, Woffla, &c., without crossing the Takkazye or any other large river, a considerable portion of which road has been trodden by myself. Such are the difficulties and dangers of penetrating into Abyssinia!

"In the course of ages Adulis became a large city and port, whence the kings of the Axumites, who were of sufficient importance to form alliances with the Byzantine emperors on equal terms, sent their fleets and armies across the Red Sea to subdue the whole coast of Arabia from Leuke Come to Sabæa.

"Whilst Adulis was thus a flourishing port, a commercial station was established at Senafé, on the edge of the table-land, a few miles south-east of Tohónda, of which place the remains still exist. Senafé appears to have supplanted Hálai (Coloë) as the intermediate station between Adulis and Axum; and its commanding position must soon have led to the opening of other communications between it and the sea, which is doubtless visible from it both to the north and to the east, its distance in a direct line to Haréna in Howákil Bay being forty-eight geographical miles, to Zulla forty-three miles, and to the south end of Annesley Bay only thirty-eight miles.

"The ascent to Senafé from the east and north, through the district of Buré, is said to be equally good with that by the way of Tohónda, the former being the direct road from the great salt plain of Harho, which supplies Abyssinia with its present currency, pieces of rock-salt. When Mr. Salt was in Abyssinia he made many inquiries about this road, through Buré, which his friend Ras Walda Selásye assured him was the best approach from the coast, it being only four days' journey to Antálo, his capital, and well supplied with both water and cattle.

"The short distance of Senafé from the sea, from which it might, if expedient, be approached in more than one direction, points it out as the best position to be occupied by an invading army on its entrance into the country. The desirability of getting away as quickly as possible, not merely from the intensely hot shores of the Red Sea, but likewise from the intervening elevations of 2,000, 3,000, and even 4,000 feet, to the fresh and bracing climate and clear atmosphere enjoyed at upwards of a mile and a half above the ocean—for Senafé is of even greater elevation than Hálai—does not require to be insisted on.

"The occupation of the passes from the salt place, and the consequent control of the supply of that valuable and necessary commodity, might form a very important strategical operation. In addition to all this, Senafé is the nearest point to Mágdala, it being less than 200 miles distant from that fortress, and all the way on the healthy tableland.

"From what is here stated, it will be seen that there is good reason for the opinion that when the Greeks of Egypt founded Adulis, at the mouth of the Hadás, they held in their hands the Key of Abyssinia; and there can be little doubt that the French were aware of this when they attempted in 1859 to acquire possession of that place and of Valentia Island (Orine), as is described in page 58 of my work, 'The British Captives in Abyssinia.' May those to whom the preparations for the contemplated expedition into Abyssinia are entrusted know how to profit by the experience of others who have preceded them!

"CHARLES BEKE."

DESCRIPTION OF THE DIFFERENT DIVISIONS OF THE PROVINCES OF TIGRE AND AMHARA IN ABYSSINIA, taken from the travels of Combes and Tamisier, and supplemented from those of Salt.

Province of Tigré.

Siré. Is bounded on the north by the Changalla, on the west by the Tacazé, on the south by Adet and Temben, on the east by Axoum. It consists generally of vast plateaux, intersected by deep valleys; a few mountains of small elevation are found in it. On the borders of the Tacazé, and to the north, great numbers of elephants and giraffes are found.

To the south of Siré are the districts of *Adet* and *Temben*, the latter is traversed by high chains of mountains, and inhabited by Mussulmans.

Bora and *Salowa* to the south, belong to the kingdom of Lasté, they are mountainous and cold. The first extends to Lake Ashangi, and to the east is the little province of *Wofela;* the rains that fall in the latter have no means of escape, and form marshes which render it unhealthy.

Salt says, Bora and Salowa form two mountainous districts northward of Lasta, and between them and the Tacazze lie the comparative low countries of Waag and Gualin, which are inhabited by Christian Agows.

To the north of Wofela is *Wojjerat*, which is much less elevated than the neighbouring regions, and might almost be classed in the low country; its inhabitants, however, having freed themselves from the Gallas, having remained Tigréens. It is one of the most wooded districts of Abyssinia, and its forests contain many wild animals, particularly elephants and rhinoceroses.

Salt says, Wojjerat is a wild country, abounding in elephants, lions, rhinoceroses, and every species of game. It is said that the rains are not so periodical as in the rest of Abyssinia, owing possibly to the forests with which it is covered.

Enderta is situated to the north of Wojjerat; Antalo, its capital, is a fine town built on the slope of a hill; its neighbourhood is fertile, beautiful, and well watered. To the east of Antalo is the district of *Désa*, which adjoins *Mantillé*, separated only by one chain of mountains from the Dobas and Wojjerat.

Salt says, to it (Enderta) are subordinate the districts of Derra, Asme, Womburta, Desa, Muntilla, and Monos; mountainous districts forming by their position, the eastern boundaries of Tigré. It also embraces within its limits the territories of Moculla, Dirbale, Gambela, Upper and Lower Gibba, Wazza, Sakarti, and Girulta.

To the south-west are the countries of *Ouombourta*, *Asma*, *Derra*.

Giralta is situate to the north-west of Antalo; it touches Temben on the west, and Haramat on the north. It is very mountainous.

Agame is one of the most important provinces of Tigré, it is bounded on the west by a great chain of mountains, stretching from Haramat to the river Belessa; to the south by Enderta, to the east by the high lands of the Taltal and the plain from which salt is procured. This province is extremely fertile, it produces every sort of grain, and even wine; but it has been much impoverished by wars; colossal mountains dominate over it.

Salt describes Agame as a rich and fertile territory, owing in a great measure to its being situated on a level, at a considerable elevation above the sea, which in the torrid climates generally ensure these advantages. It has for its eastern frontier part of the lofty ridges of mountains which extend from Senafé to Taranta, and its stronghold bordering on the Tal-tal, together with its vicinity to the salt plains, render it of great importance in the country.

Laoué, *Gunduftafte*, *Halai*, *Agguela*, and *Larai*, which are on the caravan road from Adowa and Gondar to the sea, are to the north of Agami; although their soil is fertile, yet in the dry season the springs are nearly dried up, and merchants are obliged to take another route.

The rich province of *Hamasien* is to the north-west of these districts; it is bounded on the east by the Shokos, on the north by the tribes of Bicharri, Bekla, and Boja, to the south-west by Seraweh, to the south it is prolonged to the Mareb. Hamasin and Saraweh resemble Siré in climate and productions.

The territory of *Tigré*, whose capital is Adowa, is bounded on the west by Siré, on the south-west by Temben and Adet, on the south by Giralta, on the south-east by Haramat, on the east by Agamé, and on the north by the rivers Mareb and Belessa. The mountains to the east of Adowa are very high, the range runs down the centre of the province, the general character being a range of hill forts or ambas, intersected by deep galliés, and highly cultivated plains.

Lasta is also classed with Tigré. This province which has also given its name to the kingdom of which it now forms part, is bounded on the west by the Tchera-Agous, on the north by the Ejjon-Galla, on the south-east by Angot, and on the north by Bora and Ouofila. Lasta is very mountainous, and whole armies have been known to die of cold there. The inhabitants are warlike.

Salt describes Lasta as being covered with rugged, and almost inaccessible mountains.

Province of Amhara.

Semen is the highest country in Abyssinia; its mountains are almost always covered with snow on their highest points, and ice is not unknown. The Amba-Hai, and the Amba-Béyéda are the most remarkable mountains of this country. Salt says this province may undoubtedly be considered as the highest point of land in Abyssinia, the whole range of its mountains extending in a northerly and southerly direction, about 80 miles.

Oaldubba, situated to the north-west of Semen, between the Tacazze and the Angrab, extends as far as the junction of those two rivers, and is low and unhealthy.

Oualkait is to the west of Oaldubba, it is intersected through its whole length by the two rivers, Toukour and Guangue. It is more wooded than Oaldubba. Mimosas abound.

Woggara is one of the richest provinces of Abyssinia; it is bounded on the east by the mountains of Semen, on the south by the plain of Belessa, and to the west by the territory of Tchelga; it possesses fine pastures, numerous cattle, which produce much butter and milk; it produces corn and barley, the provisions are cheap, and the inhabitants live in abundance.

Tchelga is bounded on the north by Oualkait, on the east by Woggara, on the south by Dembea, and on the west by the Abyssinian boundary, it is almost entirely inhabited by Mussulmen.

Kouara is to the west of Dembea, between Ras-el-fil (so called from its vast forests), Kouara, and Tchelga, is the province of *Dembea*, which reaches to the great lake Tzana. This province is flat, and owes its great fertility to the sludge of the lake which retires sensibly. This country possesses the best corn in Abyssinia.

The rich plain of *Bélessa* adjoins Woggara, and that part of Semen, called Meuna. This country is bounded on the west by a high chain of mountains from which terrible storms, often of rain, descend in the rainy season; on their flat tops are the wooded provinces of Ouénadéga and Mariam-Ouaha. To the south of the latter is Bégéméder.

Foggera, which extends from the River Goumara to Emfras, is a low country traversed here and there by hills, which extend from Mariam-Ouaha and Ouénadéga to the lake. It contains only miserable villages with the exception of Caroda, which owes its importance to a church; Ifag celebrated for its market, is in this province: both these places produce good wine. Between this country, the Nile, and Bégéméder, is the little district of *Daro*.

To the south is the country of *Maicha*, bounded on the east by Gojam, and on the south by the country of the Agows. This province is covered with torrents, and small veins forming ponds and marshes, which breed dangerous fevers. Corn and cattle are produced in it.

The *Damot-Agous* are shut in between Gojam, the Basso Galla, the Gonga, Kouara, Dembea, and Maicha. The country is rich in honey, butter, and cattle, the climate temperate, and soil productive.

To the west of the Damot are *Basso-Gulla*, the country is fertile.

The name of *Gojam* is sometimes given to the whole country situated in the semicircle of the Blue Nile, it is a country of extraordinary fertility, has immense meadows and magnificent pastures. The oxen are the most famous in Abyssinia, the horses were the same, but are now disappearing.

Begemeder is bounded on the west by the Nile, on the north by Foggera and Belessa, on the east by Lasta, and on the south by

Bachilo. This province is less mountainous than Semen, nevertheless great chains defend it against Lasta; towards the north it is covered by woods, to the south it is entirely destitute of trees. The horses are much esteemed and the flocks are innumerable.

Amhara, properly so called, extends between the Rivers Ouahet and Bachelot, it is bounded on the west by the Nile, which separates it from Gojam, and on the east by Lasta and Ingot.

MM. COMBES AND TAMISIER—JOURNEY FROM MASSOWAH TO ADOWA, BY THE TARANTA PASS, APRIL, 1835 (marked D and I on Map, page 39).

April 15*th*. *From Massowah to Arkiko, one hour.*—The latter the smaller town. Water of wells at Arkiko bad. In the rainy season plenty of water in the small torrent of Oues.

17*th*. *From Arkiko to Chilloki.*—Camels and mules having arrived, they left for the south through the Samhar, inhabited by the tribe of Hazortas. Two and a half hours' march up a gradual ascent, the atmosphere improving at every step, brought them to the dry torrent Catra. Gazelles and large partridges very numerous. An hour's more march brought them to high chains of rocky mountains covered with trees, which grew out of the clefts of the rocks. After traversing alternately narrow gorges and beautiful shady valleys, they arrived at an eminence called Chilloki, covered with green wood.

18*th*. *From Chilloki to Hamhamo.*—They advanced for several hours among shady mountains, covered with flowers, and arrived at the valley of Hamhamo, which appeared an enchanted garden. It is surrounded by lofty mountains, well wooded, and abounds with gazelles. At the other end they found a stream which runs the whole year, the first good water they had drunk since leaving Arkiko. On the slopes of the mountains were flocks of sheep and goats, tended by Bedouins.

19*th*. *From Hamhamo to Manta Sangla.*—The road had become rocky and difficult, the trees less numerous, the aspect of country was barren and rugged, but grand and picturesque. Chains of mountains still more elevated appeared in every direction. The pass became still narrower hemmed in between enormous mountains. Two hours after leaving Hamhamo they reached a brook; the road widened out, verdure and flowers appeared. They had entered the valley of Dobara, a green spot among majestic mountains. They followed the stream for some time and then reached another in the valley of Manta Sangla.

20*th*. *Manta Sangla to Choumfaitou.*—After traversing a stony mountainous track, they arrived at the valley of Tahtai-Tobo, watered by a brook, a beautiful spot well wooded. In another hour, following the water-course, they arrived at Haila-Tobo, by a road which, although difficult, and hemmed in between arid mountains, was admirable. They traversed the valley of Assanba, and reached Choumfaitou at the foot of Mount Taranta. They encamped under a tree full of monkeys. In the night they were disturbed by hyænas.

21*st*. *Choumfaitou to Halai.*—The baggage was shifted to bullocks to mount the Taranta Pass. An abundant stream flowed at the foot of the mountain. Two roads lead to Halai, one less difficult and wider for beasts of burden, the other shorter and steeper, only used by foot travellers. They chose the latter. The sun was hot, and they climbed with difficulty among dry trees, and aloes, and rocks. On approaching the top the trees were greener. Devoured with thirst, they reached the summit, after six hours' hard walking. After some time longer they reached Halai. The plateau of Halai is covered with resinous trees and much corn is cultivated. It is watered by a large and deep torrent, and on all sides great chains of mountains are visible. The village contains about 250 inhabitants. Large herds and flocks of cattle and sheep are found there. After the heat of the valley they found the cold intense, and were obliged to give up the idea of sleeping in the open air. They entered the house of the chief where they sat round a good fire. Their goods arrived on bullocks and they bought mules to ride. They travelled with tents.

24*th. Halai to Marda.*—They chose the most direct route to Adowa and did not go by Dixan. On starting they perceived great mountain masses on every side, for "Abyssinia is nothing but an immense mountain, cut up by profound and narrow valleys washed by impetuous torrents." After half an hour's march, they perceived on the right the hamlet of Samdi on the slope of a hill. Numerous flocks fed around. After passing the village of Deura which overlooks a vast well cultivated plain, traversed by an easy path, they went through several woods of sabines and mimosas, amid a charming landscape. They passed a chief and his attendants, some mounted on mules, some on horses, armed with lances, bucklers, and curved swords. After night-fall, they pitched their tents in the village of Marda. They had descended to Marda, and did not feel the cold there as at Taranta. It is watered by brooks of delicious water.

25*th. Marda to Sêda.*—They turned out of the road to avoid the custom-house of Gondaftafé. The path was rough and difficult; mountains on every side, shaded by numerous sabines and olives, soon replaced by dried up mimosas and the coll-qual. They passed several torrents and villages, and, after four hours' march, reached Mogouscas, where they found an abundant spring. The road became then much obstructed by stones and trees. After long ascents and descents, they reached the hamlet of Sêda, built on the slope of a mountain, having passed the great village of Achéra-Damchel on their right three-quarters of an hour before. The view from the hill above Sêda was beautiful; in front was a large and fertile valley shut in between two chains of mountains, which were broken up and distorted, and covered with rocks, presenting a great contrast to the fertility of the valley and the rich vegetation of the plain which stretched beyond.

26*th. Sêda to a desert place.*—After traversing a beautiful valley, they dismounted under a thick shade of trees, to escape the excessive heat. Peasants were feeding their flocks. About three, P.M., they started again, and having passed the *débris* of the mountains which shelter Gueurzobo, they entered the great plain, almost entirely uncultivated, which stretched before them. After traversing for a long time a very easy road, they encamped at nightfall in a desert place; rain obliged them to ditch their tent.

27*th. To Emni-Harmas.*—They soon left the plain, and, traversing hills covered with gravel and encumbered with dry trees, they reached an immense valley, watered by the great brook Séréna. They again crossed similar hills, and reached another plain, traversed by the river of Bélessa, which they followed some distance in a south-south-east direction. They turned out of their road to visit some missionaries at Emni-Harmas, a village to the south.

30*th. Emni-Harmas to a village.*—The mountains on every side were white and arid, but the villages were surrounded with trees, and the valleys covered with rich vegetation.

1st May.—They descended by a difficult path to the River Oungués, the boundary of the province of Agguéla. The plain watered by this river is carefully cultivated. After passing it they entered the mountains. The road, traced parallel to the torrent, was rocky and difficult. After a long march they rested by a brook named Kébita, near the village Zeban-Guila. They then ascended a high mountain, from whose summit they discovered a great number of villages. The country appeared less wooded than that they had traversed. They then descended into an immense and magnificent valley, covered with cultivated fields, green meadows, and beautiful trees, and traversed in its whole length by a brook, which floods it in the rains. On the right was the broken chain of black mountains, which had been indicated to them at Halai as near Adowa. They entered the mountain gorge which terminates the valley. Heavy rain fell, they lost the track and bivouacked.

2nd May.—They regained the road, and after passing a steep hill, entered a less extensive valley than the last, called Robber-Aéni (Ribieraini of Bruce). They traversed this fertile valley in a direction from east to west, and arrived at the church of Mariam-Chaouilou. They traversed the brook Assa, and arrived at Adowa, a town of 3,000 inhabitants.

MM. COMBES AND TAMISIER.—JOURNEY FROM ADOWA TO DEVRA TABOR, BY THE SELKI MOUNTAINS, JUNE, 1835 (marked K on Map page 39).

30th June.—For two hours the road was easy, through magnificent prairies and cultivation. Then they had to mount and descend very steep mountains. Violent rains fell. They reached a village, where they were refused hospitality. They then climbed a mountain, and traversed a plateau, at the end of which they arrived at Adde-Heussa.

1st July.—The road was as fatiguing as on the previous day, although the scenery was beautiful. The ground was heavy with mud, and the difficulty of the road obliged them generally to dismount from their mules. After marching five hours, they reached Devra-Guennet, a beautiful village situated on the top of a hill, with a deep torrent at its foot.

2nd July.—They had been joined by 1,000 soldiers of Oubi's army. They arrived at a spring, which in the dry season is the only one met with on this road. After climbing a high mountain, a beautiful spectacle burst on their sight. Before them was the valley of the Tacazze, and the three chains of mountains of Semen, massed one upon the other. The descent was very difficult, and in spite of the rains they found no water. The country was arid, the sun hot, and immense rocks obstructed the path. At the end of an hour they reached the village of Jibagoua. Jibagoua is without springs, and in the dry season a traveller who traversed this country alone would be in danger of perishing for want of water.

3rd July.—They went south along the sandy and dry bed of a torrent, and then by a very narrow and steep path along the edge of a precipice; then turning west they reached the Tacazé, where they found a great concourse of people. The descent had been long and fatiguing on account of its extreme steepness. The bed of the river was about 80 yards wide, and the current impetuous and tolerably deep.

This ford is formed by the Ataba, which rolls great rocks into the

Tacazé, and forms a sort of dam, which diminishes the rapidity of the current. The Tacazé separates Tigre from Amhara. In the rainy season it forms an insurmountable barrier between the two. The whole party, including the soldiers, passed it safely. The ford at the time they passed it was 5 or 6 feet deep, but it was evident that it was sometimes 9 feet deep. The river abounds in hippopotami, crocodiles, and fish, the valley in tigers and elephants. Excessive heat (95° in the shade) and periodical rains render the vegetation of the valley very luxuriant, and occasions dangerous fevers. It has been stated that this valley, at certain seasons of the year, is infested by a fly very dangerous to animals.

The road then ascended the gorge of the Ataba, crossing it several times. They then commenced to ascend a lofty mountain by a rocky and thorny path, and at the end of two hours they arrived on a fine plateau, where is the village of Torzague.

4th. They quitted Torzague, and commenced to climb again through lofty mountains, and, after three hours' march, descended by a steep and difficult path to the village of Greubeura, situated at the foot of gigantic mountains.

5th. They entered a great and fertile valley, torn by rapid and deep torrents, which fell into the Ataba. They then again struck the Ataba, and, following its banks, reached the village of Abbéna, situated at the extremity of a rich valley. On the flanks of the mountains, rising above Abbéna, they discovered a great quantity of hamlets, surrounded with trees and vegetation. The whole aspect of the country was pleasant and fertile.

6th. They followed the stream up a valley of marvellous richness. Heavy rain fell, as on almost every afternoon. They slept in a cave, and suffered much from cold.

7th. They left the river and climbed a high mountain by an "infernal" path. Its steep sides were covered by a rich vegetation. At length they arrived at the top of this prodigious mass, but on every side still higher peaks rose. This mountain is called Selki. It was very cold. They marched along the plateau for two hours, but, rain commencing, they stopped at Soana, a miserable hamlet.

8th. After an hour's march, through a boggy country, cut up by numerous torrents, and covered with abundant pasturage, they arrived at Nori. The atmosphere was charged with moisture, and the tops of the mountains covered with snow or fog.

10th. After a long and painful march, over lofty mountains, often enveloped in fog, they reached the hamlet of Amba-Ras, having left Enchetcab on their left.

11th. They followed the plateau for some time, and then descended into a valley where rain overtook them. Throughout the whole of the journey across these mountains, they had suffered much from cold and rain.

12th. After two hours' march, they passed the gate called Sancaber, shutting the finest, and, perhaps, the only road of Abyssinia. It is traced along the side of a high mountain, inaccessible from the base to the summit. Taxes are levied at this gate. They mounted by this road to a desert plateau, and then redescended into a great valley, a green meadow watered by many springs, and covered by numerous flocks. They followed it for more than an hour by an easy descent, and then arrived at the fine village of Daouarik, in the province of Ouagara. They had now left the mountains, and an open country stretched before them.

After remaining some time at Daouarik, they left it on the 27th July. After the difficult passage of the mountains of Semen, the road appeared very easy. Innumerable herds of bullocks covered the vast prairies. Villages, surrounded with trees, appeared on every side. The province of Ouagara is the richest in Abyssinia, remarkable for its prairies and fertility. They reached the village of Doougua; to the east, in the distance, they saw the lofty crests of the mountains of Semen.

2nd August.—They traversed a beautiful country, but the meadows were less in extent, and more undulating. They passed several ruined villages, and saw a good deal of cattle, but very few inhabitants; plenty of brooks and torrents. After being caught in violent rains, they reached the village of Cantiba, which is on the caravan road from Daouarik to Gondar, and at two days' journey from the latter.

3rd August.—They quitted the high road to Gondar, and followed an easy route through magnificent prairies and numerous flocks. They passed a stream called Anchoca, and in a quarter of an hour reached Dabat. The mornings were fine, but in the afternoons torrents of rain fell. The principal riches of the inhabitants consists in cattle, whose milk they sell at Gondar.

6th August.—They quitted Dabat, marching through a country similar to that of the preceding day, but rather more wooded. After an hour's march the road divided, the one to the left going to the plain of Bélessa, that to the right, which they followed, going to Duncas. To the south-east they perceived the celebrated mountain Ouechné. The country was intersected by numerous paths. They crossed the Faras-Ouaha, and reached Duncas, where there are the ruins of an old castle. Since leaving Daouarik they had found the climate milder.

8th August.—They quitted Duncas, and in an hour saw the magnificent lake of Dembea on their right. On their left was the plain of Belessa, traversed by torrents full of water. The paths were numerous; the mountains around them, covered with trees, had a sombre aspect, which contrasted with the plains they were traversing. After seven hours' march they reached Tolusguerar.

9th August.—Five hours after starting they passed a mountain covered with cascades. They then quitted the chain of mountains which occupies the whole province of Begemeder, and descended by a steep road to Derita, where they arrived early. Derita is situated on the east slope of the mountains, which rise above the plains surrounding the lake. It is on the frontier of Begemeder. There is a market every Monday at Derita, where coffee, wine, and brandy, can be bought.

19th August.—They left Derita at 11 A.M., and, after a difficult descent, arrived in the valley. The country is low and marshy, and unhealthy in the rains. They passed the brook Dendeno, and two hours afterwards traversed at the extremity of the valley the impetuous river Cheni. The fields were well cultivated, and there were numerous villages. The road was so muddy that they were obliged to halt at the hamlet of Anguot, three leagues from Derita.

20th August. — They reached the hamlet of Goub. The road through the plain was as fatiguing as that of the day before. Two hours after leaving Anguot they crossed the Rebb by a bridge of six arches, built by the Portuguese.

21st August.—At Goub they left the plain, and ascended the chain of mountains which forms the boundaries of Begemeder. The country is well wooded. The plateau, on which Devra Tabor is situated, is covered with beautiful villages, churches shaded with trees, and many streams, which flow into the Rebb. After six hours' march they reached Devra Tabor.

LETTER FROM DON ALONZO MENDEZ, PATRIARCH OF ABYSSINIA, GIVING AN ACCOUNT OF HIS JOURNEY FROM BAYLUR (AMPHILLA BAY) TO FREMONA, ON THE HIGHLANDS OF ABYSSINIA, MAY 16, 1625 (marked P on Map page 39).

WE departed Baylur on the 5th of May, afternoon, not so well furnished as we expected, for though they had promised us, and there was need of many more camels, yet we being very hasty to be gone, as apprehending the neighbourhood of Moca, they found us but fourteen, which obliged us to leave behind much of our goods, taking only the most valuable, we being now twenty-two persons, with two that joined us from the ships. Only six asses could be got, so that we rode by turns, and went afoot most part of the way, which, when it was not loose sand, was over mountains of iron mines, the stones whereof are like the dross that comes from the furnaces, and so sharp pointed that they spoilt a pair of shoes in a day, and there being no great stock of them, most of my companions were forced to make use of the pack-thread buskins we carried for the servants, and not being used to them, their feet were much galled and bloody, following the camels. Eleven days our journey lasted. Some of them, to partake of the blessing the Prophet Isaiah gives the feet of ministers of the Gospel, would not ride at all, eating very little besides rice we had with us, meeting no town to furnish us with provisions; and the heat so violent that it melted the wax in our boxes, without any shade but that of briers, which did us more harm than good, lying on the hard ground, and drinking brackish water of very ill scent, and sometimes but little of that. Yet the greatest vexation we had was the company of the camel drivers, who dealt with us most barbarously, and could never be corrected by the old furto above mentioned, who went along with us, continually craving something, and with a design to inform his King what he might demand of us. This man, the kinder we were to him, the worse he treated us, obliging us to maintain, and cook for him, and he would always be the first served; and if at any time his meat was not so soon ready as he expected, he revenged himself by not travelling that day, and playing us a thousand dog tricks, striking our men—all which we were fain to bear, for fear our goods should be left in that desert, which he would be very apt to do, because he was paid beforehand for the hire of the camels, without which he would not have stirred a foot with us.

The King of Dancali being informed of our arrival, came six days' journey, from remoter parts, to a better country, where there was good water, and sent his brother before to receive, or rather to pillage us; for soon after we met, he sent to put us in mind we should give him his present, which we could not avoid delivering in that very place, though we pleaded the things were dispersed in the several packs, that were to be opened when we came to the King, his brother's camp. To show what difference there is in men's fancies, he willingly accepted of all that was clothing, and only rejected and desired us to change him a little cabinet of diu, curiously inlaid, which is worth there five cruzados, that is, about thirteen or fourteen shillings, for a bit of cloth, worth about eight pence. The King made the same account of some curiosities of China we offered him, parting with them immediately, and being extremely fond of the clothing, though of very small value. The reason his people gave for this was that, he always living in tents, curiosities were of no use there, nor had he anything worth keeping in them.

The next day, the King sent us four mules, for the four principal fathers to come into his camp in more state; among which one fell to my share, because I was reckoned the Great Father—for so they call the superior. This name sunk that of Patriarch, or Abuna, of which the King had received some intelligence, brought him out of Ethiopia, by the Moorish Commander, and the Portuguese that came from thence. Seeing us all in the same habit, which was always that the Society wear in India, they asked for the Abuna that came from Rome, and we answered, he died at sea, meaning the Bishop of Nice, at which the King was as much concerned as we, thinking he had lost a considerable prize in him. He prepared to receive us in a hall like that the poets describe the first King of Rome had, round, enclosed and covered with hay, and so low that it obliged me to bow lower than I had intended, nor could the wind be confined in it, being open on all sides. On one of them the floor was raised about four fingers above the rest, and on it a small carpet of Lar, in Persia, worn so threadbare that it looked as if it had served all his predecessors, with a small cushion of the same antiquity, which, when he was better provided with what we gave him, he ordered to be laid for us, to sit down before him, instead of a leather we had at the first visits, and we afterwards saw it on his horse. His canopy was a piece of coarse cloth. On the right hand a chair, which was once good, with silver plates, and on the left two very large calabashes, full of a liquor he used instead of wine, and took it often before his visitors, and these were the kettledrums that went before him, when he came thither from his tent, which was a small distance, and might, for antiquity, have served Ishmael, from whom they boast they are descended.

His gravity and sedateness was well becoming a King, and he showed it in despising a small present we carried him for admittance, as his servants told us was usual, reserving the rest for another time, which he did that we might not think that sufficient; nor did he show much liking of the great present, though it was of considerable value, nor that he could find fault with it, but that there might be room for us to give more, and him to crave on, as he did during all the seventeen days we stayed there, which very much vexed us; and yet we were much obliged to him, for though he fancied we brought much more than in reality we had, and both himself and his people were very greedy, which is occasioned by the country being so poor, that for above fifty leagues I travelled through, there is not one foot fit to be sowed, and they live upon flesh and milk, and some corn brought them out of Ethiopia; yet he never ordered our goods to be searched, nor saw any of them, nor exacted any duties. The Fathers that went by the way of Mazna and Suaquem, said we should not have come off there for 150 pieces of eight.

There we began to be pinched with hunger, for though the Rectors of Bazaim and Tana had furnished us with provision enough to serve us both by sea and land, it was left at Baylur, both for want of carriage and because they told us there was plenty enough at the King's camp. But we found so little, that it was a great happiness to meet with half a peck of millet, which we eat by measure, either boiled or roasted, there being no convenience for grinding; and very often we fed upon nothing but flesh, which they sold us very dear, knowing we must eat, and they had then a good opportunity to furnish themselves with clothing, which at length began to fail, and none having faith enough to trust us, we were obliged to shorten our allowance. At our taking leave, the King would have me, as being the great Father, to ride his own horse from his tent to ours, magnifying the honour he did us therein, and telling us, that even his own brother did not mount his horse. There was no dissuading him from it, though we urged that the Fathers did not use to ride a horseback, for he was resolved the Emperor should know he did his masters that honour, so that I was obliged to mount, and went back with great noise of horse bells, and well attended.

The next day, being the 5th of June, we were dismissed with more honour than conveniency, having but one he mule, besides the beasts we brought from Baylur, so that we were little mended, except myself, who had a good mule given me by Paul Nogneyra, who would never ride in all the way, alleging he could not do it, when the fathers went afoot. Thus we travelled through uncouth lands, but with plenty of good water, the Moorish Commander and his men going along with us, as also a renegado Abyssine, who was his father-in-law.

The boundary between the kingdoms of Dancali and Tygre is a plain four days' journey in length, and one in breadth, which they call the country of salt, for there is found all that they use in Ethiopia, instead of money; being bricks, almost a span long, and four fingers thick and broad, and wonderfully white, fine, and hard, and there is never any miss of it, though they carry away never so much; and this quantity is so great that we met a caravan of it, wherein we believed there could be no less than 600 beasts of burden, camels, mules, and asses, of which the camels carry 600 of those bricks, and the asses 140 or 150, and these continually going and coming. They tell many stories concerning this Salt Field, and amongst the rest, that in some part of it, there are houses that look like stone, in which they hear human voices, and of several other creatures, and that they call such as pass that way by their names, and yet nothing can be seen. The Moorish Commander told me, that as he went by there with a Lion, Ras Cella Chistos sent to Moca, three or four of his servants vanished on a sudden, and he could never hear of them afterwards. In one place there is a mount of red salt, which is much used in physic. This is to be passed over by night, because the heat is so violent in the day, that travellers and beasts are stifled, and the very shoes parch up, as if they were laid on burning coals. We entered upon it at three in the afternoon, and it pleased God that the sun clouded, which the renegado Moor attributed to his prayers. We travelled all the night to get over the Salt Hill, only resting three times, whilst the camels' burdens were set down and loaded again; and on the 11th of June, in the morning, came to a parcel of stones, where they told us the salt was at an end. Here we all saw towards our right, a star in the sky, larger than the planet called the Morning Star, very beautiful and bright, continuing fixed in the same place whilst a man might say the Lord's Prayer and an Ave Maria. On a sudden it enlightened all the horizon, and rejoiced our hearts.

We were obliged to travel all day, that we might come to water in the evening, and had another iron mountain to pass like that of Dancali, where our Portuguese companion bid us strike off a shorter way than the caravan could go, along which we travelled afoot at least six hours, almost perished for want of drink, till a Moor we met accidentally cop-ducted us to the water, and there the caravan joined us at night. We made but a short stay here, being told that the Gallas used to resort to that water, and therefore travelled on all night to get over a great plain they continually haunt, which we found strewed with the bones of 160 persons those barbarians had butchered, and frighted us, seeing the track of their passing that way the same night, and yet we could not get over it till eight or nine the next morning, when we took to the mountains, where those people seldom go, and rested there all the remaining part of the day.

The next night we traversed another plain of Gallas, shorter than the other, and then came upon the bank of a river, along which we travelled two days, and I think it may be reckoned one of the most pleasant in the world, for the water is clear and cool, and the herbs growing along it sweet as pennyroyal, basil, and many more we know not. The banks are covered with tamarind, and those trees they call the pagod in India, besides many others, on which there were abundance of monkeys, skipping about and making faces at us. Here we met a man who brought us letters from the Fathers, and said F. Emanuel Barradas would be with us the next day at noon. The Renegado told us the camels should go no further unless we gave something more for them; and at night, that we must lie still all the next day, because their house was hard by, and they would go kill a cow in honour of St. Michael, whose festival is kept in Ethiopia on the 16th of June, and the Moors observed it. That night four or five men came up to us, sent by F. Emanuel Barradas with provisions.

The next day, at nine of the clock, we came to the foot of the mountain Sanafe, where began the command of the Moorish Captain that went for us, and so far we had hired the camels, who could not go up it, because very high and sleep. Soon after we had set up our tent, came F. Barradas, with several Ethiopians of quality, many Portuguese, mules for all the Fathers, a very fine one for the Patriarch, and abundance of provisions. Here F. Barradas advised me to put on the Episcopal robes, which when the Moorish Commander saw, he was much surprised and begged my pardon for not having known me sooner, to pay that honour that was due to my dignity, and the renegado hearing of it was so confounded, that he durst not show his face. On the 17th we ascended the mountain, which is higher than the Alps, as one of our companions said who had passed them, and thicker of cedars, cypress, and other trees, and sweet herbs, the common weed on it being extraordinary high tufts of sage, and white roses. Going down again we came into tilled grounds, full of barley and millet, which we had not seen before. The Xumo of Agamea met us at the place where we lay that night, he and the others with him bringing presents of 200 or 300 Apas, or cakes of bread each, and two or three cows, as also four, five, or six camels loaden with metheglin, all which was divided among the company, and though perhaps the presents might be the more considerable, because they were for guests who came from such remote parts, it is a settled custom in Ethiopia to entertain, and give a day's provision to all passengers, according to their quality, and if it be not done, the traveller may the next day complain of the governor of the town. The Xumo of Amba Senete, to whose house we came the third night, gave us there eight cows, and we were entertained after the Ethiopian manner, a round table being spread on the ground, and on it many apas, as broad as peck loaves, made of wheat, and a sort of pease much valued in that country, on which they lay the meat, so they are both eaten together, both flesh and dish.

Being thus attended by a great number of horsemen, richly clad, who went before skirmishing with their javelins and targets, we came to Fremona, which is a large and famous town in these parts, on the 21st of June. Thus far the Patriarch's letter.

LIEUTENANT LEFEBVRE'S JOURNEY FROM ATEBIDERA TOWARDS THE SALT PLAIN, IN JUNE, 1841 (translated with some abridgement; marked P on Map page 39).

I LEFT Atebidera on the 5th June, at 9 o'clock in the morning, taking my instruments for observations. We were six, including the guide. For four hours we did not quit the plateau of Atebidera; arrived at the edge, we descended a slope, and advanced over a silicious limestone, very arid; there was some cultivation, but the frosts are a great obstacle to the harvests. The inhabitants of the country owe their wealth to the commerce they have with the Taltal tribes.

Arrived at the extremity of the slope we came upon a precipice, which extended to the right and left as far as the eye could reach; thick vapours concealed its depth. Behind us a fine sky lighted up corn fields, and some dried bunches of bushes; no horizon being visible in front, and no ground under our feet, we seemed to have arrived at the end of the world.

The picture was striking. I wished to see it again at another time of day. The vapours having dispersed, the eye could follow a steep scarp, which terminated at a depth of 500 metres, in an immense plain; in this interval was comprised all the phases of a flourishing vegetation, from juniper, and several species of terebinths, passing by olive trees, which indicate in this climate a height of 2·200 metres,* to mimosas, which form the extreme limits beyond which no vegetation is seen. In the plain nothing is to be seen but arid sand, and, further on, a reflecting surface, from which the rays of the sun reverberate as from a calm sea under the tropics.

We passed the night in one of the villages of this frontier; we were regaled with barley-meal and water, made into balls. At 5 o'clock in the morning we began to descend to the valley of Ficho. The slope is so steep that you cannot walk without the aid of a stick; nevertheless, as the road is marked out and kept clear of rocks, in that respect preferable to the Tarenta one, camels can climb it, provided they are not loaded; but mules alone can traverse it with facility. Donkeys are excellent for the descent, but too weak for the ascent; often in the steep places the guides are obliged to help them by shoving from behind. It took us three hours and a half to reach the valley.

We then found ourselves in a ravine which encloses a rivulet, on the edge of which Ficho is built. No description can realize the sensation of heat we felt in this place; the camel is the only animal that can resist it. Many go, it is true, as far as the Salt Plain, but their masters take care to provide themselves with grass and fresh straw before quitting the highlands; besides, it is only a fatigue of some days, after which the animals find repose in a better climate. In the hut of the chief who

* A metre is 3 feet 3¼ inches.

received us we perceived a mule, kept at great cost, which did not prevent it from being a complete specimen of osteology.

The town of Ficho may contain 1,500 inhabitants; its huts, placed on the slope of the ravine, at a few paces from the brook, are badly constructed, and judging from their miserable appearance, it would not be imagined that that is one of the richest countries in Abyssinia. One would say that the drying influence of the sun has also touched the brains of the inhabitants, for they seem to have no ideas but that of gain. Drinking is their only pleasure, and they are constantly drunk. Frequent disputes are the result, from which arises the general habit of never appearing anywhere, even at their own doors, without their buckler. Our host did not lay it down, even in his own house, and poured us out to drink with one hand, holding his weapon with the other. This custom may also have arisen from the permanent state of hostility in which the Christian population lives with the Taltals. An Abyssinian, who comes to settle at Ficho, makes his fortune there in two years, but what sentiment but the love of gain could make such a residence endurable.

It is not only from Atebi that people come to the market of Ficho; many neighbouring towns, especially Addigrate, the capital of Agamè, send their caravans there. The merchants of Dessa and Oikamessal also come there by a valley which cuts that of Ficho perpendicularly above the town of Endolote.

At a certain period of the year these caravans go for the salt to the place where it is procured, and it is by their transit that Ficho becomes rich; but, when the great heats come, in the months of June, July, August, and September, the inhabitants of the plateau, not daring to expose themselves to the danger of such a climate, abandon the traffic to the Taltals, who are used to the climate; the latter only take the salt as far as Ficho.

On our arrival, we saw the Taltals of the different tribes to the west of the plain of salt; the names of the principal ones are Rorôme, Ararat, Bori. These people bring to market camels, which they exchange for thalers, grain, and cotton cloth. The value of a camel at Ficho varies from three to seven thalers.

Our host entertained us after dinner with an account of the commerce in salt which is carried on on the banks of Lake Alelbad. He said that this lake often changes its shape and place, which he expressed in these terms: the lakes moves. Often, he added, on going to a place which the evening before was quite solid, you suddenly break through, and disappear in the abyss. But what is more frightful is the overflow of the waters; sometimes the lake rises like a mountain, and falls again into the plain like a deluge; entire caravans, men and beasts are engulphed. There are, however, precursory signs, of which mounted men only can take advantage, by flying at the utmost speed of their animals; occasionally some of them have thus escaped, and it is from them these terrible details are procured.

The salt is found in the neighbourhood of the lake, in horizontal beds, two inches thick, which are got out with wooden levers; the pieces are then cut into the proper size for sale with little hatchets. A little to the north three volcanoes are found, at the foot of which sulphur is found mixed with salt; from these volcanoes is often heard a dull sound, which the Tantals call the Devil's thunder.

All this appeared to me very curious, and I asked for a guide to the place, but I was told the season was too unfavourable; an Abyssinian, much more a white man, could not support the heat; the Taltals themselves often fell, struck by congestion of the brain. Notwithstanding these reasons I insisted, and a Tantal was given me to take me on the morrow as far as I could go.

On the 7th June, directly after sunrise, I commenced the journey with my guide, and descended the valley of Ficho. It widened out as we advanced, and I found in some places alluvial banks, on which mimosas grow, and several shrubs, of which camels eat the leaves. An hour and a-half after leaving Ficho, we arrived at a point where our road cut that of Dessa; the two form nearly a right angle. There the mountains separate, and form a sort of basin. We descended for an hour more, then the slope became less steep, the alluvial banks less extensive; there is more water, and the vegetation becomes fresher. The valley continued to widen, and on the left of the brook was a rich tract, about a mile broad and two miles long, on which grew several fine sycamores, and other species of trees of large size, unknown in Europe. These woods were very thick; I was obliged to dismount to penetrate them. After ten minutes march we found ourselves among abandoned cotton plantations, not far from which were some fruit trees, and behind them a little church of Greek architecture, built by the Ethiopians, which was easily discernible by the cement, which is composed of clay, mixed with chopped straw, instead of lime. An Abyssinian colony had been founded here by Sebagadis, who wished to build also a fortress there to hold the Taltals in check. At the time that we visited it, the place was deserted; the banana and pomegranate trees grew together without hindrance. Their fruits served still to refresh the merchant, who was bold enough to leave the caravan, and risk an encounter with a Taltal. . . . . .

After having eaten some fruits, we tried to advance to Mankel Kelié, from which the salt plain and lake may be seen; but I was suddenly taken with vertigo, and fell insensible. I recovered my senses after some moments, and ordered myself to be placed immediately on my mule, sustained by two men, and led back. I forbad them to stop, however much I might suffer. I fainted again on the road, and in that state arrived at Ficho.

At sunrise I was tied on to my mule again to reascend the plateau. . . . . The next day I resumed the journey to Atebidera, where I arrived at nine o'clock in the morning.

Among the information which I had obtained from my host at Ficho, there was one portion which I should like to have verified, namely, to know whether it was possible to go from that place to Amphilah in four days, and to Massowah in five. This journey, which had never been made, would have been very curious; but the season was not favourable, and I was too well persuaded of this to wish to repeat the experience.

LIEUTENANT LEFEBVRE'S JOURNEY FROM MASSOWAH TO ADOWA IN JUNE, 1839 (translated with some abridgement; marked B and I on Map, page 39).

6th June.—We quitted Massowah. On landing on the continent we found eight camels and six mules under the guidance of some Chohos.

Although our pace was quicker than that of the caravans of Egypt, it took us nearly three hours to reach Arkiko. The sons of the Naib begged us to stop with them until the heat of the day was passed. We acquiesced, the heat being very great, the thermometer marking 45° (113° Fahrenheit) in a thatched house. The sandy soil of the coast refracts the rays of the sun with a force which renders a march by day very painful. The eye sees no vegetation, but some stunted acacias, which, by encumbering the road, add to the difficulties of the march.

The Naib wanted us to take the road by Dixan in preference to that by Halai, which is the shortest. We were indifferent by which road we entered Abyssinia, but the narratives of the travellers who had preceded us had impressed us so unfavourably with the Naib that we determined to go by Halai. We marched during the night, and at sunset we stopped at Oueia, which most travellers have described as a charming place. Probably the insupportable heat of the sun, the burning aridity of the soil, and the fatigues of the road, had rendered them easy to please, for the oasis of Oueia is composed of a bunch of palm trees, a little scorched grass, and a spring of brackish water. We may add that this magnificence is set in a valley bounded by schistous and sterile mountains, where nothing attests the passage of man.

As soon as the heat was passed, we resumed our route, and encamped in the evening at Hamhamo, taking care to light great fires to keep off wild beasts. Here commence the gorges formed by the spurs of the Ethiopian plateau. From this point the road follows the bed of a torrent, moistened by some rills of water, which never dry up. The temperature becomes endurable, but the nature of the ground, and its steepness, make the march difficult. It took us three days to reach Toubbo, the last station before arriving at the foot of Taranta.

As our guides insisted on our going by Dixan, and not by Halai, I left my companions with the baggage, whilst I went on to examine the two roads.

I advanced into the middle of interlaced gorges, in which the traveller may easily lose his way. In these vast solitudes I experienced an undefinable impression, and felt crushed by the greatness of nature.

As we rose, the country changed its aspect, and became more habitable. We met from time to time herds of cattle, whose shepherds saluted my guide respectfully. Some asked for snuff, to which I added the snuff-box, and received in exchange jars of milk, and an infinity of blessings.

At the first sight of Taranta it seemed impossible that caravans could climb its steep and slippery slopes; but the heavily laden oxen climb with wonderful strength and agility over rocks apparently impracticable, and thanks to my mule I soon found myself at the top of the ascent. An air almost cold succeeded to the intolerable heat of the preceding days, and the kolkoual, the tree peculiar to Abyssinia, which Bruce was the first to describe, showed me that I had reached the edge of the scene of my explorations.

We advanced for some time longer over a tolerably well cultivated plateau to Halaye. From thence the eye embraces a wide view, stopped only by the curtain of high mountains which form the north-west side of the basin of Adowa. These serrated peaks cut the horizon in a thousand fantastic forms. At my feet was spread out Tigré, a succession of valleys, of which my eye followed all the windings. An infinity of brooks traversed the country. I stopped for some minutes to gaze on this magnificent picture, which the rising sun lighted up with a mixture of gold and silver.

On approaching Halai, I had been accosted by a very intelligent adventurer called Betléem, mentioned by MM. Combes and Tamisier in the interesting account of their journey. He conducted me to the chief of the village, the *choum guedé*, and gave me the necessary information, which enabled me to send at once to my companions to join me.

Next day I was told that armed men, descending from Dixan, had arrested my companions, encamped at the junction of the two roads, and wanted to compel them to take the one leading to their town. I immediately started with a band of the Chohos of Halai.

After two hours' journey we perceived the soldiers of Dixan. After some threatening movements the two parties joined, and agreed that, whichever road we took, they would divide the profits arising from the hire of the beasts of burden for our baggage.

Next day it required all our moral force to reduce to just proportions their ridiculous demands. They wished first to open our boxes, pretending they were too heavy for the oxen, but in reality to see what they contained. The proprietors of the oxen appeared disposed to make us pass a second night at the foot of Taranta. Nevertheless, after three hours' discussion, we came to terms, and about four o'clock started for Halaye. Night obliged us to encamp on the first step.

As soon as we reached Halaye we prepared to start for Adowa, the capital of Tigré, where we proposed to pass the rainy season. This town, being a centre of easy communications with the other countries of Abyssinia, we could there obtain the necessary information for continuing our explorations. We started in a south-south-west direction. During the four days' march which separate Halaye from Adowa we met with the most favourable reception from the natives, with the exception of those of Eguela. This district is, nevertheless, one of the richest of Tigré, but, being on the caravan route, it has too many opportunities for practising hospitality to put it often in practice.

This absence of goodwill obliged us to sleep in the open air, on stones symmetrically arranged by the caravans, a precaution rendered indispensable by a very unpleasant species of ant named *Dekondeki*, which gets into the clothes and hair of people who sleep on the earth, and whose bite is insupportable.

For Lefebvre's itinerary of this route, see page 192.

LIEUTENANT LEFEBVRE'S JOURNEY FROM ADOWA TO ADDIGRAHT AND ANTALO, OCTOBER, 1840 (translated with some abridgement; marked L 2 and O on Map, page 39).

The country we had to traverse had the reputation of being little hospitable, and the civil war which had desolated it prevented any regular markets being held. We were obliged, therefore, to carry provisions with us, as is the custom among Abyssinians of distinction. They generally consist of a little corn and theff; barley-meal for mixing with cold water when an oven cannot be procured; meal of peas and beans, mixed with red and black pepper; butter, honey, pimento sauce, and powdered meat.

To prepare this meat, it is cut into extremely thin shreds, which are dried until they are as hard as wood; they are then pounded in a mortar with black and red pepper, cinnamon and cloves. All the meat of a bullock, thus treated, is hardly a load for a man. To eat it, a pinch or two is boiled in water, and a little butter is added: soon the whole swells, thickens, and forms an agreeable broth, far preferable to the salt meat of ships.

We left Adowa on the 5th October, at 7 o'clock in the morning. We took the road to Entitcho by the village of Memessa, and made our first halt at the wells of Mégara Tsameri, at the foot of a mountain which closes the valley. From thence, mounting the plateau of Guendepta, and leaving Mount Semayata behind us, we descended on the other side into the valley of Maye Kerbahar, and slept at Addikrasse.

At seven o'clock in the morning we resumed our march across a country furrowed by numerous brooks, of which the principal rise in the mountain of Guendepta. At noon we reached the first hills of the district of Entitcho. We stopped in the village, and I passed part of the night in making astronomical observations. In the evening Schaffner was taken ill, and I left him to the hospitalities of MM. Ferret and Galinier, who lived close by. On the morrow I continued my journey.

For two hours I went along the crest of the mountains of Entitcho, which join those of Darha and Amba Saneyti, and form, with those of Debra Damo, a rich basin, which comprehends the districts of Seriro and Béezet. After having descended the eastern slope of the mountain Entitcho, I entered first the prairie of Seriro, traversed by a brook called Gabata. I then mounted by a slight incline to the country of Béezet, and passed the night in the village of that name, situated at the foot of the mountain Alékoua. Next day I entered the high country of Agamé, whose slopes are thickly wooded. After a fatiguing day's march I arrived at the village of Addi Baria, whose elevation is 3,000 metres; the air was very brisk, and the light cotton stuff, which forms the ordinary costume of the Abyssinians, was replaced by woollen clothes.

Here we experienced the first refusal of hospitality. . . . . .

The next day, until noon, I remained on the tops of the hills, and then, descending, traversed a plain of great extent, which on one side descends abruptly towards the Taltal country by a vast precipice, from which rise some mountains, whose blue tops are hardly elevated above the level of the plain and form the first of the two descents which have to be passed before reaching the sea.

The plain belongs to the district of Addigrate. It appeared to be thickly inhabited, and everything breathed an air of wealth; the neighbourhood of the salt mines and the presence of a population at once commercial and agricultural, made themselves felt. . . . . .

The next day, Sunday, I made my last stage. Following the range of hills which bounds the plain to the south, and leaving the town of Addigrate some miles to the north, I arrived about two o'clock in front of the camp, which was established in the midst of an uncultivated country, at a point from whence a ravine debouched leading to the Taltal country. The tent of the Commander-in-Chief was placed on a little eminence in the form of a fort, and all round, on the plain, was arranged the tents of the principal officers against which the straw huts of the soldiers were arranged in circles. These huts formed an empty space, where the horses, mules, and cattle were tethered, and part of which served as a threshing floor . . . . . .

I passed some days in the camp, and I employed them exclusively in geographical labours. I was on the point of leaving when I received a visit from two men belonging to Balgada Arèa, who pressed me, on the part of their master, to visit Enderta and Ouodgérata, which they told me were the most curious and hospitable provinces of Abyssinia. I did not want pressing to decide me to profit by this opportunity, and prepared to set out.

Addigraht to Antalo.

I went the same day to take leave of Oubié, who presented me with two cows, and on the morrow I quitted the camp, taking a westerly direction towards the plains of Haramat. I stopped at a little village named Ouéléle, but, learning that the neighbourhood was infested by the bands of Guébra Rafael, I determined to alter my route and regain the chain of Agamé and follow the high plateaux of Atibidera, Ouomberta, and Dessa, which is the ordinary route to Enderta of the salt merchants.

I first reached a range of high hills, composed of white sandstone and quartz. Their sides contain cavities, veritable eagle nests, which the agile Abyssinians alone can reach. If, by chance, a little fresh water filters through the rock, the cavern becomes an excellent fortification, and may become a refuge for a fugitive or for a band of robbers. I found a narrow valley, called Soussobé Gabia, or the market of Soussobé, and, following it for some time, arrived at the foot of the mountain on which Atebidera is built. We were arranging our camp for the night when we were accosted by a man who had been one of the brave soldiers of Cassaye. He invited me to pass the night in his village, and to accept for supper a cow and some honey. My host was called the Chalaka Chékoulabe; formerly all the country we had traversed during the day belonged to him; now, completely despoiled, he lives from the fruits of his excursions amongst the Taltals, who are born enemies of the Abyssinians. Having learnt that I was going to Antalo by Atebidera, he persuaded me to take by preference the road by Denguellet, which was shorter, and promised to give me a guide. We parted very good friends. On leaving him I passed by a narrow gorge which led to the Amba of the Chalaka; the road was bordered by olive trees, junipers, and plants called Taddo, employed to ferment hydromel. Enormous blocks of quartz were seen scattered among the sandstones and clay schists. The extremity of the defile led us to the country of Denguellet, and I had only to pass a few hills before arriving at the vast plains of Enderta.

I stopped at the entrance of the defile of Dongollo to breakfast under a grotto covered with jasmine. We were near a hill twenty metres high; a church, called Mariam Corver, has been cut out of the rock composing it, and the inhabitants of the country pretend that it was made by God. Having passed the Guenfel river, which issues from this defile, we entered the plain of Aouza. In the middle of excellent pasture I saw cattle of a stronger nature than those of Lower Tigré, and the great horns of some of them proved that they had been crossed with Taltal bulls, which are remarkable for the strength of their muscles and the disproportionate size of their horns, which attain a size of six feet long by six inches in diameter. In the evening I arrived at Aouza, a little town near which the River Guenfel forms a cascade and falls into a valley, along whose sides calcareous rock, with fossils of the Jurassic series, crop out.

Aouza has a population of about 1,200 souls. The houses are terraced instead of being conical and covered with straw, as is generally the case in Abyssinia.

Eight kilometres south of Aouza the River Agoula, which separates the plateau of Tera from that of Ouomberta, is found. In the valley I saw five Taltals in ambush, armed with lances and bucklers. Some moments after we had a more pacific rencontre; it was a great number of merchants leading donkeys and some camels loaded with salt, which announced to us the neighbourhood of the towns bordering the plains where it is procured. For four hours we had been marching in a country which appeared uninhabited.

On entering Ouomberta we began to find some cottages. We passed from thence into the province of Dessa. And hoped to arrive that evening at Antalo, for we perceived the mountains which command that town; we therefore hurried on across the great interminable plains. Nevertheless night overtook us before we had reached a suitable habitation. We began to hear the cry of the hyænas. At each instant our guide hoped to arrive, and after wandering about we ended by floundering in a marsh, not being able to distinguish anything at five paces from us. At length our feet struck firmer ground, a path presented itself to our sight, we followed it as our only resource. Two or three times we lost it, either in crossing brooks or mounting rocks, an operation in which my mule acquitted itself very well. At length we perceived a light, and soon we were winding in a labyrinth of houses. We were in the village of Elkèle.

I quitted Elkèle at daybreak, and went towards Antalo, where I expected to find Balgada Arèa, but in the plain of Afgole I met several horsemen, who were hastening to the camp of their master, who was going to direct, in person, an expedition against a revolted chief. I stopped, then, at the village of Afgole, and sent to the camp of Arèa to demand an interview. About three o'clock, not seeing the messenger return, I started, and arrived at Antalo at the end of an hour. . . . . .

I returned to Antalo, where I remained a day. On the morrow I passed the village of Afgole, and followed a valley, which, after two hours' march, brought me to Tchèleukot.

This town is built in the middle of a basin, surrounded by high hills, which were cultivated in the time of Ras Ouelda Salassé, but which now are overrun by thorny plants. A large brook, which flows round the town, makes the environs fertile. Each house is provided with a garden, and in some are planted olive trees, junipers, cedars, and vines. A charming *coup d'œil* is the result. Tchèleukot has two churches; one is constructed with a cylindrical wall, according to the Abyssinian method. . . . . .

In the time of the Ras the population of Tchèleukot was considerable, but it does not now exceed 3,000 souls. It is one of the Guèdam, or towns of refuge.

LIEUTENANT LEFEBVRE'S ROUTE FROM ADOWA TO MASSOWAH, BY KAYAKHOR AND AILET, IN JANUARY, 1840 (translated from the original with some abridgement; marked H on Map, page 39).

I TOOK the Seraé road, and stopped the first day at Béeza, at the house of the Choum Ato Ouelda Raphaël. He gave me a guide to the Mareb. The road leading there was bounded by a series of hills which belong entirely to the District of Béeza; on their rounded summits are pleasant groups of houses. The last hill, before arriving at the plains in the midst of which the river runs, is covered with trees; it is called Amba Christophe; the plain is thickly wooded, but with small trees, except at the edge of the water, where we saw several fine timber trees, all peculiar to Abyssinia. As we were in the dry season, we were not afraid of pitching our tents there. The river was dry and we were obliged to dig in the sand to obtain water. The inhabitants came to warn us that two troops of elephants had been seen in the neighbourhood, and that lions and hyænas were numerous. (They saw the elephants, but were dissuaded by their guides from firing, on account of the small bore of their fowling-pieces. The guides told them that, unless mounted on horseback, it was not safe to attack them unless there was a ravine or other place where they could take refuge).

The next day, having started at daybreak, we arrived early at Gondet. I sent to salute the Choum, Ato Akilas, and to ask him for lodgings in the village. He allotted me a house next his, and gave me two young rams, a pot of honey, a sheep, and a fillet of beef. . . . . . . Gondet has only two classes of inhabitants, husbandmen and hunters. The first, being obliged to carry on their work in the plains of Mareb, often at great distances from their habitations, are always armed to defend their cattle against wild beasts, and the other class, among whom must be ranked the nobles of the country, is, by the nature of its occupations, continually exposed to the greatest dangers, which render it hardy and warlike. . . . . . .

I took from one of the summits of Gondet a general *coup d'œil* of the mountains of Tigré, from Axoum as far as Okoulé Gouzaye, towards the Taranta chain. At my feet was the valley of the Mareb, of which the level is here 1,100 metres. The plateau of Seraé, which it surrounds completely, has a mean elevation of 2,000 metres. It results from this that the vegetation in the first of these two regions is quite tropical, whilst the plateau has a European climate. It is only opposite Chiré, above Medebaye Tabor, that the Mareb is never dry. This is why in that place the pursuit of wild animals is so productive; tormented by flies and the other insects of the low countries which surround Abyssinia, these animals ascend to the high lands, but they are obliged to stop at the limit of the waters.

. . . . . . I devoted the two following days to completing my works at Gondet, and then took leave of my friend Akilas. I entered a very narrow valley following a north-north-easterly direction, which, after an hour's march, took me to the foot of a steep slope, where there is only a narrow path worn by mules. A fine plain, composed of a rich and fertile soil, recompensed me for this difficult ascent; I could perceive, as far as the town of Addi Hoala, no prairie; everything was in crops; but the remainder of the route, as far as Addi Cassemo, on the contrary, contained numerous prairies, watered by many brooks, where I saw magnificent flocks.

On leaving Addi Cassemo, we traversed a desert which isolated travellers avoid for fear of banditti. We then reached the village Enna Hoyola, on approaching which we heard the sounds of music which announced the celebration of a fête. . . . . . .

The next day we started, and soon arrived at Amba Zareb, on our left were the mountains of Amacene (Hamasen), and at our feet a level country, a sort of desert which leads to the village of Châha, called also Maye Tada, that is to say white water, because the water found there traverses a stratum of tufa, which whitens it. The houses of Châha have flat roofs covered with rammed earth; they are so arranged as to enclose a large space into which all the houses open, and to which there is but one access. This arrangement is adopted to facilitate their defence against the attacks of the Chohos, who often push their excursions as far as this part of the frontier of Tigré.

From thence we again descended into the valley of the Mareb, and we passed that river at a place only one day's distance from its source; after having ascended the opposite slope, we arrived on the plain of Eguela Goura. The caravans often halt there, both going and returning; they can buy corn there, which is not usually the case elsewhere in Abyssinia, except at Gondar. At half-past eleven in the morning we were in the town of Eguela Goura, at the house of the Choum Aptaye, who had the reputation of being a very rich man. Some persons told us that he had jars full of gold, which was undoubtedly absurd; but there is no doubt that he possessed much grain, for, without reckoning the overflowing granaries of his vast house, I perceived on the outside a great quantity of full sacks, and every moment loaded bullocks augmented their number. . . . . .

The Choum gave me the guide I asked for. As we were in the dry season, I left with him my baggage and mules, and my two servants to look after them; I then started for Kaïkor, the last Christian town on this frontier. On the road I met the Debtera Sahalo, who brought me a letter from Dr. Petit. . . . . .

The plain of Eguela Goura continues until a short way from Kaïkor; it is then terminated by a steep slope, which leads to another step, where there is a village whose inhabitants act as guides between that place and the sea. They make the journey, which takes three days, for a quarter of a thaler to a poor man, but they take more from merchants and caravans, sometimes as much as five thalers; but their greatest gain proceeds from the sale of corn, butter, and other eatables for the use of travellers. They are honest, trustworthy, and courageous, and the interval between the Ethiopian plateau and Kaïkor is perhaps the safest part of Abyssinia for merchants.

We passed the night at our guide's house, and on the morrow we left Kaikor travelling north-east across a wooded plain, in which there was a great number of gazelles and pintadoes. A hyæna passed us, taking no more notice than a tame dog would have done. We then descended towards another step, although the slope was steep, and the road encumbered with rocks; we saw numerous traces of elephants. At the foot of the descent a narrow valley commences; it is traversed in its whole length by a cool and shady stream, and numerous herds feed in the rich grass by its side. Some tribes of Chohos have built here and there their huts. In the midst of this grass a clump of trees affords a delicious shade.

It is here, at the foot of an enormous sycamore, that caravans usually stop during the heat of the day. This place is called Ayederesso. After having reposed there some instants, we followed, for about two hours, the valley which runs north and south; we then traversed the chain which bounds it on the east, and we descended again into another lower valley which runs north-north-west. Another brook traversed this valley with a little grass on its banks; we stopped there to pass the night. It was about four o'clock; we profited by the little time which remained before sunset to collect some wood, for we were obliged to keep up a fire on account of wild beasts. . . .

On quitting this station, we crossed a chain on our left, and descended again into a valley, lower than the preceding ones, and which, at the end of two hours, led us into a sort of basin, where other valleys terminate and empty their water. This place is called Medeumnar, a name which the Tigreans give to all junctions of many waters. In the country of the Gallas this name is changed into that of Djeumma.

We began to find many Choho villages; they should rather be called camps, for they are composed of osiers, arranged in a circle, and covered with leather and rush mats . . . . . .

Some time after having passed these villages, we halted to breakfast; at that moment we perceived a herd of antelopes, of which we killed one . . . We were going to put our pieces of meat on a species of Euphorbia, but the inhabitants hastened to prevent us, because the contact alone of this plant is poisonous. Several analogous species in this country have very active poisonous qualities.

We were at length in the valley of Ailat, where, during the rainy season, almost all the flocks belonging to the different tribes of Chohos are collected; the inhabitants of the frontier of Serae and Amacène also bring theirs down here when the rains cease in the highlands, and commence in the low countries. We had never seen anything to compare with the number of cattle which we saw on our road, not even in the Pampas of South America. The plain of Ailat appeared to be the finest country in the world; but, in the dry season the soil, which we then saw covered with verdure, is converted into a fine dust, impregnated with saltpetre, which penetrates the skin, and sometimes causes sores. The leaves of the trees disappear, the branches are blackened; everything has the appearance of having been ravaged by fire. The water is all dried up, except in some fetid marshes. All the inhabitants hasten to quit the desolate country, and to remove their flocks to the plateau, leaving the lions and other wild beasts in tranquil possession of the place.

It was at the village Ailate that my colleagues were to meet me; at four o'clock in the afternoon we were very near it, but my mule was tired, and I could go no further. I accepted the hospitality that was offered us in a camp of Chohos; it was composed of about a thousand huts, disposed in a circle on four rows; two doors only gave admittance into the circular interior; they were left open in the day, but were shut at night after the cattle had entered.

. . . .

On the morrow, 7th February, we reached Ailate, after an hour's march . . . . . .

I employed the rest of the day in getting some geographical information, and in visiting the hot springs, which are four miles south-southwest of Ailate. These springs, whose temperature is 56° (133° Fahr.), issue from four orifices, distant about a metre from one another, among talcose schists traversed by veins of white quartz and rose feldspar.

LIEUTENANT LEFEBVRE'S ROUTES.

1. First Route from Messoah to Adowa.

First day. Seven hours for a mule.

Oye-Négousse.—The road traverses an arid and desert country.

Second day. Nine hours.

Medeummar.—Country varies in aspect in the dry and wet seasons: it is burnt up in the first, but in the second covered with foliage and verdure; the yellow dust gives way to fresh prairies, where innumerable flocks pasture.

Third day. Nine hours.

Aye-Deresso.—In the valleys formed by the spurs of the Ethiopian plateau. The air begins to become cool, and water is found on the greater part of the road.

Fourth day. Five hours.

Kayé-Kor.—Kayé-Kor is the first place where water is found after leaving Aye-Deresso. It is reached by a wooded plain, inhabited by ferocious animals.

Fifth day. Five hours.

Châah.—On leaving Kayé-Kor the Ethiopian plateau is reached. The road traverses a cultivated and thickly peopled country. Châah is on the other side of the Nareb on the right bank.

Sixth day. Four hours.

Koudo-Fellassi.—The first part of the road is through a desert and wooded country, haunted by bandits and Chohos, who attack travellers when they are not on their guard; the second part is by the fertile plateau of Serae.

Seventh day. Nine hours.

Addi Hoala.—Over plains watered by numerous brooks.

Eighth day. Four hours.

Goudei.—On the descent leading from the plateau to the Mareb.

Ninth day. Four hours.

Addis Addi.—The Mareb is repassed: a low country, wooded, of high temperature.

Tenth day. Seven hours.

Chaâguené.—On cultivated hills, where only a few villages.

Eleventh day. Five hours.

Adowa.—Through green well-watered valleys.

2. Second Route from Messowah to Adowa.

First day. Six hours.

Oueia.—Road dry, burnt up, so hot that it can only be travelled over by night.

Second day. Five hours.

Hamhamo.—Same as the first day.

Third day. Seven hours.

Toubo.—Through narrow valleys, at the bottom of which are torrents which never dry.

Fourth day. Seven hours.

Dixan or Halaye.—The mountain of Tarenta is climbed; Halaye and Dixan are first Christian towns met with after leaving the coast.

Fifth day. Nine hours.

Eguerzobo.—Through a very mountainous country, whose population is one of the least hospitable in Christian Abyssinia.

Sixth day. Nine hours.

Equela.—The Rivers Tserana and Belessa are crossed and the chain of Logota is passed.

Seventh day. Nine hours.

Adowa.—The two rivers Mémenè and Onguaye are crossed: a rich country.

3. Route from Massowah to Atebi.

In the first four days Halaye is reached (see second route).

Fifth day. Seven hours.

Segonete.—On the edge of the Ethiopian plateau; the second part of the route is made on a lower level; several unimportant brooks are crossed.

Sixth day. Nine hours.

Bihate.—Through an arid and desert country.

Seventh day. Seven hours.

Addigrate.—A higher elevation is reached, rich in pasture. Addigrate is the capital of Agamé.

Eighth day. Six hours.

Agoddi.—Through a low country, at first rich and fertile, then dry and destitute of vegetation.

Ninth day. Eight hours.

Atebi.—Over elevated plateaux, where barley is the only cereal cultivated.

4. Route from Messoah to Antalo.

Atebi is reached in nine days. (See previous route.)

Tenth day. Four hours.

Aikamessal.—Following the course of the brook Alecti which further on is increased by receiving the waters of the Province of Ouomberta, and takes the name of the River Agoula. Aikamessal is one of the points by which the caravans descend to the plain of salt.

Eleventh day. Six hours.

Dessa.—Over a tolerably fertile plateau, but thinly inhabited. Dessa is also an assembling point from which caravans descend to the plain of salt.

Twelfth day. Five hours.

Kouchaine-Tcheleukot.—Continuation of the same plateau.

Thirteenth day. Six hours.

Antâlo.—The two little rivers Guembéla and Antabate are crossed. Country inhabited and covered with cultivation.

5. Route from Messoah to Abi-Addi.

Seventh day. Adowa.

Eighth day. Six hours.

Zoungui.—Through an uneven country belonging to the Kollas (low country).

Ninth day. Eight hours.

Meretta.—The first part of the road is almost a desert and nothing but some groups of mimosas are seen. The river Ouéri is crossed, shut in between deep banks; the country assumes a less desolate aspect on approaching Meretta, of which the position is very picturesque and surrounded by numerous hamlets.

Tenth day. Six hours.

Abi-Addi.—Through a country thickly inhabited and frequented by a great number of merchants who trade in salt with Abi-Addi. This town serves as a depôt for this article for the provinces of Sémien.

6. Route from Messoah to Areza.

The three first days are employed in traversing the Choho country.

Third day. Seventeen hours.

Ouaky.—First station on the Ethiopian plateau.

Fourth, fifth, and sixth days. Eighteen hours.

Aréza.—Through low country, furrowed by numerous ravines. Aréza is in the middle of the great hunting-grounds for the elephant, rhinoceros and buffalo.

7. Route from Messoah to Debab Gouna.

Seventh day.

Adowa.

Eighth day. Four hours.

Axoum.—Almost always across a plain; country rich and fertile, principally cultivated with theff and corn. With the exception of some hills the country is entirely level.

Ninth day. Six hours.

Tambouhh.—The brook Maye-Tchout is crossed, which waters fine prairies and fields of theff; from thence the hills of Akabsiré are reached, behind which is the valley of Guerzela; after having crossed them a descent is made into the valley of Tamboukh.

Tenth day. Four hours.

Belasse.—After passing the fine prairies of Seleuloah, a little chain is passed which borders the plain of Chiré on the north, and the district of Belasse is entered.

Eleventh day. Eight hours.

Maye-Témine.—Through a plain. Country of cereals and pastures.

Twelfth day. Six hours.

Debabgouna.—Through a plain, the same as the preceding.

8. Route from Messoah to Eguela-Goura.

Fourth day.

Kayé-Kor.—First station on the plateau.

Fifth day. Three hours.

Eguela-Goura.—On leaving Kayé-Kor a steep ascent is made; but the rest of the route is through a plain, formed of fields of cereals and a small number of meadows.

9. Route from Messoah to Ficho and the Salt Plain.

Ninth day.

Atebi.—(See the route from Messoah to Atebi).

Tenth day. Three hours.

Assote.—Road through a plain.

Eleventh day. Three hours.

On leaving Assote the eastern slope of the Ethiopian plateau is descended by a very steep road.

Twelfth day. Four hours.

Wells of Saba.—Through the bed of a torrent strewed with rolled flints.

Thirteenth day. Eight hours.

Mankel-Kelié.—A slope is descended, and a vast plain, for a great part desert, is entered. Near Mankel-Kelié are some cultivated spots and a few scattered huts inhabited by the Taltal tribe of Borôme.

Fourteenth day. Nine hours.

Gara.—The road descends sensibly, and soon there is nothing but a burning desert without any species of vegetation. The halt is made near the spring of Gara, from which the water issues boiling.

10. Second Route.

Ficho may be reached by a much shorter line through valleys from the shore; for the Salt Lake is only at a very short distance from the coast; but this route is only practicable for natives of the country.

11. First Route from Messoah to Addi-Abo.

Eighth day.

Axoum.—(See the itinerary from Messoah to Debabgouna).

Ninth day.

Maye-Touaro.—This town is at the extremity of the valley of Tambouhh, at the place where it debouches into the plain of Seleuloah.

Tenth day.

Medebayetaber.—Country hilly and generally wooded, with numerous ravines, and a great number of brooks, which fall into the River Mareb.

Eleventh day.

Kayé-Beit.—Low country, well wooded. It is difficult to travel without a guide in this country.

Twelfth day.

Addi-Abô.—On approaching Addi-Abô, a great number of villages and well cultivated fields are met. Nevertheless a part of the population devotes itself to hunting which is very productive.

12. Second Route.

On leaving Maye-Touaro a road may be followed by the hills of Koyeta and the district of Addi-Onfito, and on the fourth day Addi-Abô is reached. The people met with on this road are more hospitable than those of the districts of Medebaye Taber and Kayé-Beit.

13. Route from Messoah to Aosseba.

Twelfth day.

Kouihaine-Tcheleukot.—(See the route from Messoah to Antalo).

Thirteenth day.

Aosseba.—The country traversed consists of plateaux rich in pasture and cereals. Aosseba is on the southern frontier of Enderta; the inhabitants have frequent relations with the Taltals and Gallas, from whom they buy ivory and some rhinoceros horns.

14. Route from Messoah to Assakelti.

Thirteenth day.

Antalo.—(See the route from Messoah to Antalo).

Fourteenth day. Four hours.

Addirake.—The plain of Antalo and the River Bouillé is traversed;

the province of Ouodgérate, of which Addirake is one of the principal villages, is then entered.

Fifteenth day. Five hours.

Beit-Maria.—After traversing a plain covered with pasturage, and bounded by high mountains, a narrow valley, watered by a large brook coming from the high peak of Aladjié, is ascended.

Sixteenth day. Five hours.

Sessate.—Still in the same valley, the ascent of which is continued; the high chain of Ouodgérate is then crossed by the pass of Aladjié. It is on its opposite slope that the village of Sessate is built.

Seventeenth day. Eight hours.

Tsaâfti.—After having traversed the valley of Atsalla, the chain, which bounds it on the opposite side to Aladjié, is crossed, and a second valley, named Aiba, is entered, behind which is a large plain, from whence a descent is made into the basin of Dôba, in the middle of which is situated the market of Tsaâfti.

Eighteenth day. Six hours.

Assakelti.—The road leads through several interlaced valleys, which are bordered by hills with round summits, on which are grouped hamlets of an agreeable aspect. Several districts are thus traversed, of which the most important is that of Mehhane, followed by that of Aya, on leaving which a rugged mountain is climbed, which leads to a slope of the mountain of Mossobo. There is situated the village of Assakelti, the ordinary residence of the governor of Achangui.

15. Route from Messoah to Debra Abbaye.

Eleventh day.

Maye-Teméne.—(See the route from Messoah to Debabgouna).

Twelfth day. Five hours.

Maye-Chebéni.—Road through plain.

Thirteenth day. Five hours.

Tembela.—Country hilly for the first part of the road; a plain covered with cotton and fields of maize is then reached.

Fourteenth day. Eight hours.

Debra-Abbaye.—Country wooded and cut up by ravines, where the principal cultivation is that of cotton. The market of Debra-Abbaye is one of the most important for ivory and furs.

16. Route from Messoah to Dembelasse.

Sixth day.

Koudofelassi.—(See the route from Messoah to Adowa).

Seventh day.

Dembelasse.—A descent is made by the western slope of the plateau of Sérаé. The inhabitants of Dembelasse are almost all hunters, and have the reputation of being cruel and inhospitable.

17. Route from Messoah to Samere.

Antalo is passed; then, in one day, the plain of Sahharte is crossed; one of the angles of which touches the town of Samеré. This market supplies with grain a great part of Tigré; excellent iron is also brought there by the inhabitants of the neighbouring province of Bora.

18. Routes from Messoah to Sokota.

The first passes by Achangui, from which Sokota is reached in a day and a-half; it is the road by the high country.

The second passes by Sameré and the valley of the Tellaré (this road crosses an unhealthy country where the temperature is very high). The market of Sokota is on the road of the merchants who visit the Galla provinces to the east, and it is also the central point of the commerce with Ouadela and the Ouello tribes who possess the finest wools of Abyssinia.

19. Route from Adowa to Gondar.

Fourth day.

Maye-Teméne.—(See the route from Messoah to Debabgouna).

Fifth day.

Maye-Chebéni.—Prairies.

Sixth day.

Maye-Temkate.—The road is through plains.

Seventh day.

Maye-Aini.—The Taccazé, which runs in a profound fissure, is crossed; Maye-Aini is on the side opposite to Maye-Temkate, in the country of Berra Ouesseya.

Eighth day. Four hours.

Maye-Téclite.—As far as Maye-Kessate the country is level and the soil appears fertile; but it is nevertheless nearly deserted, probably on account of the Chankallas, who sometimes make incursions there and massacre the isolated inhabitants. Between Maye-Aini and Maye-Kessate the River Sarentia, one of the affluents of the Taccazé, is met with. On leaving Maye-Kessate many hills are crossed, where the population begins to collect; half-way the River Ounguiha is crossed.

Ninth day. Eight hours.

Debeubaheur.—The country traversed is formed by the prolongation of the spurs of the chain of Sémiéne, and these spurs form between them narrow valleys, in the middle of which flow the Rivers Enzo, Bouheia, Ansia, Zarima. After crossing this last river a steep road is ascended for two hours until Debeubaheur is reached which is on the highest step of the mountain of Lamalmon.

Tenth day. Five hours.

Debarek.—The ascent of Lamalmon is completed and a country of plains is entered, forming part of the province of Ouoguéra. Before arriving at Debeubaheur the district of Oulkefite is reached.

Eleventh day. Seven hours.

Chimberazéguéne.—The road is through a plain covered with pasture, the country is well peopled; climate cold.

Twelfth day. Eight hours.

Tsâkdebeur.—The same country: but the plateau lowers whilst advancing south-west, and is covered by numerous brooks, which often, in winter,* stop the communication.

Thirteenth day. Six hours.

Gondar.—The road continues over plains until the descent of Bambélo. Before arriving at Gondar the River Maguelche is crossed by a bridge.

* Query, summer, which is the rainy season.

20. First Route from Adowa to Basso.

Thirteenth day.

Gondar.—First route.

Fourteenth day.

Feurkabeur.—After descending from Gondar and traversing the River Kaâh and Dembea-Goumara, the road goes along the edge of Lake Tsana.

Fifteenth day.

Ifag.—The road continues along the lake following a plain, bounded on the east by a chain with flat summits, on which are built the towns of Amba-Mariam, Emfraze, &c. Before reaching Ifag the River Arnogarno is crossed.

Sixteenth day.

Madéramariam.—After quitting Ifag the plain of Foguéra is crossed which is watered by the River Reb. A chain of hills is then crossed and the country is broken as far as Madéramariam. The little River Goumara is met with on the road.

Seventeenth day.

Chimé Guiorguis.—The road follows green valleys; halfway the River Gota is crossed.

Eighteenth day.

Andabeit.—The road through plains. You encamp on the border of a ravine which leads to the Nile.

Nineteenth day.

Mota.—The Nile is traversed, and Mota reached on the opposite bank.

Twentieth day.

Oüofite.—Road through a plain. The Rivers Ázouari and Teguidar are crossed.

Twenty-first day.

Debraouerk.—Road through a plain: some hills before arriving at Debraouerk. The Rivers Ennate, Tché, Guelguel-Tché, Idane, and Feza are passed.

Twenty-second day.

Yebeurte.—Through a plain. The country is furrowed by numerous brooks, which descend from the chain of Telba-Ouaha.

Twenty-third day.

Debeite.—Some hills are crossed, which are the ramifications of the chain of Telba-Ouaha. On leaving Debeite the country is generally woody.

Twenty-fifth day.

Dagate.—The Rivers Soa, Mouga, Betchet, and Bagana, are crossed.

Twenty-seventh day (siemong).

Yedje-Oulie.—The River Yeda is crossed.

Twenty-eighth day.

Bassô.—The River Tchamoga is crossed.

21. Second Route from Adowa to Basso.

First day.

Djenda.—Country of plains; thickly peopled, and fertile.

Second day.

Takoussa.—The same.

Third day.

Danguelbeur.—The province of Agâo Mèdeur is reached.

Fourth day.

Atchafeur.—Over an elevated plateau, where pasturage abounds; excellent iron is found.

Fifth day.

Metcha.—Country of plains, in the middle of which rise the sources of the Blue Nile.

Sixth day.

Bourié.—On the slope of the plateau of Agâo Médeur.

Seventh day.

Denguelbeur.—A narrow pass is traversed, which leads to the high lands.

Eighth day.

Atchafeur.—Country intersected with ravines. Atchafeur is the second town of this name in the province of Agaô Médeur.

Ninth day.

Dembetcha.—On the chain of Telba-Ouaha.

Tenth day.

Godera.—Near this town the River Felane rises.

Eleventh day.

Bassô.—The Rivers Fetane and Beur are crossed.

22. Route from Adowa to Charia.

First day.

Zoungui.—The districts of Addi-Kéré and Serhi are traversed. The soil is rich and well cultivated, although mountainous, and cut by ravines.

Second day.

Aouzienne.—The River Ouéri is passed, whose bed is profoundly shut in; the rest of the road is over a plain.

Third day.

Aouza.—The Rivers Selheu and Gueba are passed; the rest of the road is flat.

Fourth day.

Aréna.—The River Agoula is crossed.

Fifth day.

Antalô.—Road level. The great brook Guembela is crossed.

Sixth day.

Beit Maria.—The River Bouillè is crossed.

Seventh day.

Sessate.—The high chain of Ouodgerate is crossed.

Eighth day.

Tsaâfli.—The Rivers Atsala and Aiba are crossed.

Ninth day.

Achangui.—One of the seven districts of the province of Dôba. In the middle is a lake two leagues in diameter.

Tenth day.

Late.—An elevated chain, which borders lake Achangui to the south, is crossed, and you halt on the reverse of that of Aina, of which Tsera-Guedel forms one of the most elevated steps.

Twelfth day.

Oualdia.—Capital of the province of Yedjou. The high lands of Lasta have been left behind, and a valley with a temperate climate entered. The Rivers Ala and Toukour are crossed.

Thirteenth day.

Guerado.—Mussulman district of the province Yedjou. The Rivers Edéfe, Ouaha, Mersa, and Guerado, are passed.

Fourteenth day.

Sirba.—The districts of Sekala and Zéletcherk are traversed, and the province of Outchalé entered.

Fifteenth day.

Cossaro.—The River Méllé is crossed, and the valleys of Djari and Katti ascended.

Sixteenth day.

Entcharô.—The slope of the chain which borders the valley of Katti to the east is followed.

Seventeenth day.

Madjetié.—The valley Ouárakallo is followed, in which flows the River Borkenna.

Eighteenth day.

Chéréfa.—The River Toukour is crossed, and you descend from the plateau of Guemza to the torrent of Lembelete, of which you ascend the bed as far as Doulloute. There you descend into the valley of Moukeméda, which is watered by the River Nazaro. The eastern slope of the high chain, of which the foot has been followed since Oualdia, takes here the name of Guedme.

Nineteenth day.

Arogouratti.—The plain of Neguesso is traversed, in which run the Rivers Djaou, Saour, and Gacha-belebde. Arogouratti is situated on a spur of the great chain in the district of Mengneuste.

Twentieth day.

Goudje-Amba.—Road crosses the plain of *Rôbi.* Before ascending to Goudje-Amba, which is on the summit of an elevated peak, the River Rôbi is crossed.

Twenty-first day.

Tcheuno.—Several spurs of the plateau of Choa are crossed. Between the two last is the valley of Tchenno, watered by the River Aouadi.

Twenty-second day.

Aliyô-Amba.—After having crossed the spur which forms the southern side of the valley of Tchenno, several torrents which descend from the mountains of Ankober are crossed. Aliyô-Amba is the point where exchanges are made between the caravans which come from the interior of Abyssinia and those which bring, by way of Tedjoura, the products of India or Europe.

23. Route from Gondar to Madjetie.

First day. Four hours.

Ambamariam.—The Rivers Sodié and Arnogamo are crossed.

Second day. Five hours.

Derita.—The heights of Ouaina Dega may be followed, or the plain which borders the lake.

Third day. Seven hours.

Debratabor.—The River Reb is crossed.

Fourth day. Six hours.

Estié.—An elevated plateau, from which Lake Tsana is visible, is followed for some time, and a part of Godjam and Lasta.

Fifth day. Eight hours.

Tchetchéo.—Flat country, where horses and cattle abound. The River Tchetchéo is crossed.

Sixth day. Six hours.

Nebit.—Country watered by a great number of brooks, of which the greater part are affluents of the Taccazé.

Seventh day. Seven hours.

Daonte.—Country of high plateaux, where very fine woollens are found, which form an important object of commerce with the rest of Abyssinia.

Eighth day. Eight hours.

Sekala.—The high plateaux are descended by a spur, which abuts on the valley Yedjou.

Ninth day. Five hours.

Sirba.—The road follows the fertile plain of Zététcherk, which is a dependence of Yedjou, and, after surmounting several hills, the market of Sirba is reached.

Tenth day. Seven hours.

Kossarô.—The River Mellé is crossed; then, ascending the valleys of Djari and Katti, the pass of Kossarô is reached in a cold and wet country.

Eleventh day. Four hours.

Combolcha.—A descent is made from the pass of Kossarô into the basin of Combolcha, which is surrounded on all sides by high mountains.

Twelfth day.

Koterne.—You debouch by a defile between the Aini-Amba and Tgof into the Valley Ouasékallo, watered by the River Borkenna. The town of Koterne is situated on an isolated mamelon.

Thirteenth day. Seven hours.

Madjétié.—The River Borkenna and several of its affluents are crossed. The country is flat and well wooded.

24. Route from Gondar to Mélaksanko.

Seventh day.

Daonte.—(See the road from Gondar to Madjetié.)

Eighth day. Eight hours.

Melaksanka.—The Bachelo is crossed.

25. Route from Adowa to Gouraguie.

Twenty-second day.

Aliyô-Amba.—(See the itinerary from Adowa to Aliyô-Amba.)

Twenty-third day. Eight hours.

Dililla.—Several very elevated steps are ascended, and the province of Boulga is entered, on an elevated plateau at the foot of Mount Méguezazé.

Twenty-fourth day. Nine hours.

Bora-Addo.—You continue to follow the high lands of Boulga, which are rich in cultivation and pasture.

Twenty-fifth day. Six hours.

Aouache.—Fertile and well-peopled plains. Country for cavalry.

Twenty-sixth day. Seven hours.

Soddo.—The country of Soddo composed of fertile hills, with clumps of forest trees, is crossed.

Twenty-seventh day. Nine hours.

Absala.—Road across a wooded country, where a very high tree, called Zegba, may be specially remarked.

Twenty-eighth day. Eight hours.

Ahimeléle.—This country is covered with coffee trees.

Twenty-ninth day. Seven hours.

Lake Zouaye.—This lake occupies the centre of the province of Gouraguié. Seven islands may be counted on it, of which the principal is Debrasina.

26. Route from Aliyo-Amba to Saka.

First day. Eight hours.

Angolola.—The torrent Airara is crossed, and the ascent climbed which leads to the plateau of Choa. The road afterwards is almost entirely flat. The country is generally cultivated, and produces abundance of corn and barley, but not a single tree is seen. Halfway an affluent of the River Berésa is crossed, and, two hours before arriving at Angolola, that river itself is crossed.

Second day. Eight hours.

Fintchoa.—The River Tchatcha is passed, and you march through a thinly inhabited country, although the soil is fertile.

Third day. Seven hours.

Messeur-Medeur.—Country of plains. Some unimportant brooks are crossed.

Fourth day. Nine hours.

Roguié.—You descend a step, and almost immediately cross the Guermana, an affluent of the Aouache. As far as the hills of Roguié the road passes in the midst of meadows and fields, where corn, theff, and some vegetables, are grown. The country is watered by the River Hakaki, several affluents of which are met with.

Fifth day. Five hours.

Endodé.—You descend a step, which leads to the valley of Aouache, leaving to the east the mountain of Fouri, and to the west that of Endotto. At the foot of this step the country is covered with magnificent pasturage, which extends as far as Endodé.

Sixth day. Nine hours.

Bélcho Ori.—The Aouache is crossed, and a well-peopled country is entered. Barley and corn is principally cultivated.

Seventh day. Ten hours.

Oualisso.—The road continues across a plain, but the country becomes wooded, and it is furrowed by a great number of watercourses, which fall into the River Ouahabé.

Eighth day. Ten hours.

Saka.—The River Ouahabé, running to the south-east, is crossed. The country lowers towards the south, and forms several basins, separated by low hills, at the foot of which the coffee tree is cultivated. Halfway the Guibé is crossed, which the Abyssinians say is larger than the Nile.

27. Route from Aliyo-Amba to the Country of Aoussa.

First day. Six hours.

Tchénno.—Country of hills. The River Kaléna is crossed.

Second day. Eight hours.

Mafoude.—The River Aouadi is crossed. By the defile of Kéraba a tolerably high chain is passed, which forms one of the spurs of the plateau of Choa. Behind this spur is another parallel one, which must be crossed before reaching the valley of Mafoude.

Third day. Four hours.

Mengueuste.—The hills of Mafoude are descended, at the foot of which is the River Rôbi; a vast plain, partly desert and well wooded, is then entered. The elephant, and all carnivorous animals, are very common in it.

Fourth day. Seven hours.

Chéréfa.—The hills of Mengueuste are descended, and the plain of Néguesso entered, where maize and cotton are cultivated. This plain is watered by the Djaou and three of its affluents, which all descend from the plateau of Choa. After having traversed Djaou you ascend to Chéréfa.

Fifth day. Eight hours.

Saramba.—On leaving Chéréfa, you descend into a fertile, well-peopled plain, whose inhabitants are Mussulmans; several hills are then crossed, and the plain of Mouka Méda entered, in the middle of which flows the River Nazaro; it is at the western edge of this plain, at the foot of the chain of Guedme, that the town called Saramba is situated.

Sixth day. Six hours.

Madjétié.—The River Nazaro is crossed a second time, and you arrive, by the ascent of Doulloute, at the head of the valley of Sembeléte, which is then descended until it debouches into the plain of Chafa. The River Toukour is then crossed, and an ascent made to the plateau of Guemza. Madjétié is situated on the first step.

Seventh day. Seven hours.

Rékié.—Several large brooks, which descend from the plateau of Guemza, water the plain of Chafa, which is traversed during this day. A short time before reaching Rékié the River Borkenna is crossed.

Eighth day. Eight hours.

Leide.—Country hilly and intersected by ravines.

Ninth day. Six hours.

Ilala.—Country cultivated, but inhabited by Gallas who are exceedingly inhospitable.

Tenth day. Seven hours.

Kaâb.—The aspect of the country does not change.

Eleventh day. Six hours.

Taâ.—The country is inhabited by exceedingly ferocious tribes of Gallas. These tribes are nomade; their wealth consists in flocks and camels.

Twelfth day. Seven hours.

Bakarsa.—The banks of the River Mellé are followed, forming two strips of verdure, rather narrow, beyond which the soil is very arid.

Thirteenth day. Six hours.

Agamti.—The banks of the river Mellé are still followed.

Fourteenth day. Six hours.

Mellé.—The River Mellé is crossed, and a halt made on the left bank.

Fifteenth day. Seven hours.

Arabatessa.—The country, generally sandy and desert, is strewn with oases tolerably well peopled.

Sixteenth day. Seven hours.

Tchéfi.—Country the same.

Seventeenth day. Six hours.

Aoussa.—The Aouache is crossed. The banks of the River Aouache are wooded. They are frequented by wandering tribes, but they make no stay there, because the air is unhealthy. Aoussa is the capital of a little province of the Adal country. The neighbourhood is cultivated, but to the productions of the soil the inhabitants of Aoussa join the profits of commerce. Aoussa is one of the great marts of exchange between the high lands of Abyssinia and several ports of the Red Sea, Zeila, Tedjoura, Béloul and Eide.

28. Route from Entcharo to Aoussa.

From Entcharo, Leide is reached in five hours. (For the rest of the route see the itinerary from Aliyo-Amba to Aoussa).

29. Route from Aoussa to Tedjoura.

First day. Five hours.

The Aouache is crossed and Bila reached.

Second day. Seven hours.

Kourkoura is reached.

Third day. Six hours.

To Arho.

Fourth day. Seven hours.

To Dourgourgoura.

Fifth day. Seven hours.

To Kaballé.

Sixth day. Eight hours.

To Rahéta.

Seventh day. Eight hours.

To Arguita.

Eighth day. Eight hours.

To Heigounoal.

Ninth day. Nine hours.

To Tedjoura.

The whole of this route is inhabited by ferocious people against whom it is necessary to be always on your guard.

BRUCE'S JOURNEY FROM MASSOWAH TO GONDAR IN NOVEMBER, 1769 (marked D, I, and J on Map, page 39).

According to Achmet's desire, we left Arkeeko the 15th November, 1769, taking our road southward, along the plain, which is not here above a mile broad, and covered with short grass nothing different from ours, only that the blade is broader. After an hour's journey, I pitched See Maps, pages 113 168.

my tent at Laberhey, near a pit of rain water. The mountains of Abyssinia have a singular aspect from this, as they appear in three ridges. The first is of no considerable height, but full of gullies and broken ground, thinly covered with shrubs; the second, higher and steeper, still more rugged and bare; the third is a row of sharp, uneven-edged mountains, which would be counted high in any country in Europe. Far above the top of all towers that stupendous mass, the mountain of Taranta, I suppose one of the highest in the world, the point of which is buried in the clouds, and very rarely seen but in the clearest weather; at other times abandoned to perpetual mist and darkness, the seat of lightning, thunder, and of storm.

Taranta is the highest of a long steep ridge of mountains, the boundary between the opposite seasons. On its east side, or towards the Red Sea, the rainy season is from October to April; and on the western, or Abyssinian side, cloudy, rainy, and cold weather prevails from May to October.

On the 16th, in the evening, we left Laberhey; and, after continuing about an hour along the plain, our grass ended, the ground becoming dry, firm, and gravelly, and we then entered into a wood of acacia-trees of considerable size. We now began to ascend gradually, having Gedem, the high mountain which forms the bay of Arkeeko, on our left, and these same mountains which bound the plain of Arkeeko to the west on our right. We encamped this night on a rising ground called Shillokeeb, where there is no water, though the mountains were everywhere cut through with gullies and watercourses made by the violent rains that fall here in winter.

The 17th we continued along the same plain, still covered thick with acacia-trees. They were then in blossom, had a round yellow flower, but we saw no gum upon the trees. Our direction had hitherto been south. We turned westerly through an opening in the mountains, which here stand so close together as to leave no valley or plain space between them but what is made by the torrents in the rainy season, forcing their way with great violence to the sea.

The bed of the torrent was our only road; and, as it was all sand, we could not wish for a better. The moisture it had strongly imbibed protected it from the sudden effects of the sun, and produced, all along its course, a great degree of vegetation and verdure. Its banks were full of rack-trees, capers, and tamarinds, the two last bearing larger fruit than I had ever before seen, though not arrived to their greatest size or maturity.

We continued this winding according to the course of the river, among mountains of no great height, but bare, stony, and full of terrible precipices. At half-past eight o'clock we halted, to avoid the heat of the sun, under shade of the trees before mentioned, for it was then excessively hot, though in the month of November, from ten in the morning till two in the afternoon. We met this day with large numbers of Shiho, having their wives and families along with them, descending from the tops of the high mountains of Habesh, with their flocks to pasture, on the plains below near the sea, upon grass that grows up in the months of October and November, when they have already consumed what grew in the opposite season on the other side of the mountains.

At two o'clock in the afternoon we resumed our journey through a very stony, uneven road, till five o'clock, when we pitched our tent at a place called Hamhammon, on the side of a small green hill some hundred yards from the bed of the torrent. The weather had been perfectly good since we left Masuah; this afternoon, however, it seemed to threaten rain; the high mountains were quite hid, and great part of the lower ones covered with thick clouds; the lightning was very frequent, broad and deep-tinged with blue; and long peals of thunder were heard, but at a distance. This was the first sample we had of Abyssinian bad weather.

The river scarcely ran at our passing it; when, all of a sudden, we heard a noise on the mountains above, louder than the loudest thunder. Our guides upon this flew to the baggage, and removed it to the top of the green hill; which was no sooner done than we saw the river coming down in a stream about the height of a man and breadth of the whole bed it used to occupy. The water was thick tinged with red earth, and ran in the form of a deep river, and swelled a little above its banks, but did not reach our station on the hill.

Hamhammon is a mountain of black stones, almost calcined by the violent heat of the sun. This is the boundary of the district: Samhar, inhabited by the Shiho from Hamhammon to Taranta, is called Hadaffa; it belongs to the Hazorta.

This nation, though not so numerous as the Shiho, are yet their neighbours, live in constant defiance of the Naybe, and are of a colour much resembling new copper, but are inferior to the Shiho in size, though very agile. All their substance is in cattle; yet they kill none of them, but live entirely upon milk. They, too, want also an original word for bread in their language, for the same reason, I suppose, as the Shiho. They have been generally successful against the Naybe, and live either in caves or in cabannes, like cages, just large enough to hold two persons, and covered with an ox's hide. Some of the better sort of women have copper bracelets upon their arms, beads in their hair, and a tanned hide wrapt about their shoulders.

The nights are cold here even in summer, and do not allow the inhabitants to go naked, as upon the rest of the coast; however, the children of the Shiho, whom we met first, were all naked.

The 18th, at half-past five in the morning, we left our station on the side of the green hill at Hamhammon. For some time our road lay through a plain so thick set with acacia trees that our hands and faces were all torn and bloody with the strokes of their thorny branches. We then resumed our ancient road in the bed of the torrent, now nearly dry, over stones which the rain of the preceding night had made very slippery.

At half-past seven we came to the mouth of a narrow valley, through which a stream of water ran very swiftly over a bed of pebbles. It was the first clear water we had seen since we left Syria, and gave us then unspeakable pleasure. It was in taste excellent. The shade of the tamarind-tree, and the coolness of the air, invited us to rest on this delightful spot, though otherwise, perhaps, it was not exactly conformable to the rules of prudence, as we saw several huts and families of the Hazorta along the side of the stream, with their flocks feeding on the branches of trees and bushes, entirely neglectful of the grass they were treading under foot.

At two o'clock we continued our journey among large timber trees till half-past three, along the side of the rivulet, when we lost it. At half-past four we pitched our tent at Sadoon, by the side of another stream, as clear, as shallow, and as beautiful as the first; but the night here was exceedingly cold, though the sun had been hot in the daytime. Our desire for water was by this time considerably abated. We were everywhere surrounded by mountains, bleak, bare, black, and covered

Q 2

with loose stones, entirely destitute of soil; and besides this gloomy prospect we saw nothing but the heavens.

On the 19th, at half-past six in the morning, we left Sadoon, our road still winding between mountains in the bed, or torrent of a river, bordered on each side with rack and sycamore trees of a good size. I thought them equal to the largest trees I had ever seen; but upon considering, and roughly measuring some of them, I did not find one 7½ feet diameter, a small tree in comparison of those that some travellers have observed, and much smaller than I expected; for here every cause concurred that should make the growth of these large bodies excessive.

At half-past eight o'clock, we encamped at a place called Tubbo, where the mountains are very steep, and broken very abruptly into cliffs and precipices. Tubbo was by much the most agreeable station we had seen; the trees were thick, full of leaves, and gave us abundance of very dark shade. There was a number of many different kinds so closely planted that they seemed to be intended for natural arbours. Every tree was full of birds, variegated with an infinity of colours, but destitute of song; others, of a more homely and more European appearance, diverted us with a variety of wild notes, in a style of music still distinct and peculiar to Africa, as different in the composition from our linnet and goldfinch, as our English language is to that of Abyssinia.

We left Tubbo at three o'clock in the afternoon, and we wished to leave the neighbourhood of the Hazorta. At four we encamped at Lila, where we passed the night in a narrow valley, full of trees and brushwood, by the side of a rivulet. These small but delightful streams, which appear on the plain between Taranta and the sea, run only after October. When the summer rains in Abyssinia are ceasing, they begin again on the east side of the mountains; at other times no running water is to be found here, but it remains stagnant in large pools, whilst its own depth, or the shade of the mountains and trees, prevent it from being exhaled by the heat of the sun till they are again replenished with fresh supplies, which are poured into them upon return of the rainy season. Hitherto we had constantly ascended from our leaving Arkeeko, but it was very gradually, indeed almost imperceptibly.

On the 20th, at six o'clock in the morning, we left our station at Lila, and about seven we began to ascend the hills or eminences which serve as the roots or skirts of the great mountain Taranta. The road was on each side bordered with nabca or jujeb trees, of great beauty, and sycamores perfectly deprived of their verdure and branches.

We saw to-day plenty of game. The country here is everywhere deprived of the shade it would enjoy from these fine trees by the barbarous axes of the Hazorta. We found everywhere immense flocks of antelopes; as also partridges of a small kind, that willingly took refuge upon trees; neither of these seemed to consider us as enemies. The antelopes let us pass through their flocks, only removing to the right or to the left, or standing still and gazing upon us till we passed. But as we were then on the confines of Tigré, or rather on the territory of the Bahamagash, and as the Hazorta were in motion, everywhere removing towards the coast, far from the dominions of the Abyssinians to which we were going, a friend of their own tribe, who had joined us for safety, knowing how little trust was to be put in his countrymen when moving in this contrary direction, advised us by no means to fire, or give any unnecessary indication of the spot where we were, till we gained the mountain of Taranta, at the foot of which we halted at nine in the morning.

At half-past two o'clock in the afternoon we began to ascend the mountain, through a most rocky, uneven road, if it can deserve the name, not only from its incredible steepness, but from the large holes and gullies made by the torrents, and the huge monstrous fragments of rocks which, loosened by the water, had been tumbled down into our way. It was with great difficulty we could creep up, each man carrying his knapsack and arms; but it seemed beyond the possibility of human strength to carry our baggage and instruments. Our tent, indeed, suffered nothing by its falls; but our telescopes, time-keeper, and quadrant, were to be treated in a more delicate and tender manner.

Our quadrant had hitherto been carried by eight men, four to relieve each other; but these were ready to give up the undertaking upon trial of the first few hundred yards. A number of expedients, such as trailing it on the ground (all equally fatal to the instrument) were proposed. At last, as I was incomparably the strongest of the company, as well as the most interested, I, and a stranger Moor, who had followed us, carried the head of it for about 400 yards over the most difficult and steepest part of the mountain, which before had been considered as impracticable by all.

We found it impossible to pitch our tents, from the extreme weariness in which our last night's exertion had left us. But there was another reason also; for there was not earth enough covering the bare sides of Taranta to hold fast a tent-pin; but there were a variety of caves near us, and throughout the mountain, which had served for houses to the old inhabitants; and in these found a quiet and not inconvenient place of repose, the night of the 20th November.

All the side of the mountain of Taranta, which we had passed, was thick-set with a species of tree which we had never before seen, but which was of uncommon beauty and curious composition of parts; its name is kol-quall. Though we afterwards met it in several places of Abyssinia, it never was in the perfection we now saw it in Taranta.

On the 21st, at half-past six in the morning, having encouraged my company with good words, increase of wages, and hopes of reward, we began to encounter the other half of the mountain, but, before we set out, seeing that the ass of the stranger Moor, which was bit by the hyæna, was incapable of carrying his loading further, I desired the rest, every one, to bear a proportion of the loading till we should arrive at Dixan, where I promised to procure him another which might enable him to continue his journey.

This being ended, I soon perceived the good effect. My baggage moved much more briskly than the preceding day. The upper part of the mountain was indeed steeper, more cragged, rugged, and slippery than the lower, and impeded more with trees, but not embarrassed so much with large stones and holes. Our knees and hands, however, were cut to pieces by frequent falls, and our faces torn by the multitude of thorny bushes. I twenty times now thought of what Achmet had told me at parting, that I should curse him for the bad road shown to me over Taranta; but bless him for the quiet and safety attending me in that passage.

The middle of the mountain was thinner of trees than the two extremes; they were chiefly wild olives which bear no fruit. The upper part was close covered with groves of the oxy cedars, the Virginia or berry-bearing cedar, in the language of the country called Arze. At last we gained the top of the mountain, upon which is situated a small village called Halai, the first we had seen since our leaving Masuah. It is chiefly inhabited by poor servants and shepherds keeping the flocks of men of substance living in the town of Dixan.

All sorts of cattle are here in great plenty; cows and bulls of exquisite beauty, especially the former; they are, for the most part, completely white, with large dew-laps hanging down to their knees; their heads, horns, and hoofs, perfectly well turned; the horns wide like our Lincolnshire kine; and their hair like silk.

The plain on the top of the mountain Taranta was, in many places, sown with wheat, which was then ready to be cut down, though the harvest was not yet begun. The grain was clean, and of a good colour, but inferior in size to that of Egypt. It did not, however, grow thick, nor was the stalk above fourteen inches high. The water is very bad on the top of Taranta, being only what remains of the rain in the hollows of the rocks, and in pits prepared for it.

Being very tired, we pitched our tent on the top of the mountain. The night was remarkably cold, at least appeared so to us, whose pores were opened by the excessive heat of Masuah; for at mid-day the thermometer stood 61°, and at six in the evening 59°; the barometer, at the same time, 18½ inches French. The dew began to fall strongly, and so continued till an hour after sunset, though the sky was perfectly clear, and the smallest stars discernible.

On the 22nd, at eight in the morning, we left our station on the top of Taranta, and soon after began to descend on the side of Tigré, through a road the most broken and uneven that ever I had seen, always excepting the ascent of Taranta. After this we began to mount a small hill, from which we had a distinct view of Dixan.

The cedar trees, so tall and beautiful on the top of Taranta, and also on the east side, were greatly degenerated when we came to the west, and mostly turned into small shrubs and scraggy bushes. We pitched our tent near some marshy ground for the sake of water, at three quarters past ten, but it was very bad, having been for several weeks stagnant. We saw here the people busy at their wheat harvest; others, who had finished theirs, were treading it out with cows or bullocks. They make no use of their straw; sometimes they burn it, and sometimes leave it on the spot to rot.

We set out from this about ten minutes after three, descending gently through a better road than we had hitherto seen. At half-past four in the evening, on the 22nd of November, we came to Dixan. Halai was the first village, so is this the first town in Abyssinia, on the side of Taranta. Dixan is built on the top of a hill, perfectly in form of a sugar loaf; a deep valley surrounds it everywhere like a trench; and the road winds spirally up the hill till it ends among the houses.

It was on November 25th, at ten in the morning; we left Dixan, descending the very steep hill on which the town is situated. It produces nothing but the kol-quall tree all around it. We passed a miserable village called Hadhadid, and, at eleven o'clock, encamped under a daroo tree, one of the finest I have seen in Abyssinia, being 7½ feet diameter, with a head spreading in proportion, standing alone by the side of a river which now ran no more, though there is plenty of fine water still stagnant in its bed. This tree and river is the boundary of the territory, which the Naybe farms from Tigré, and stands within the province of Bahamagash, called Midré Bahar.

The 26th, at seven in the morning, we left our most pleasant quarters under the daroo tree, and set forward with great alacrity. About a quarter of a mile from the river we crossed the end of the plain Zarai, already mentioned. Though this is but three miles long, and one where broadest, it was the largest plain we had seen since our passing Taranta, whose top was now covered wholly with large, black, and very heavy clouds, from which we heard and saw frequent peals of thunder and violent streams of lightning. This plain was sown partly with wheat, partly with Indian corn; the first was cut down, the other not yet ripe. Two miles farther we passed Addicota, a village planted upon a high rock; the sides towards us were as if cut perpendicular like a wall. Here was one refuge of the Jesuits when banished Tigré by Facilidas, when they fled to the rebel John Akay. We after this passed a variety of small villages on each side of us, all on the top of hills; Dorcatta and Embabuwhat on the right, Azaria on the left.

At half-an-hour past eleven we encamped under a mountain, on the top of which is a village called Hadaur, consisting of no more than eighty houses, though, for the present, it is the seat of the Bahamagash.

On the 27th, we left Hadaur, continuing our journey down a very steep and narrow path between two stony hills; then ascended one still higher, upon the top of which stands the large village of Goumbubba, whence we have a prospect over a considerable plain all sown with the different grain this country produces, wheat, barley, teff, and tocuffo; simsim (or sesame), and nook; the last is used for oil.

We passed the village of Dergate, then that of Regticat, on the top of a very high hill on the left, as the other was on our right. We pitched our tent about half-a-mile off the village called Bananda. It was the 29th we left our station at Bananda, and had scarcely advanced a mile when we were overtaken by a party of about twenty armed men on horseback.

The first part of our journey to-day was in a deep gully; and, in half-an-hour we entered into a very pleasant wood of acacia-trees, then in flower. In it likewise was a tree, in smell like a honeysuckle, whose large white flower nearly resembles that of a caper. We came out of this wood into the plain, and ascended two easy hills; upon the top of these were two huge rocks, in the holes of which, and within a large cave, a number of the blue fork-tailed swallows had begun their nests. These, and probably many if not all the birds of passage, breed twice in the year, which seems a provision against the losses made by emigration perfectly consonant to divine wisdom. These rocks are by some said to be the boundaries of the command of the Bahamagash on this side; though others extend them to the Balezat.

We entered again a straggling wood, so overgrown with wild oats that it covered the men and their horses. The plain here is very wide; it reaches down on the west to Serawé, then distant about twelve miles; it extends from Goumbubba as far south as Balezat. The soil is excellent; but such flat countries are very rare in Abyssinia. This, which is one of the finest and widest, is abandoned without culture, and is in a state of waste.

After passing the wood, we came to the river, which was then standing in pools. I here, for the first time, mounted on horseback, to the great delight of my companions from Baranda, and also of our own, none of whom had ever before seen a gun fired from a horse galloping, excepting Yasine and his servant, now my groom, but neither of these had ever seen a double-barrelled gun. We passed the plain with all the diligence consistent with the speed and capacity of our long-eared convoy; and having now gained the hills, we bade defiance to the

Serawé horse, and sent our guard back perfectly content, and full of wonder at our fire-arms, declaring that their master, the Bahamagash, had he seen the black horse behave that day, would have given me another much better.

We entered now into a close country, covered with brushwood, wild oats, and high bent-grass; in many places rocky and uneven, so as scarce to leave a narrow part to pass. Just in the very entrance a lion had killed a very fine animal called Agazan; it is of the goat kind.

At noon we crossed the river Balezat, which rises at Ade Shiho, a place on the S. W. of the province of Tigré; and, after no very long course, having been once the boundary between Tigré and Midré Bahar, (for so the country of the Bahamagash was called), it falls into the Mareb, or ancient Astusaspes. It was the first river then actually running that we had seen since we passed Taranta; indeed, all the space is but very indifferently watered. This stream is both clear and rapid, and seems to be full of fish. We continued for some time along its banks, the river on our left, and the mountains on our right, through a narrow plain, till we came to Tomumbusso, a high pyramidal mountain, on the top of which is a convent of monks, who do not, however, reside there, but only come hither upon certain feasts, when they keep open house, and entertain all that visit them. The mountain itself is of porphyry.

There we encamped by the river's side, and were obliged to stay this and the following day, for a duty, or custom, to be paid by all passengers.

On the 1st December we departed from Balezat, and ascended a steep mountain, upon which stands the village Noguet, which we passed about half-an-hour after. On the top of the hill were a few fields of teff. Harvest was then ended, and they were treading out the teff with oxen. Having passed another very rugged mountain, we descended and encamped by the side of a small river, called Mai Kol-quall, from a number of these trees growing about it. This place is named the Kella, or castle, because, nearly at equal distances, the mountains on each side run for a considerable extent straight and uneven, in shape like a wall with gaps at certain distances, resembling embrasures and bastions. This rock is otherwise called Damo, anciently the prison of the collateral heirs male of the royal family.

The river Kol-quall rises in the mountains of Tigré, and after a course nearly N.W. falls into the Mareb. It was at Kella we saw, for the first time, the roofs of the houses made in form of cones; a sure proof that the tropical rains grow more violent as they proceed westward. About half-a-mile on the hill above is the village Kai-bara, wholly inhabited by Mahometan Gibbertis; that is, native Abyssinians of that religion.

It was in the afternoon of the 4th that we set out from Kella; our road was between two hills covered with thick wood. On our right was a cliff, or high rock of granite, on the top of which were a few houses that seemed to hang over the cliff rather than stand upon it. A few minutes after three o'clock we passed a rivulet, and a quarter of an hour afterwards another, both which run into the Mareb. We still continued to descend, surrounded on all sides with mountains covered with high grass and brushwood, and abounding with lions. At four we arrived at the foot of the mountain, and passed a small stream which runs there.

We had seen no villages after leaving Kella. At half-past four o'clock we came to a considerable river called Angueah, which we crossed, and pitched our tent on the farther side of it. It was about fifty feet broad, and three in depth. It was perfectly clear, and ran rapidly over a bed of white pebbles, and was the largest river we had yet seen in Habesh. In summer there is very little plain ground near it but what is occupied by the stream; it is full of small fish, in great repute for their goodness.

This river has its name from a beautiful tree, which covers both its banks. This tree, by the colour of its bark and the richness of its flower, is a great ornament to the banks of the river. A variety of other flowers fill the whole level plain between the mountain and the river, and even some way up the mountains. In particular, great variety of jessamine, white, yellow, and parti-coloured. The country seemed now to put on a more favourable aspect; the air was much fresher and more pleasant, every step we advanced after leaving Dixam; and one cause was very evident; the country where we now passed was well watered with clear running streams; whereas, nearer Dixam, there were few, and all stagnant.

The 5th, we descended a small mountain for about twenty minutes, and passed the following villages, Zabangella, about a mile N.W.; at a quarter of an hour after, Moloxito, half-a-mile farther S.E.; and Mansuetemen, three quarters of a mile E.S.E. These villages are all the property of the Abuna, who has also a duty upon all merchandize passing there; but Ras Michael had confiscated these last villages on account of a quarrel he had with the last Abuna, Af-Ya-goube.

We now began first to see the high mountains of Adowa, nothing resembling in shape to those of Europe, nor, indeed, any other country. Their sides were all perpendicular rocks, high like steeples, or obelisks, and broken into a thousand different forms.

At half-past eight o'clock we left the deep valley, wherein runs the Mareb, W.N.W.; at the distance of about nine miles above it is the mountain, or high hill, on which stands Zarai, now a collection of villages, formerly two convents built by Lalibala; though the monks tell you a story of the queen of Saba residing there, which the reader may be perfectly satisfied she never did in her life.

The Mareb is the boundary between Tigré and the Bahamagash, on this side. It runs over a bed of soil; is large, deep, and smooth; but upon rain falling it is more dangerous to pass than any river in Abyssinia, on account of the frequent holes in its bottom. We then entered the narrow plain of Yeeha, wherein runs the small river, which either gives its name to, or takes it from it. The Yeeha rises from many sources in the mountains to the west; it is neither considerable for size nor its course, and is swallowed up in the Mareb.

The harvest was in great forwardness in this place. The wheat was cut, and a considerable share of the teff in another part; they were treading out this last-mentioned grain with oxen. The dora and a small grain called telba (of which they make oil), was not ripe.

At eleven o'clock we rested by the side of the mountain whence the river falls. All the villages that had been built here bore the marks of the justice of the Governor of Tigré. They had been long the most incorrigible banditti in the province. He surrounded them in one night, burnt their houses, and extirpated the inhabitants; and would never suffer anyone since to settle there. At three o'clock in the afternoon we ascended what remained of the mountain of Yeeha, came to the plain upon its top, and, at a quarter before four, passed the village of that name, leaving it to the S.E., and began the most rugged and dangerous descent we had met with since Taranta.

At half-past five in the evening we pitched our tent at the foot of the hill, close by a small, but rapid and clear stream, which is called Ribieraini. This name was given it by the banditti of the villages before mentioned, because from this you see two roads; one leading from Gondar, that is, from the westward; the other from the Red Sea to the eastward. One of the gang that used to be upon the outlook from this station, as soon as any caravan came in sight, cried out, "Ribieraini," which in Tigré signifies they are coming this way; upon which notice everyone took his lance and shield, and stationed himself properly to fall with advantage upon the unwary merchant; and it was a current report, which his present greatness could not stifle, that, in his younger days, Ras Michael himself frequently was on these expeditions at this place. On our right was the high, steep, and rugged mountain of Samayat, which the same Michael, being in rebellion, chose for his place of strength, and was there besieged and taken prisoner by the late King Yasons.

The rivulet of Ribieraini is the source of the fertility of the country adjoining, as it is made to overflow every part of this plain, and furnishes a perpetual store of grass, which is the reason of the caravans choosing to stop here. Two or three harvests are also obtained by means of this river; for, provided there is water, they sow in Abyssinia in all seasons. We perceived that we were now approaching some considerable town, by the great care with which every small piece of ground, and even the steep sides of the mountains, were cultivated, though they had ever so little soil.

On Wednesday, the 6th December, at eight o'clock in the morning, we set out from Ribieraini, and in about three hours travelling on a very pleasant road, over easy hills and through hedge-rows of jessamine, honey-suckle, and many kinds of flowering shrubs, we arrived at Adowa, where once resided Michael Suhul, Governor of Tigré. It was this day we saw, for the first time, the small long-tailed green paroquet, from the hill of Shillodee, where, as I have already mentioned, we first came in sight of the mountains of Adowa.

Adowa is situated on the declivity of a hill, on the west side of a small plain surrounded everywhere by mountains. Its situation accounts for its name, which signifies pass, or passage, being placed on the flat ground immediately below Ribieraini; the pass through which everybody must go on their way from Gondar to the Red Sea.

This plain is watered by three rivulets which are never dry in the midst of summer; the Assa, which we cross just below the town when coming from the eastward; the Mai Gogua, which runs below the hill whereon stands the village of the same name formerly, though now it is called Fremona, from the monastery of the Jesuits built there; and the Ribieraini, which, joining with the other two, falls into the River Mareb, about 22 miles below Adowa. There are fish in these three streams, but none of them remarkable for their size, quantity, or goodness. The best are those of Mai Gogua, a clear and pleasant rivulet, running very violently and with great noise. This circumstance and ignorance of the language has misled the reverend father Jerome, who says that the water of Mai Gogua is called so from the noise that it makes, which, in common language, is called guggling. This is a mistake, for Mai Gogua signifies the river of owls.

There are many agreeable spots to the south-east of the convent, on the banks of this river, which are thick-shaded with wood and bushes. Adowa consists of about 300 houses, and occupies a much larger space than would be thought necessary for these to stand on, by reason that each house has an inclosure round it of hedges and trees; the last chiefly the wanzey. The number of these trees so planted in all the towns, screen them so, that, at a distance, they appear so many woods.

But what deservedly interested us most was the appearance of our kind and hospitable landlord, Janni. He had sent servants to conduct us from the passage of the river, and met us himself at the outer door of his house. I do not remember to have seen a more respectable figure. He had his own short white hair, covered with a thin muslin turban, a thick well-shaped beard, as white as snow, down to his waist. He was clothed in the Abyssinian dress, all of white cotton, only he had a red silk sash, embroidered with gold, about his waist, and sandals on his feet, his upper garment reached down to his ankles. He had a number of servants and slaves about him of both sexes; and, when I approached him, seemed disposed to receive me with marks of humility and inferiority, which mortified me much, considering the obligations I was under to him, the trouble I had given, and was unavoidably still to give him. I embraced him with great acknowledgments of kindness and gratitude, calling him father; a title I always used in speaking either to him or of him afterwards, when I was in higher fortune, which he constantly remembered with great pleasure.

He conducted us through a courtyard planted with jessamine to a very neat, and, at the same time, large room, furnished with a silk sofa; the floor was covered with Persian carpets and cushions. All round flowers and green leaves were strewed upon the outer yard, and the windows and sides of the room stuck full of evergreens in commemoration of the Christmas festival that was at hand. I stopped at the entrance of this room; my feet were both dirty and bloody; and it is not good breeding to show or speak of your feet in Abyssinia, especially if anything ails them, and, at all times, they are covered. He immediately perceived the wounds that were upon mine. Both our clothes and flesh were torn to pieces at Taranta, and several other places, but he thought we had come on mules furnished us by the Naybe, for the young man I had sent to him from Kella, following the genius of his countrymen, though telling truth was just as profitable to him as lying, had chosen the latter, and seeing the horse I had got from the Bahamagash, had figured in his own imagination a multitude of others, and told Janni that there were with me horses, asses, and mules in great plenty; so that when Janni saw us passing the water, he took me for a servant, and expected for several minutes to see the splendid company arrive, well mounted upon horses and mules caparisoned.

He was so shocked at my saying that I performed this terrible journey on foot, that he burst into tears, uttering a thousand reproaches against the Naybe for his hardheartedness and ingratitude, as he had twice, as he said, hindered Michael from going in person and sweeping the Naybe from the face of the earth. Water was immediately procured to wash our feet. And here began another contention, Janni insisted upon doing this himself, which made me run out into the yard, and declare I would not suffer it. After this the like dispute took place among the servants. It was always a ceremony in Abyssinia to wash the feet of those that come from Cairo, and who are understood to have been pilgrims at Jerusalem.

This was no sooner finished than a great dinner was brought exceedingly well-dressed. But no consideration or entreaty could prevail upon my kind landlord to sit down and partake with me. He would stand, all the time, with a clean towel in his hand, though he had plenty of servants; and afterwards dined with some visitors who had come out of curiosity to see a man arrived from so far. Among these was a number of priests; a part of the company which I liked least, but who did not show any hostile appearance. It was long before I cured my kind landlord of these respectful observances, which troubled me very much; nor could he wholly ever get rid of them, his own kindness and good heart, as well as the pointed and particular orders of the Greek patriarch Mark constantly suggesting the same attention.

In the afternoon, I had a visit from the governor, a very graceful man, of about sixty years of age, tall and well favoured. He had just then returned from an expedition to the Tacazzé, against some villages of Ayto Tesfos, which he had destroyed, slain 120 men, and driven off a number of cattle. He had with him about sixty muskets, to which I understood, he had owed his advantage. These villages were about Tübalaque, just as you ascend the farther bank of the Tacazzé. He said he doubted much if we should be allowed to pass through Woggora, unless some favourable news came from Michael; for Tesfos of Samen, who kept his government after Joas's death, and refused to acknowledge Michael, or to submit to the king, in conjunction with the people of Woggora, acted now the part of robbers, plundering all sorts of people that carried either provisions or any thing else to Gondar, in order to distress the king and Michael's Tigré soldiers, who were then there.

The church of Mariam is on the hill S.S.W. of the town, and east of Adowa; on the other side of the river, is the other church, called Kedus Michael. About nine miles north, a little inclined to the east, is Bet Abba Garima, one of the most celebrated monasteries in Abyssinia. It was once a residence of one of their kings; and it is supposed that, from this circumstance ill understood, former travellers have said the metropolis of Abyssinia was called Germé.

Adowa is the seat of a very valuable manufacture of coarse cotton cloth, which circulates all over Abyssinia instead of silver money; each web is sixteen peek long of $1\frac{3}{4}$ width, their value a pataka; that is, ten for the ounce of gold. The houses of Adowa are all of rough stone, cemented with mud instead of mortar. That of lime is not used but at Gondar, where it is very bad. The roofs are in the form of cones, and thatched with a reedy sort of grass, something thicker than wheat straw. The Falasha or Jews enjoy this profession of thatching exclusively; they begin at the bottom, and finish at the top.

Excepting a few spots taken notice of as we came along from Ribieraini to Adowa, this was the only part of Tigré where there was soil sufficient to yield corn; the whole of the province besides is one entire rock. There are no timber trees in this part of Tigré unless a daroo or two in the valleys, and wanzeys in towns about the houses. At Adowa, and all the neighbourhood, they have three harvests annually. Their first seed time is in July and August; it is the principal one for wheat, which they then sow in the middle of the rains. In the same season they sow tocusso, teff, and barley. From the 20th November, they reap first their barley; then their wheat, and last of all their teff. In room of these they sow, immediately upon the same ground, without any manure, barley, which they reap in February; and then often sow teff, but more frequently a kind of veitch, or pea, called Shimbra; these are cut down before the first rains, which are in April. With all these advantages of triple harvests, which cost no fallowing, weeding, manure, or other expensive processes, the farmer in Abyssinia is always poor and miserable.

The cattle roam at discretion through the mountains. The herdsmen set fire to the grass, bent, and brushwood before the rains, and an amazing verdure immediately follows. As the mountains are very steep and broken, goats are chiefly the flocks that graze upon them.

The province of Tigré is all mountainous; and it has been said, without any foundation in truth, that the Pyrenees, Alps, and Apennines, are but mole-hills compared to them. I believe, however, that one of the Pyrenees above St. John Pied de Port, is much higher than Lamalmon; and that the mountain of St. Bernard, one of the Alps, is full as high as Taranta or rather higher. It is not the extreme height of the mountains in Abyssinia that occasions surprise, but the number of them, and the extraordinary forms they present to the eye.

On the 17th we set out from Adowa, resuming our journey to Gondar; and after passing two small villages, Adega Net and Adega Daid, the first about half a mile on our left, the second about three miles distant on our right, we encamped at sunset near a place called Bet Hannes, in a narrow valley, at the foot of two hills, by the side or a small stream.

On the 18th, in the morning, we ascended one of these hills, through a very rough stony road, and again, came into the plain, wherein stood Axum, once the capital of Abyssinia, at least as it is supposed. For my part, I believe it to have been the magnificent metropolis of the trading people, or Troglodyte Ethiopians, called properly Cushites, for the reason I have already given, as the Abyssinians never built any city, nor do the ruins of any exist at this day in the whole country. But the black or Troglodyte part of it, called in the language of scripture Cush, in many places have buildings of great strength, magnitude, and expense, especially at Azab, worthy the magnificence and riches of a state which was from the first ages the emporium of the Indian and African trade, whose sovereign, though a Pagan, was thought an example of reproof to the nations, and chosen as an instrument to contribute materially to the building of the first temple which man erected to the true God.

The ruins of Axum are very extensive; but, like the cities of ancient times, consist altogether of public buildings. In one square, which I apprehend to have been the centre of the town, there are forty obelisks, none of which have any hieroglyphics upon them.

It was the 20th of January, at seven o'clock in the morning, we left Axum; our road was at first sufficiently even, through small valleys and meadows; we began to ascend gently, but through a road exceedingly difficult in itself, by reason of large stones standing on edge, or heaped one upon another; apparently the remains of an old large causeway, part of the magnificent works about Axum.

The last part of the journey made ample amends for the difficulties and fatigue we had suffered in the beginning. For our road on every side was perfumed with variety of flowering shrubs, chiefly different species of jessamine; one in particular of these called Agam (a small four-leaved flower) impregnated the whole air with the most delicious odour, and covered the small hills through which we passed in such profusion, that we were at times almost overcome with its fragrance. The country all round had now the most beautiful appearance, and this was heightened by the finest of weather, and a temperature of air neither too hot nor too cold.

At 11 o'clock of the 20th, we pitched our tent in a small plain by the banks of a quick clear running stream; the spot is called Mai

Shum. There are no villages, at least that we saw, here. A peasant had made a very neat little garden on both sides of the rivulet, in which he had sown abundance of onions and garlic, and he had a species of pumpkin, which I thought was little inferior to a melon. This man guessed by our arms and horses that we were hunters, and he brought us a present of the fruits of his garden, and begged our assistance against a number of wild boars.

On the 21st we left Mai Shum at seven o'clock in the morning, proceeding through an open country part sown with teff, but mostly overgrown with wild oats and high grass. We afterwards travelled among a number of low hills, ascending and descending many of them, which occasioned more pleasure than fatigue. The jessamine continued to increase upon us, and it was the common bush of the country.

We now descended into a plain called Selech-lecha, the village of that name being two miles east of us. The country here has an air of gaiety and cheerfulness superior to anything we had ever yet seen. Poncet was right when he compared it to the most beauteous part of Provence. We crossed the plain through hedgerows of flowering shrubs, among which the honeysuckle now made a principal figure, which is of one species only, the same known in England; but the flower is larger and perfectly white, not coloured on the outside as our honeysuckle is. Fine trees of all sizes were everywhere interspersed; and the vine, with small black grapes of very good flavour, hung in many places in festoons, joining tree to tree, as if they had been artificially twined and intended for arbours.

After having passed this plain we again entered a close country through defiles between mountains thick covered with wood and bushes. We pitched our tent by the water-side judiciously enough as travellers, being quite surrounded with bushes, which prevented us from being seen in any direction.

On the 22nd we left Selech-lecha at seven o'clock in the morning, and at eight passed a village of two hundred yards on our left without seeing any one; but, advancing half-a-mile farther, we saw a number of armed men, from sixty to eighty, and we were told they were resolved to oppose our passage unless their comrades, taken the night before, were released. The people that attended us on the part of Welleta Michael as our escort, considered this as an insult, and advised me by all means to turn to the left to another village immediately under the hill, on which the house of Welleta Michael, mother to Welleta Gabriel, their governor was situated, as there we should find sufficient assistance to force these opponents to reason. We accordingly turned to the left, and, marching through thick bushes, came to the top of the hill above the village, in sight of the governor's house, just as about twenty men of the enemy's party reached the bottom of it.

The governor's servants told us that now was the time if they advanced to fire upon them, in which case they would instantly disperse, or else they would cut us off from the village. But I could not enter into the force of this reasoning, because if this village was strong enough to protect us, which was the cause of our turning to the left to seek it, these twenty men putting themselves between us and the village took the most dangerous step for themselves possible, as they must unavoidably be destroyed; and if the village was not strong enough to protect us, to begin with bloodshed was the way to lose our lives before a superior enemy. I therefore called to the twenty men to stop where they were, and send only one of their company to me; and upon their not paying any attention I ordered Yasine to fire a large blunderbuss over their head, so as not to touch them. Upon the report they all fled, and a number of people flocked to us from other villages; for my part, I believe some who had appeared against us came afterwards and joined us. We soon seemed to have a little army, and in about half-an-hour a party came from the governor's house with twenty lances and shields, and six firelocks, and presently after the whole multitude dispersed. It was about ten o'clock when under their escort, we arrived at the town of Siré, and pitched our tent in a strong situation in a very deep gulley on the west extremity of the town.

The province of Siré, properly so called, reaches from Axum to the Tacazzé. The town of Siré is situated on the brink of a very steep, narrow valley, and through this the road lies which is almost impassable. In the midst of this valley runs a brook bordered with palm trees, some of which are grown to a considerable size, but bear no fruit; they were the first we had seen in Abyssinia.

The town of Siré is larger than that of Axum; it is in form of a half-moon fronting the plain, but its greatest breadth is at the west end; all the houses are of clay, and thatched; the roofs are in form of cones, as indeed are all in Abyssinia. Siré is famous for a manufacture of coarse cotton cloths, which pass for current money through all the province of Tigré, and are valued at a drachm, the tenth part of a wakea of gold, or near the value of an imperial dollar each; their breadth is a yard and a quarter. Besides these, beads, needles, cohol, and incense at times only, are considered as money.

Although Siré is situated in one of the finest countries in the world, like other places it has its inconveniences. Putrid fevers, of the very worst kind, are almost constant here; and there did then actually reign a species of these that swept away a number of people daily.

On the 24th, at seven o'clock in the morning, we struck our tent at Siré, and passed through a vast plain. All this day we could discern no mountains, as far as eye could reach, but only some few detached hills standing separate on the plain, covered with high grass, which they were then burning, to produce new with the first rains. The country to the north is altogether flat, and perfectly open; and though we could not discover one village this day, yet it seemed to be well inhabited, from the many people we saw on different parts of the plain, some at harvest and some herding their cattle. The villages were probably concealed from us on the other side of the hills.

At four o'clock we alighted at Maisbinni at the bottom of a high, steep, bare cliff of red marble, bordering on purple, and very hard. Behind this is the small village of Maisbinni; and in the south another still higher hill, whose top runs in an even ridge like a wall. At the bottom of this cliff, where our tent was pitched, the small rivulet Maisbinni rises, which, gentle and quiet as it then was, runs very violently in winter, first north from its source, and then winding to S.W. it falls in several cataracts near a hundred feet high, into a narrow valley, through which it makes its way into the Tacazzé. Maisbinni, for wild and rude beauties, may compare with any place we had ever seen.

This day was the first cloudy one we had met with, or observed this year. The sun was covered for several hours, which announced our being near the large river Tacazzé.

On the 25th, at seven in the morning, leaving Maisbinni, we continued on our road, shaded with trees of many different kinds. At half an hour after eight we passed the river, which at this place runs west; our road this day was through the same plain as yesterday, but broken and full of holes. At ten o'clock we rested in a large plain called Dagashaha; a hill in form of a cone stood single about two miles north from us; a thin straggling wood was to the S.E., and the water, rising in spungy, boggy, and dirty ground, was very indifferent; it lay to the west of us.

Dagashaha is a bleak and disagreeable quarter; but the mountain itself, being seen far off, was of great use to us in adjusting our bearings, the rather that, taking our departure from Dagashaha, we came immediately in sight of the high mountain of Samem, where Lamalmon, one of that ridge, is by much the most conspicuous; and over this lies the passage or high road to Gondar. We likewise see the rugged, hilly country of Salent, adjoining to the foot of the mountains of Samen. We observed no villages this day from Maisbinni to Dagashaha; nor did we discern, in the face of the country, any signs of culture or marks of great population. We were, indeed, upon the frontiers of two provinces which had for many years been at war.

On the 26th, at six o'clock in the morning, we left Dagashaha. Our road was through a plain and level country, but, to appearance, desolated and uninhabited, being overgrown with high bent grass and bushes, as also destitute of water. We passed the solitary village Adega, three miles on our left, the only one we had seen. At eight o'clock we came to the brink of a prodigious valley, in the bottom of which runs the Tacazzé, next to the Nile the largest river in Upper Abyssinia. It rises in Angot (at least its principal branch) in a plain champaign country, about 200 miles S.E. of Gondar, near a spot called Sonami Midre. It has three springs' heads, or sources, like the Nile; near it is the small village Gourri.

It must be confessed that during the inundation these things wear a contrary face. It carries in its bed near one third of all the water that falls in Abyssinia; and we saw the mark the stream had reached the preceding year, eighteen feet above the bottom of the river, which we do not know was the highest point that it arrived at. But three fathoms it certainly had rolled in its bed; and this prodigious body of water, passing furiously from a high ground in a very deep descent, tearing up rocks and large trees in its course, and forcing down their broken fragments scattered on its stream, with a noise like thunder echoed from a hundred hills, these very naturally suggest an idea that from these circumstances it is very rightly called the Terrible. But then it must be considered that all rivers in Abyssinia at the same time equally overflow; that every stream makes these ravages upon its banks; and that there is nothing in this that peculiarly affects the Tacazzé, or should give it this special name; at least such is my opinion, though it is with great willingness I leave every reader in possession of his own, especially in etymology.

At half an hour past eight we began a gradual descent, at first easily enough, till we crossed the small brook called Maitemquet, or the water of baptism. We then began to descend very rapidly in a narrow path, winding along the side of the mountain, all shaded with lofty timber trees of great beauty. About three miles farther we came to the edge of the stream at the principal ford of the Tacazzé, which is very firm and good; the bottom consists of small pebbles, without either sand or large stones. The river here at this time was fully 200 yards broad, the water was perfectly clear, and running very swiftly; it was about three feet deep. This was the dry season of the year, when most rivers in Abyssinia ran now no more.

The banks of the Tacazzé are all covered at the water's edge with tamarisks, behind which grow high and straight trees, that seem to have gained additional strength from having often resisted the violence of the river. Few of these even ever lose their leaves, but are either covered with fruit, flower, or foliage the whole year; indeed, abundantly with all three during the six months fair weather.

Beautiful and pleasant, however, as this river is, like everything created, it has its disadvantages. From the falling of the first rains in March till November, it is death to sleep in the country adjoining to it, both within and without its banks; the whole inhabitants retire and live in villages on the top of the neighbouring mountains; and these are all robbers and assassins who descend from their habitations on the heights to lie in wait for, and plunder, the travellers that pass. Notwithstanding great pains have been taken by Michael, his son, and grandson, governors of Tigré and Siré, this passage had never been so far cleared but every month people are cut off.

The plenty of fish in this river occasions more than an ordinary number of crocodiles to resort hither. These are so daring and fearless that when the river swells, so as to be passable only by people upon rafts, or skins blown up with wind, they are frequently carried off by these voracious and vigilant animals. There are also many hippopotami, which in this country are called Gomari. I never saw any of these in the Tacazzé, but at night we heard them snort, or groan, in many parts of the river near us. There are also vast multitudes of lions and hyænas in all these thickets. We were very much disturbed by them all night. The smell of our mules and horses had drawn them in numbers about our tent, but they did us no further harm, except obliging us to watch. I found the latitude of the ford by many observations, the night of the 26th, taking a medium of them all, to be 13° 42′ 45″ north.

The river Tacazzé is, as I have already said, the boundary of the province of Siré. We now entered that of Samen.

On the 27th of January, a little past six in the morning, we continued some short way along the river side, and at forty minutes past six o'clock came to Ingerohha, a small rivulet rising in the plain above, which, after a short course through a deep valley, joins the Tacazzé. At half-past seven we left the river and began to ascend the mountains which forms the south side of the valley, or banks of that river. The path is narrow, winds as much and is as steep as the other, but not so woody. What makes it, however, still more disagreeable is, that every way you turn you have a perpendicular precipice into a deep valley below you. At half-past eight we arrived at the top of the mountain, and at half-past nine halted at Tabulaqué, having all the way passed among ruined villages, the monuments of Michael's cruelty or justice; for it is hard to say whether the cruelty, robberies, and violence of the former inhabitants did not deserve the severest chastisement.

We saw many people feeding cattle on the plain, and we again opened a market for flour and other provisions, which we procured in barter for cohol, incense, and beads. None but the young women appeared.

Our tent was pitched at the head of Ingerohha, on the north of the plain of Tabulaqué. This river rises among the rocks at the bottom of a little eminence, in a small stream, which from its source runs very swiftly, and the water is warm. The peasants told us that in winter in time of the rains, it becomes hot and smoked. It was in taste, however, good; nor did we perceive any kind of mineral in it. Tabulaqué, Anderassa, and Mentesegla, belong to the Shum of Addergey, and the

Viceroy of Samen, Ayto Tesfos, the large town of Hauza, is about eight miles south and by east of this.

On the 28th, at forty minutes past six o'clock in the morning, we continued our journey, and at half-past seven saw the small village Motecha on the top of the mountain, half a mile south from us. At eight we crossed the river Aira; and at half-past eight the river Tabul, the boundary or the district of Tabulaqué, thick covered with wood, and especially a sort of cane or bamboo, solid within, called there Shemale, which is used in making shafts for javelins, or light darts thrown from the hand, either on foot or on horseback, at hunting or in war.

We alighted on the side of Anderassa, rather a small stream, and which had now ceased running, but which gives the name to the district through which we were passing. Its water is muddy and ill-tasted, and falls into the Tacazzé, as do all the rivers we had yet passed. Dagashaha bears N.N.E. from this station. A great dew fell this night; the first we had yet observed.

The 29th, at six o'clock in the morning, we continued our journey from Anderassa through thick woods of small trees, quite overgrown, and covered with wild oats, reeds, and long grass, so that it was very difficult to find a path through them. We were not without considerable apprehension from our nearness to the Shangalla, who were but two days' journey distant from us to the W.N.W., and had frequently made excursions to the wild country where we now were. Hauza was upon a mountain south from us; after travelling along the edge of a hill, with the river on our left hand, we crossed it; it is called the Bowiha, and is the largest we had lately seen.

At nine o'clock we encamped upon the small river Angari, that gives its name to a district which begins at the Bowiha where Anderassa ends. The river Angari is much smaller than the Bowiha; it rises to the westward in a plain near Montesegla; after running half a mile it falls down a steep precipice into a valley, then turns to the N.E., and, after a course of two miles and a half farther, joins the Bowiha a little above the ford.

The small village Angari lies about two miles S.S.W., on the top of a hill Hauza (which seems a large town formed by a collection of many villages), is six miles south, pleasantly situated among a variety of mountains, all of different and extraordinary shapes; some are straight like columns, and some sharp in the point and broad in the base, like pyramids and obelisks, and some like cones. All these, for the most part, inaccessible, unless with pain and danger, to those that know the paths, are places of refuge and safety in time of war, and are agreeably separated from each other by small plains producing grain. Some of these, however, have at the top water and small flats that can be sown, sufficient to maintain a number of men independent of what is doing below them. Hauza signifies delight, or pleasure, and probably such a situation of the country has given the name to it. It is chiefly inhabited by Mahometan merchants, is the entrepôt between Masuah and Gondar, and there are here people of very considerable substance.

The 30th, at seven in the morning, we left Angari, keeping along the side of the river. We then ascended a high hill, covered with grass and trees, through a very difficult and steep road; which ending, we came to a small and agreeable plain, with pleasant hills on each side, this is called Mentesegla. At half-past seven we were in the middle of three villages of the same name, two to the right and one on the left, about half-a-mile distance. At half-past nine we passed a small river called Daracoy, which serves as the boundary between Addergey and this small district Mentesegla. At a quarter-past ten, we encamped at Addergey, near a small rivulet called Mai-Lumi, the river of limes or lemons, in a plain scarce a mile square, surrounded on each side with very thick wood in form of an amphitheatre. Above this wood are bare, rugged, and barren mountains. Midway in the cliff is a miserable village, that seems rather to hang than to stand there, scarce a yard of level ground being before it to hinder its inhabitants from falling down the precipice. The wood is full of lemons and wild citrons from which it acquires its name. Before the tent, to the westward, was a very deep valley, which terminated this little plain in a tremendous precipice.

The river Mai-Lumi, rising above the village, falls into the wood, and there it divides itself in two; one branch surrounds the north of the plain, the other the south, and falls down a rock on each side of the valley, where they unite, and, after having run about a quarter of a mile farther, are precipitated into a cataract of 150 feet high, and run in a direction south-west into the Tacazzé. The river Mai-Lumi was, at this time, but small, although it is violent in winter; beyond this valley are five hills, and on the top of each is a village.

We did not leave Addergey till near ten o'clock in the forenoon of the 4th of February. We continued our journey along the side of a hill, through thick wood and high grass; then descended into a steep narrow valley, the sides of which had been shaded with high trees, but in burning the grass, the trees were consumed likewise, and the shoots from the roots were some of them above eight feet high since the tree had thus suffered that same year; the river Angueah runs through the middle of this valley; after receiving the small streams before mentioned, it makes its way into the Tacazzé. It is a very clear, swift running river, something less than the Bowiha.

Hauza was from this S.E. eight miles distant. Its mountains, of so many uncommon forms, had a very romantic appearance. At one o'clock we alighted at the foot of one of the highest, called Debra Toon, about half-way between the mountain and village of that name, which was on the side of the hill about a mile N.W. Still farther to the N.W. is a desert, hilly district, called Adebarea, the country of the slaves, as being the neighbourhood of the Shangalla, the whole country between being waste and uninhabited.

The mountains of Waldubba, resembling those of Adebarea lay north of us about four or five miles. Waldubba, which signifies the Valley of the Hyæna, is a territory entirely inhabited by the monks.

The water is both scarce and bad at Debra Toon, there being but one spring or fountain, and it was exceedingly ill-tasted. We did not intend to make this a station, but having sent a servant to Hauza to buy a mule in room of that which the hyæna had eaten, we were afraid to leave our man, who was not yet come forward, lest he should fall in with the Shum of Addergey, who might stop the mule for our arrears of customs.

The pointed mountain of Dagashaha continued still visible; I set it this day by the compass, and it bore due N.E. We had not seen any cultivated ground since we passed the Tacazzé.

The 5th, at seven o'clock in the morning, we left Debra Toon, and came to the edge of a deep valley bordered with wood, the descent of which is very steep. The Anzo, larger and more rapid than the Angueah, runs through the middle of this valley; its bed is full of large, smooth stones, and the sides composed of hard rock, and difficult to descend; the stream is equally clear and rapid with the other. We ascended the valley on the other side, through the most difficult road we had met with since that of the valley of Siré. At ten o'clock we found ourselves in the middle of three villages, two to the right and one on the left; they are called Adamara, from Adama, a mountain; on the east side of which is Tchober. At eleven o'clock we encamped at the foot of the mountain Adama, in a small piece of level ground, after passing a pleasant wood of no considerable extent. Adama, in Amharic signifies pleasant; and nothing can be more wildly so than the view from this station.

Tchober is close at the foot of the mountain, surrounded on every side except the north by a deep valley covered with wood. On the other side of this valley are the broken hills which constitute the rugged banks of the Anzo. On the point of one of these, most extravagantly shaped, is the village Shahagaanah, projecting as it were over the river; and behind these the irregular and broken mountains of Salent appear, especially those around Hauza, in forms which European mountains never wear; and still higher, above these, is the long ridge of Samen, which run along in an even stretch till they are interrupted by the high conical top of Lamalmon, reaching above the clouds, and reckoned to be the highest hill in Abyssinia, over the steepest part of which, by some fatality, the reason I do not know, the road of all caravans to Gondar must lie.

As soon as we passed the Anzo, immediately on our right, is that part of Waldubba, full of deep valleys and woods, in which the monks used to hide themselves from the incursions of the Shangalla, before they found out the more convenient defence by the prayers and superior sanctity of the present saints. Above this is Adamara, where the Mahometans have considerable villages, and by their populousness and strength, have greatly added to the safety of the monks, perhaps not altogether completed yet by the purity of their lives. Still higher than these villages is Tchober, where we now encamped.

On the left hand, after passing the Anzo, all is Shahagaanah, till you come to the river Zarima. It extends in an east and west direction, almost parallel to the mountains of Samen, and in this territory are several considerable villages; the people are much addicted to robbery and rebellion, in which they were engaged at this time. Above Salent is Abergali, and above that Tamben, which is one of the principal provinces in Tigré, commanded at present by Kefla Yasous, an officer of the greatest merit and reputation in the Abyssinian army.

On the 6th, at six o'clock in the morning, we left Tchober, and passed a wood on the side of the mountain. At a quarter-past eight we crossed the river Zarima, a clear stream running over a bottom of stones. It is about as large as the Anzo. On the banks of this river, and all this day, we passed under trees larger and more beautiful than any we had seen since leaving the Tacazzé. After having crossed the Zarima, we entered a narrow defile between two mountains, where ran another rivulet; we continued advancing along the side of it, till the valley became so narrow as to leave no room but in the bed of the rivulet itself. It is called Mai-Agam, or the water or brook of jessamine, and falls into the Zarima, at a small distance from the place wherein we passed it. It was dry at the mouth (the water being there absorbed and hid under the sand), but above, where the ground was firmer, there ran a brisk stream of excellent water, and it has the appearance of being both broad, deep, and rapid in winter. At ten o'clock we encamped upon its banks, which are here bordered with high trees of cummel, at this time both loaded with fruit and flowers. There are also here a variety of other curious trees and plants; in no place, indeed, had we seen more, except on the banks of the Tacazzé; Mai-Agam consists of three villages; one, two miles distant, east-and-by-north, one at same distance N.N.W.; the third at one mile distant, S.E. by S.

On the 7th, at six o'clock in the morning, we began to ascend the mountain; at a quarter-past seven, the village Lik lay east of us; Muras, a country full of low but broken mountains, and deep narrow valleys, bears N.W.; and Walkayt, in the same direction, but farther off. At a quarter-past eight, Gingerohha, distant from us about a mile S.W.; it is a village situated upon a mountain that joins Lamalmon. Two miles to the N.E. is the village Taguzait, on the mountain which we were ascending. It is called Guza by the Jesuits, who strangely say that the Alps and Pyrenees are inconsiderable eminences to it. Yet, with all deference to this observation, Taguzait, or Guza, though really the base of Lamalmon, is not a quarter of a mile high.

Ten minutes before nine o'clock we pitched our tent on a small plain called Dippebaha, on the top of the mountains, above a hundred yards from a spring, which scarcely was abundant enough to supply us with water, in quality as indifferent as it was scanty. The plain bore strong marks of the excessive heat of the sun, being full of cracks and chasms, and the grass burnt to powder. There are three small villages so near each other that they may be said to compose one. Near them is the church of St. George, on the top of a small hill to the eastward, surrounded with large trees.

Since passing the Tacazzé, we had been in a very wild country, left so, for what I know, by nature, at least now lately rendered more so by being the theatre of civil war. The whole was one wilderness without inhabitants, unless at Addergey. The plain of Dippebaha had nothing of this appearance; it was full of grass, and interspersed with flowering shrubs, jessamine and roses, several kinds of which were beautiful, but only one fragrant. The air was very fresh and pleasant, and a great number of people passing to and fro animated the scene.

We met this day several monks and nuns of Waldubba; I should say pairs, for they were two and two together. They said they had been at the market of Dobarké on the side of Lamalmon, just above Dippebaha. Both men and women, but especially the latter, had large burdens of provisions on their shoulders, bought that day, as they said, at Dobarké, which shewed me they did not wholly depend upon the herbs of Waldubba for their support. The women were stout and young, and did not seem by their complexion to have been long in the mortifications of Waldubba. I rather thought that they had the appearance of healthy mountaineers; and were, in all probability, part of the provisions bought for the convent; and by the sample one would think the monks had the first choice of the market, which was but fit, and is a custom observed likewise in Catholic countries. The men seemed very miserable, and ill-clothed, but had a great air of ferocity and pride in their faces. They are distinguished only from the laity by a yellow cowl, or cap, on their head. The cloth they wear round them is likewise yellow, but in winter they wear skins dyed of the same colour.

On the 8th, at three quarters past six o'clock in the morning, we left Dippebaha, and at seven had two small villages on our left; one on the S.E., distant two miles, the other on the south, one mile off. They are called Wora, and so is the territory for some space on each side of them; but beyond the valley all is Shahagaanah to the root of Lamalmon. At a quarter past seven, the village of Gingerohha was three miles on our right; and we were now ascending Lamalmon, through a very narrow road, or rather path, for it scarcely was two feet wide anywhere. It was a spiral winding up the side of the mountain, always on the very brink of a precipice. Torrents of water, which in winter carry prodigious stones down the side of this mountain, had divided this path into several places, and opened to us a view of that dreadful abyss below, which few heads can (mine at least could not) bear to look down upon.

On the 9th February, at seven o'clock, we took leave of the friends whom we had so newly acquired at Lamalmon, all of us equally joyful and happy at the news. We began to ascend what still remained of the mountain, which, though steep and full of bushes, was much less difficult than that which we had passed. At a quarter-past seven we arrived at the top of Lamalmon, which has from below the appearance of being sharp-pointed. On the contrary, we were much surprised to find there a large plain, part in pasture, but more bearing grain. It is full of springs, and seems to be the great reservoir from whence arise most of the rivers that water this part of Abyssinia. A multitude of streams issue from the very summit in all directions; the springs boil out from the earth in large quantities, capable of turning a mill. They plough, sow, and reap here at all seasons; and the husbandman must blame his own indolence, and not the soil, if he has not three harvests. We saw, in one place, people busy cutting down wheat; immediately next to it others at the plough, and the adjoining field had green corn in the ear; a little farther it was not an inch above the ground.

Lamalmon is on the N.W. part of the mountains of Samen. That of Gingerohha, with two pointed tops, joins it on the north, and ends these mountains here, and is separated from the plain of St. Michael by a very deep gulley. Neither Lamalmon nor Gingerohha, though higher than the mountains of Tigré, are equal in height to some of those of Samen. I take those to the S.E. to be much higher, and, above all, that sharp-pointed hill Amba Gideon, the present residence of the Governor of Samen, Ayto Tesfos. This is otherwise called the Jews Rock, famous in the history of this country for the many revolts of the Jews against the Abyssinian kings.

The mountain is everywhere so steep and high that it is not enough to say against the will, but without the assistance of those above, no one from below can venture to ascend. On the top is a large plain, affording plenty of pasture, as well as room for ploughing and sowing for the maintenance of the army, and there is water at all seasons in great plenty, and even fish in the streams upon it; so that, although the inhabitants of the mountain had been often besieged for a considerable time together, they suffered little inconvenience from it, nor ever were taken unless by treason, except by Christopher de Gama and his Portuguese, who are said, by their own historians, to have stormed this rock, and put the Mahometan garrison to the sword. No mention of this honorable conquest is made in the annals of Abyssinia, though they give the history of this campaign of Don Christopher in the life of Claudius, or Atzenaf Segued.

On the top of the cliff where we now were, on the left hand of the road to Gondar, we filled a tube with quick-silver, and purged it perfectly of outward air; it stood this day at 20¼ English inches. Dagashaha bears N.E. by E. from our present station upon Lamalmon. The language of Lamalmon is Amharic, but there are many villages where the language of the Falasha is spoken. These are the ancient inhabitants of the mountains, who still preserve the religion, language, and manners of their ancestors, and live in villages by themselves. Their number is now considerably diminished, and this has proportionally lowered their power and spirit. They are now wholly addicted to agriculture, hewers of wood and carriers of water, and the only potters and masons in Abyssinia. In the former profession they excel greatly, and, in general, live better than the other Abyssinians, which these in revenge attribute to a skill in magic, not to superior industry. Their villages are generally strongly situated out of the reach of marching armies, otherwise they would be constantly rifled, partly from hatred and partly from hopes of finding money.

On the 10th, at half-past seven in the morning, we continued along the plain on the top of Lamalmon; it is called Lama; and a village of the same name bore about two miles east from us. At eight o'clock we passed two villages called Mocken, one W. by N. at one mile and a half, the other S.E. two miles distant. At half-past eight we crossed the River Macara, a considerable stream running with a very great current, which is the boundary between Woggora and Lamalmon. At nine o'clock we encamped at some small villages called Macara, under a church named Yasous. On the 11th of February, by the meridian altitude of the sun at noon, and that of several fixed stars proper for observation, I found the latitude of Macara to be 13° 6′ 8″. The ground was everywhere burnt up, and though the nights were very cold, we had not observed the smallest dew since our first ascending the mountain. The province of Woggora begins at Macara; it is all plain, and reckoned the granary of Gondar on this side, although the name would denote no such thing, for Woggora signifies the stony or rocky province.

The mountains of Lasta and Belessen bound our view to the south; the hills of Gondar on the S.W., and all Woggora lies open before us to the south, covered, as I have said before, with grain. But the wheat of Woggora is not good, owing probably to the height of that province. It makes an indifferent bread, and is much less esteemed than that of Woggora and Dembea, low, flat provinces, sheltered with hills, that lie upon the side of the lake Tzana.

On the 12th we left Macara at seven in the morning, still travelling through the plain of Woggora. At half-past seven saw two villages called Erba Tenfa, one of them a mile distant, the other half a mile on the N.W. At eight o'clock we came to Woken, five villages not two hundred yards distant from one another. At a quarter past eight we saw five other villages to the S.W. called Warrar, from one to four miles distant, all between the points of east and south. The country now grows inconceivably populous, vast flocks of cattle of all kinds feed on every side, having large and beautiful horns, exceedingly wide, and bosses upon their backs like camels; their colour is mostly black.

At a quarter past eight we passed Arena, a village on our left. At nine we passed the River Girama, which runs N.N.W. and terminates the district of Lamalmon, beginning that of Giram. At ten the church of St. George remained on our right, one mile from us; we crossed a river called Shimbra Zuggan, and encamped about two hundred yards from it. The valley of that name is more broken and uneven than any part we had met with since we ascended Lamalmon. The valley called also Shimbra Zuggan is two miles and a half N. by E. on the top of a hill surrounded with trees. Two small brooks, the one from S.S.E. the other from S.E. join here, then fall into the rivulet.

The 13th, at seven in the morning, we proceeded still along the plain; at half-past seven came to Arradara, and afterwards saw above twenty other villages on our right and left, ruined and destroyed from the lowest foundation by Ras Michael in his late march to Gondar. At half-past eight the church of Mariam was about a hundred yards on our left. At ten we encamped under Tamamo. The country here is full of people; the villages are mostly ruined, which, in some places, they are rebuilding. It is wholly sown with grain of different kinds, but more especially with wheat. For the production of this they have everywhere extirpated the wood, and now labour under a great scarcity of fuel. Since we passed Lamalmon, the only substitute for this was cows' and mules' dung, which they gather, make into cakes, and dry in the sun. From Addergey hither, salt is the current money, in large purchases such as sheep or other cattle; cohol and pepper for smaller articles, such as flour, butter, fowls, &c. At Shimbra Zuggan they first began to inquire after red Surat cotton cloth, for which they offered us thirteen bricks of salt; four peeks of this red cloth are esteemed the price of a goat. We began to find the price of provisions augment in a great proportion as we approached the capital.

This day we met several caravans going to Tigré, a certain sign of Michael's victory; also vast flocks of cattle driven from the rebellious provinces, which were to pasture on Lamalmon, and had been purchased from the army. Not only the country was now more cultivated, but the people were cleanlier, better dressed, and apparently better fed than those in the other parts we had left behind us. Indeed from Shimbra Zuggan hither there was not a foot excepting the path on which we trod that was not sown with some grain or other.

On the 14th, at seven o'clock in the morning, we continued our journey. At ten minutes past seven we had five villages of Tamamo three miles on our left; our road was through gentle rising hills, all pasture ground. At half past seven the village of Woggora was three miles on our right, and at eight the church of St. George a mile on our left, with a village of the same name near it; and ten minutes after Angaba Mariam, a church dedicated to the virgin, so-called from the small territory Angaba, which we are now entering. At fifty minutes past eight we came to five villages called Angaba, at small distances from each other. At nine o'clock we came to Kossogué, and entered a small district of that name. The church is on a hill surrounded with trees. On our left are five villages, all called Kossogué, and, as it were on a line, the farthest at three miles distance; near ten we came to the church of Argiff in the midst of many ruined villages. Three miles on our left hand are several others called Appano.

After having suffered with infinite patience and perseverance the hardships and danger of this long and painful journey, at forty minutes past ten we were gratified at last with the sight of Gondar, according to my computation about ten miles distance. The king's palace (at least the tower of it) is distinctly seen, but none of the other houses, which are covered by the multitude of wanzey trees growing in the town, so that it appears one thick, black wood. Behind it is Azazo, likewise covered with trees. On a hill is the large church of Tecla Haimanout, and the river below it makes it distinguishable; still farther on is the great lake Tzana, which terminates our horizon.

At forty-five minutes past ten we began to ascend about two miles through a broken road, having on our right in the valley below the river Tchagassa; and here begins the territory of that name. At fifty-five minutes past ten, descending still the hill, we passed a large spring of water, called Bambola, together with several plantations of sugar canes which grow here from the seed. At eleven o'clock the village Tchagassa was about half a mile distant from us on our right on the other side of the river. It is inhabited by Mahometans, as is Waalia, another small one near it. At twelve o'clock we passed the river Tchagassa over a bridge of three arches, the middle of which is Gothic, the two lesser Roman. This bridge, though small, is solid and well cemented, built with stone by order of Facilidas, who probably employed those of his subjects who had retained the arts of the Portuguese, but not their religion.

The Tchagassa has very steep, rocky banks. It is so deep, though narrow, that without this bridge it scarce would be passable. We encamped at a small distance from it, but nearer Gondar. Here again we met with trees (small ones indeed), but the first we had seen since leaving Lamalmon, excepting the usual groves of cedars. It is the Virginia cedar or oxy-cedros, in this country called Arz, with which their churches are constantly surrounded.

On the 15th, at ten minutes past seven, we began to ascend the mountain; and at twenty minutes after seven passed a village on our left. At seven and three-quarters we passed Tiba and Mariam, two churches, the one on our right, the other on our left, about half a mile distant; and near them several small villages inhabited by Falasha, masons, and thatchers of houses employed at Gondar. At half-past eight we came to the village Tocutcho, and in a quarter of an hour passed the river of that name, and in a few minutes rested on the river Angrab, about half a mile from Gondar.

Tchagassa is the last of the many little districts which together compose Woggora, generally understood to be dependent on Samen, though often, from the turbulent spirit of its chiefs, struggling for independency, as at the present time, but sure to pay for it immediately after. In fact, though large, it is too near Gondar to be suffered to continue in rebellion; and being rich and well cultivated, it derives its support from the capital as being the mart of its produce. It is certainly one of the fruitfullest provinces in Abyssinia, but the inhabitants are miserably poor, notwithstanding their threefold harvests. Whereas in Egypt, beholden to this country alone for its fertility, one moderate harvest gives plenty everywhere.

Woggora is full of large ants and prodigious swarms of rats and mice which consume immense quantities of grain; to these plagues may be added still one, the greatest of them all, bad government, which speedily destroys all the advantages they reap from nature, climate, and situation.

STEUDNER'S JOURNEY FROM CHANKAR (SOUTH-WEST OF GONDAR) TO MAGDALA, 1862 (marked S on Map, page 39).

THE plain of Dembea is void of trees, shrubs, and stones, and skulls of cattle are used for the outlets of the channels, constructed to irrigate the fertile soil. We continued three-quarters of an hour to Serava, a village of some size. On the road we met immense herds of cattle belonging to the Sellan. They are Christians and migratory herdsmen. A tract along the lake as far as Fogara is reserved for their herds, and it is not permitted to cultivate it. It bears a luxuriant vegetation of grass, and is called Aravĭĕ. The Sellan live in hemispherical straw huts, similar to those of the Bogos. They are under a chief who assigns every herd its pasturing ground. During the rainy season they, with their herds, are in the low lands towards Metamma, and on the first of Maskarem (10th September) they come to the plateau along the lake. Between Chankar and Serava we crossed the Dirma, which enters the lake near the Eastern Cape of Gorgora, which is called Debra Sina.

24th February.—We only travelled two hours this day direct to the east, as far as the village Adisgie. The very fertile plain is partly cultivated especially with Carthumus (Shuf). Extensive flats of luxuriant grass, alternate with very fertile fields. The plain of Dembea is entirely without trees; the many villages lie in dense woods or thickets formed of *Donax* trees (shambuko), 30 to 35 feet high, among which is seen occasionally a willow or a tree like *Veronia*. The houses are built of shambuko, the only building material of the district, and covered with grass. The dried dung of cows forms the only fuel, and throws out much heat. The water, at this time of the year, is scooped from wells 10 to 14 feet deep.

The Dembea plain is very healthy, as the wind blows twice a-day from the lake, but the districts of Dingel Ber and Foggara suffer from fevers. The nights in Dembea are very cold when compared to the temperature during the day. Water is always found 8 to 10 feet below the surface, but the soil, where it is not covered with high, thick grass, is traversed by fissures 1 to 3 feet deep which render riding difficult. The ground is perfectly level and free from stones, and the shambuko woods, in which lie the villages and churches, form the only breaks in the plain.

Beyond Adisgie we crossed the Mageteh, and 25 minutes after we had left the village a small rivulet, the Guasa, and further on, at the eastern corner of the lake, the Woin Arab. The two latter come from Amba Chara. As far as Ambo we rode along the grassy Dembea plain, generally close to the shore of the lake, which was enlivened by thousands of geese, ducks, and black swans. . . . . A few villages of the despised Woito, who live all round and subsist principally from fishing and hunting the hippopotamus, which they eat to the horror of all orthodox Abyssinians, are close to the shore. The huts of the Woito resemble those of the Sellan. . . . Here I saw the first boats. They are made of a very light sort of cane ("*Dongola*"), tied together. Some of these canes are as thick as an arm. They are perhaps a *papyrus*, and grow in Gurafa and Alafa. These boats are called *Tanquas*. They are 12 to 20 feet long, are pointed and turned up at both ends, and have a flat bottom. They at once fill in part with water, but do not capsize or sink. They are only used along shore.

After a ride of two hours and three-quarters, first over the plain of Dembea, then through tracts covered with dense shrubs of acacias, rolwal, stunted dates, &c., alternating with fine meadows, we reached the Arno-Garno, below the junction of the Arno and Garno. . . . . In three-quarters of an hour we reached the village Emfras, consisting of a few huts, hidden amongst splendid fig and wonsa trees. The Shum of this place is known as one of the greatest rogues of this neighbourhood.

26th.—We started early, crossed the hilly district of Tisba. After ascending for nearly two hours amidst the most luxuriant shrubs, we reached a plain, also covered with shrubs. Mount Efag rose above it, towards the south. After three hours' ride from Emfras, we reached the market place of Efag, which lies on the south foot of Mount Efag, and dismounted in the shade of gigantic juniper trees, surrounding a church. The market held here is the most important of Southern Abyssinia, with the exception of that of Baso, in Gojam. We found here on sale 300 to 400 oxen, as many asses and mules, 80 to 100 horses, (one of which caused universal admiration, and was sold for 11 thalers, the ordinary price for a good horse varying between 2 and 5 thalers). There was on sale much cotton and coffee (35 Nötte cost 1 thaler). 28 pieces of salt cost 1 thaler. . . . . The grape disease has destroyed nearly all the vines. . . We continued another half-hour over the plain of Foggara, to a large village.

27th.—On the next morning, riding over the extremely fertile but bare plain of Foggara, we reached the Reb in three-quarters of an hour. Its bed is cut deep into the fertile soil. We waded through the river. An old Portuguese bridge, still passable, is at some distance from our ford. Our road led us for three hours and a quarter through the fertile plain to the sinuous bed of the Sellien Woha (Date Water). The plain is bare, excepting some isolated shrubs of acacias. Another hour through a similar country brought us to Ambo, where there are several mineral springs in the bed of the Ferren Wuha. . . One quarter of an hour after we had left Ambo, we passed the isolated rock Amora Géddel, *i.e.*, Eagle's Eyrie, which rises at least 250 feet above the plain, and is inaccessible, and then continued at a rapid pace towards the high slope of the Debra Tabor plateau. As the sun disappeared in the lake towards the west, we had ascended the highest of the rocky terraces, all of which bore a luxuriant vegetation. The air up here was cold, and there was a fog. Our guide, in the darkness and in the uniformly undulating country, lost his way. At one time we blundered through cultivated fields, then through dense shrubs. We crossed the beds of torrents . . . . and at length, at 9 p.m., we reached the missionary station Gaffatt (near Debra Tabor). . . . .

8th March.—We left Gaffatt on 8th March. We rode through several well cultivated mountain valleys, showing splendid wheat fields between wooded hills, on which stood churches. After one hour's ride we continued along a bad road for one hour and a quarter to the Reb, which here rushes along its rocky bed between olive trees. We passed the village and district of Gulqúal. The road gradually ascends, through shrubs of rose trees and *hypericum* to the bare plateau of Guna. Twice more we had to cross the Reb, about 2 or 3 feet deep, before reaching the plateau. The latter is bare, cultivated in places, and has an altitude of 9,800 to 10,000 feet. Aloes, clover, *erica* and *hypericum* are met with, but generally speaking the plateau is barren. . . . It rained . . temperature 48° F. . . We were obliged to dismount, for our mules could hardly keep their footing on the fat, slippery soil. We crossed several streams flowing north-east and east towards the Takazze. After an uninterrupted march of five hours and a-half, we ascended another 500 feet, and half-an-hour afterwards we reached Dèttera, a village in the Sittim district (10,500 feet). . . It rained and thundered until late at night. . . .

9th March.—As the ground was still quite wet, we only got away at half-past seven, and even then riding was not very pleasant, and we preferred to lead our mules. . . We continued in a south-east direction over parallel ranges of hills, 200 to 400 feet high, as far as a rather considerable rivulet, running along its rocky bed to the Takazze, where we halted after a march of three hours. On the left (north-east) deep valleys descended towards the Takazze. . . We had here a fearful shower of rain and hail. . . We were again obliged to lead our animals over the slippery ground. . The road ascends gradually as far as Chechcho, and we were probably again 10,500 feet above the sea, when we reached the eastern edge of the plateau. The descent along the steep but broad road, in its present state, will always be stored in our memories. At this place a small "Amba" rises on the narrow back of a range, between deep valleys. It bears the promising name of Nefas Motcha, that is, "Windy Road," and fully deserves that name. At the Amba there resides one of the missionaries, who directs the construction of a road from Nefas Motcha to Zebit. We were hardly able to climb the Amba, though it only rises 200 feet above the road.

10th March.—On the 10th we started late, and rode along the newly made road, first towards the north-north-east, then east-north-east along the mountain slope, and crossed, after two hours, a narrow mountain pass, beyond and above the church Medhanie Alem. The slopes of the mountains and rocky precipices rise up to 10,500 feet, and still bear a dense vegetation of acacias, olives, &c. The new road only goes a short distance beyond the narrow pass, and soon we ascended up a steep zigzag path, through splendid groves of olives, about 400 feet, to the edge of the plateau of Zebit, which attains an elevation of 11,000 feet. The soil of the plateau is excellent, and shrubs of hypericum, olives, and celastus grow. Half an hour east of the plateau there is the small village of Zebit, with large stores of grain, guarded by a Shum and some soldiers. . . The rocks from Gaffat to this place are volcanic; trachytes and fine basalts are met with. The plateau of Zebit is not cultivated at all, but in the adjoining valleys and on the terraces of the mountain slopes there are many fields of barley, whilst teff is grown in the lower valleys. We met numerous herds of cattle. The breeding of mules and asses flourishes. It rained and hailed almost throughout the day.

11th March.—From Zebit we marched three hours and a half over the level plateau to Gergera. The plateau is almost without cultivation or water, and there are shrubberies of roses, celastrus, olives, etc. In some places the plateau is scarcely more than a quarter of an hour in width. At Gergera we descended to the plateau of Wadela, which is well cultivated in some parts. We only passed one small village of five or six huts, as most of the villages are built on the terraces of the slopes, away from the road. After a ride of nearly two hours (from Gergera) towards the east-south-east, we reached Wokiéta village, situate on a terrace of the northern slope of the Wadela plateau.

12th March.—Next morning we ascended by a very bad road the upper terrace of Wadela, the real plateau, which we had only left to camp near Wokiéta. The plateau sinks very gradually, and some isolated flat hills excepted, it is level. It is covered with grass to the neighbourhood of Betèhor, and there are scarcely any fields. On approaching Betèhor district, churches and hamlets are perceived on the isolated hills rising on the plateau. The bottoms, where water remains longer, are well cultivated. On the plateau there is only short grass. . . Juniper and olive trees are found near the churches. We rode three hours in the forenoon to the rivulet of Yannicha Gĕdus Mikael, where we stayed one hour. There was rather much water in the rivulet. Like the following rivulets it flows to the Iidda.

We now crossed a flat, aloe-covered hill, along the eastern foot of which flows the considerable rivulet Bansh Etie to the Iidda. There were ducks and geese, also an Ibis. We rode in a north-east direction over the undulating plateau, past the church Beit Yohannis, to a group of houses in the Betehor district, where we arrived after an afternoon ride of two hours, and pitched our tent on the ground saturated by this day's rain. . . . These plateaux are scooped out towards their centre, rising thence towards the edges. This day, for instance, we descended to 9,700 feet, and then gradually ascended again 300 feet towards the edge.

13th March.—After an hour's march we reached a rivulet, swollen by the rains, and rushing over the perpendicular columns of basalt forming its bed, towards the Djidda. We sought a long time for a ford to wade through this torrent, the name of which we could not ascertain. It probably is the Wons Bahr, which we passed on our journey back near where it falls into the valley. After another three quarters of an hour we reached the upper edge of the Iidda valley, near the church of Betèhor. The Iidda valley is 2,500 to 2,800 feet deep. Wadela, similarly to Talanta, slopes down to the river in two terraces. The upper terrace, 800 to 900 feet high, is passed on a road not very badly constructed, and thus reaches the lower and broader terrace, where are many acacias and Kolkwal trees. This terrace is about half an hour wide, and at its edge we met several hundred monkeys. From the ruins of a church we descended 1,800 feet, over a very steep zigzag path, to the principal terrace. Our mules, not being able to pass, went by a broader path along the terrace which leads down to a narrow valley joining the Iidda; and though they travelled a longer distance, they reached the river nearly as quickly as we did. The lower slopes are very steep, and full of channels formed by the water rushing down. They bear a luxuriant growth of shrubs, but there are no trees. The lowest slope is quite perpendicular, like the topmost one. Descending along the edge of the lowest terrace took us a long hour, whilst we had done the upper one in half an hour. The rocks consist of trachytic lava and basalt, as well as greywacke. . . . The bed of the Iidda at this spot is 120 to 150 paces wide, and filled with boulders varying in size from a fist to a man's head, and the river rushes through these in three branches, each ten to twelve paces wide, and two feet deep. Owing to the rapid current, and to the depth of some places, the river cannot be forded everywhere, and where we crossed it, though the depth did not exceed eighteen inches or two feet, our people and animals had to make some effort to maintain themselves. The opposite side of the valley corresponds in all respects to that of Wadela, down which we had come. After a short rest in the shade of fine sycamores, by the river side, we ascended a very steep, broad, zigzag edge—the first, principal terrace—and we passed the night in the village Averkut, a short distance east of the road. The village is surrounded by fields; the church of the village, Chaôt gumma Giyorgis, stands in a grove of Kolkwal trees.

14th March.—We started at 8 A.M.; rode for half an hour over the plateau, through fields and acacia shrubs, and then ascended the upper terrace by a zigzag path. We reached the upper plateau after a march of one hour and a half. It is bare, has a rich black soil, and we saw many fields from which the harvest had been taken home. Talanta Baha, a church on the plateau, was south of us. We left it on the right. We rode south-south-east over the plateau of Talanta; and one hour and three-quarters after we had left the edge of the Iidda valley, we reached that of the Beshilo. The upper edges of the Beshilo valley are much further apart than those of the Iidda. At our feet we perceived the valley of the Beshilo, 3,000 to 3,500 feet deep, and beyond it rose the many plateaux of Woro Haimanot, of small elevation, their edges cut off perpendicularly; and between them rose the Amba of the Negus, Magdala, and the plateau of Tanta towards the south-east. . . . The upper descent is very steep, and there are many Kolkwal trees, whilst on the lower slopes there are acacias and shrubs. The stony path at first leads past columns of basalt, down a steep declivity. It then intersects a layer of white sandstone, and after having crossed the narrow lower terrace it leads down a steep zigzag path, full of rolling stones, to the river bed. We found the bodies of many animals who had succumbed to fatigue, or fallen down the precipices, along the road. In some parts the passage is rendered possible by rude bridges, formed of the trunks of trees, and covered with brushwood and earth. The bed of the river is about 150 paces wide, and full of boulders of basalt, amongst which the river, at present 30 paces wide, flows along with a strong current. We rested in the bed of the river from one to three o'clock, and then rode up the bed of a tributary stream for two hours in a south direction. . . . We ascended steadily, but gradually, as far as a hill in front of Magdala, and ascended it by a steep zigzag path. There are here several small villages on the advanced hills around the fortress, between narrow ravines, four to five hundred feet deep, and densely covered with Kolkwal. We pitched our tent close to two huts, on the summit of the hill.

15th March, 1862.—On the following day we ascended a steep, stony path to Magdala. A mere chance and the strength of my mule here saved me from a fall down the precipice . . . On reaching the lower terrace of the mountain, we found ourselves on a small plateau, above which rises perpendicularly on the highest terraces the fortress proper. We now rode in the midst of rocks. This is the advanced fortress, for the citadel, if I may be permitted to use that term, rises further south on the same plateau. On that part of the plateau which lies between the advanced fortress and the Amba, serving as citadel, there are a few houses called Islam-gie. This portion, however, also forms part of the fortress, and it can be defended easily, as it rises on most points perpendicularly from the valley to a height of six to eight hundred feet. The advanced northern Amba is not as high as the main fortress which commands it, as well as the plateau. The main fortress rises several hundred feet above the plateau of Islam-gie, and on that side there are some inconsiderable works in masonry, to fill up some gaps which might possibly be escaladed. A few other points are defended by abatis, but for the rest it is a fortress by nature. As we were not permitted to pass the main Amba, we were not able to reach the principal road leading to Tanta, we descended a narrow, almost perpendicular path, on the eastern side of the Amba, for several hundred feet, to the spurs of the mountain, and thence, by less steep paths down to the bottom of the Woro Haimanot valley, which is about 1,500 feet below the plateau of the fortress. I do not think a European mule could have got down this path, even without a burden. In the bottom of the valley we found some water in a hole of the dry bed of the river, and there we rested for some time . . . The ascent to Tanta, on the other side of the valley, though fatiguing owing to the stony path, was mere child's play compared to the descent from Magdala.

Tanta is not a village, but a fortified plateau, forming part of the fortress, on which a number of huts have been erected for stores, &c. . . . In front of us, beyond the valley of Woro Haimanot, we had the entire fortress of Magdala, commanded by the guns of Tanta. It ascends precipitously from the valley, and a narrow rocky ridge connects it towards the south-west with the plateau of Tanta. This ridge bounds the Woro Haimanot valley on the south, and the main road between the principal Amba of Magdala and Tanta leads along it. Towards the north we saw the Amba Kuahit, beyond the Beshilo . . . Tanta is a plateau, which rises in several terraces. The few accessible spots are defended by masonry walls, with the exception of two, and these also can be rendered inaccessible in a short time. It forms part of the fortress of Magdala, and being several hundred feet higher, it commands it. On the south it is connected with a large plateau, extending south and south-west, and a broad ditch separates this plateau from Tanta. This ditch is generally filled with earth, but in time of danger it is cleared out. At this spot the plateau is hardly 200 paces wide. Short grass covers the plateau, and there are some shrubs on the hills rising on its edges. . . . The rocky precipices consist of phonolite and trachyte; there are veins of pitch stones—sometimes taken to be mineral coal. A large church has been excavated in the rocks of one of the hills. In its subterranean passages there were kept, at the time of our stay, about 100 cows. The magazines stand on the western edge of the plateau, opposite Magdala. They contained about 6,000 ardeb of corn, furnished in the course of this year by the tributary Gallas . . . East of Tanta the deep Ambĕla Sieda valley, with a broad, dry river bed, extends north to the Beshilo . . . The weather, on the whole, was favourable. We only had a few thunder-storms, with heavy showers of rain and hail. Every morning we had heavy dew.

LINE OF ADVANCE OF EXPEDITION.

The operations of the reconnoitring party, under Colonel Merewether, are given in the following extracts from his letter of the 12th November, 1867, from which it appears that Ansley Bay has been chosen as the landing place for the Expedition, and that the probable line of advance will be by Sanafe, and perhaps, also, by the Haddas to Tohonda, or Tekonda:—

"We have just returned from a most interesting and important reconnaisance up the pass from Koomoglee* to within five miles by road from Senafee, a distance of 41 miles. There were some very bad places in one part; but the road has been made now by the sappers easy for passage of cavalry, infantry, mules, and camels; and it will, I think, prove the chief line of route, as leading at once to a good position on the highlands of Abyssinia in the direction we have to go, and to a spot within easy reach.

"To-morrow we start up the Haddes to examine that, to as near as we can get to Tekonda, without actually entering it, or compromising the inhabitants by opening communications with them.

"Sir Robert Napier's excellent proclamation was sent out on the 6th inst., and I hope for the best results from it. Directly the ruler of Tigré, now Prince Kassai, a rebel against Theodorus, shows he intends acting in a friendly manner towards us, there will be no impropriety in visiting both Tekonda and Senafee; but until he does it would not be just to the people of those places to make them run

* Evidently the place marked Kumayli on map.

the risk of encountering his displeasure before we were in a position to protect them.

"I have been very vexed not to find a suitable plateau short of the Abyssinian highlands, but I was misled by the richness of the Agametta plateau, west of Massowah, and have only now learnt, what no one seemed to have been able to tell me before, that as you go south of the latitude of Massowah the lower hills become more purely volcanic; indeed, in some places entirely, so that vegetation diminishes *pari passu*.

"The troops that have landed are, I am happy to say, in excellent health and spirits. The 3rd Light Cavalry had been losing horses from fever, but to-day on my arrival here I was glad to find the disease disappearing. It was clearly owing to the effects of the sea voyage from India, and being cooped up on board ship."

PRINTED BY HARRISON AND SONS, ST. MARTIN'S LANE.

ROUTE MAP OF ABYSSINIA.

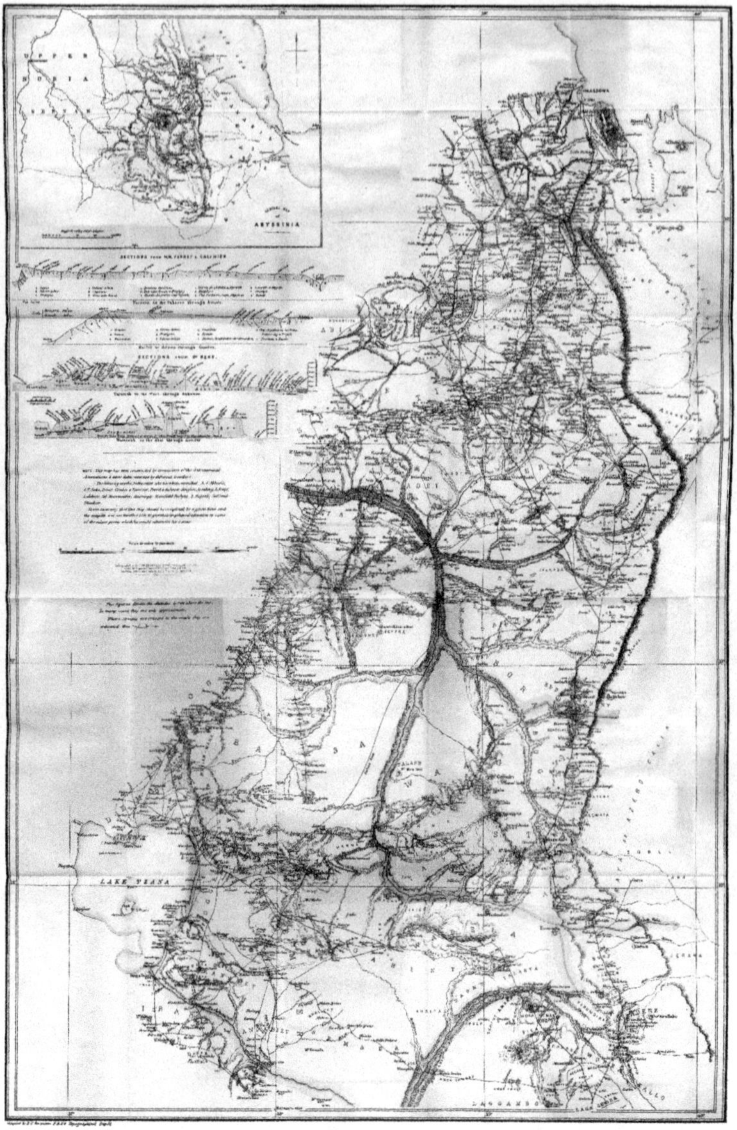

Made in the USA
Monee, IL
04 November 2026

67391555R10144